LONGMAN
WORLD GUIDE TO
MAMMALS

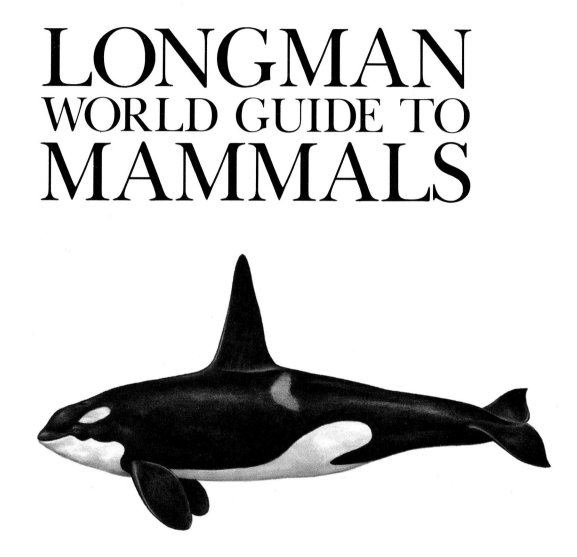

Consultant Editor Dr Philip Whitfield

Longman

Longman World Guide to Mammals
was conceived, edited and designed by
Marshall Editions Limited,
71 Eccleston Square, London SW1V 1PJ

Published by
Longman Group Limited,
Longman House, Burnt Mill, Harlow,
Essex CM20 2JE, England
and Associated Companies throughout the world

First published 1985

British Library Cataloguing in Publication Data

Longman world guide to mammals.—(Longman
  wildlife library)
  1. Mammals—Dictionaries
  I. Whitfield, Phillip, 1944–
  599'.003'21   QL701.2

ISBN 0-582-89211-2

Consultant Editor: **Dr Philip Whitfield,**
Zoology Department, King's College, University of
London

Consultant: **Dr D. M. Stoddart,**
Zoology Department, King's College, University of
London

Artists: **Graham Allen**
        Dick Twinney (pp. 83, 85, 125–57, 191, 193)

Editor: **Jinny Johnson**
Text Editor: **Gwen Rigby**
Art Director: **John Bigg**
Researcher: **Pip Morgan**
Production: **Barry Baker**
           Janice Storr

Contents

Filmset in Goudy Old Style by
Filmtype Services Limited,
Scarborough, North Yorkshire

Origination by Gilchrist Brothers Limited, Leeds

Printed in Spain by TONSA, San Sebastian.

Introduction

The *Longman World Guide to Mammals* sets out to provide a comprehensive catalogue of the staggering range of mammals. It is clearly impossible to be comprehensive at the species level, so we have looked at a higher level in the classification hierarchy and organized the book at family level, at which it is possible to be comprehensive.

The classification of animals into groups, often seems a mystifying or intimidating exercise, as does the scientific naming of animals that goes with it. But both are merely an attempt to organize the creatures into recognizable groups which show their relationships. The Latin- or Greek-based names are enormously useful because of their stability: the scientific name of a creature remains the same all over the world, but it may have dozens of common names. An animal species is a group of animals that can, at least potentially, successfully breed with one another. This is a natural grouping based on the intrinsic attributes and activities of the animals themselves. Each species is given a unique, two-part name in which the second component is specific to that species. The tiger, for example, is called *Panthera tigris*. *Panthera* is its generic name — the genus *Panthera* contains 5 different species of big cat — while *tigris* is its specific name, which refers only to the tiger. The generic and specific names are always printed in italic.

The species names are only the first two rungs of a taxonomical ladder of hierarchy. Many more rungs are required in order to encompass the patterns of similarity which exist. In ascending order, the most commonly used levels are species, genus, family, order and class.

To return to the tiger, this animal and all other cats belong to the Cat Family, Felidae, and all have certain physical and behavioural characteristics in common. This family is grouped with other families of related animals, such as dogs, viverrids and mustelids, in the Order Carnivora, and this and all other orders of mammals belong to the Class Mammalia. It is at the vital, family level that this book is comprehensive. A résumé of the major characteristics of each family is accompanied by a number of representative examples of the family, so that although each species cannot be shown, a close relative of it will be. Each order is also mentioned. Some of the larger families, such as Old World Rats and Mice, Muridae, are split up into many subfamilies, since this facilitates the description of the various groups and avoids gross generalization.

In something so complex as taxonomy, there is bound to be argument. Where there is particular controversy over the placing of a particular species, this is mentioned in the text.

The tiger is a species of cat, and everyone knows — or think they know — what a tiger looks like. However, species do not consist of identical individuals. Enormous genetic diversity exists within a species as a cursory examination of our own species, *Homo sapiens*, reveals. In many animal species it is possible to identify groupings, known as subspecies or races. These often represent geographically localized forms of a species that show characteristic differences from one another. The tiger has half a dozen geographically distinct subspecies, which vary in size and in fur coloration and patterning. Subspecies of a species can still interbreed. All subspecies, however, are human-defined and they are essentially arbitrary, unlike the species themselves, which correspond more or less precisely to actual interbreeding groups of animals.

This book is a catalogue and, as with any catalogue, part of its organization emphasizes similarities. Similar animals are grouped into their family assemblages and this format enables the shared habits and structures of related animals to be easily grasped. Ultimately though, there is another way of responding to the patterns of animal organization delineated here. Instead of emphasizing the shared characteristics, one can marvel at the almost infinite inventiveness of the life-styles and physical structure of mammals that is a joy in itself. There is space and opportunity enough in this book to savour the amazing diversity of mammals, and to find fascination in the contrasts, even between members of the same family.

Philip Whitfield

Conservation Status

Many of the world's animal species are in danger of extinction or are becoming rare, often as a direct result of human activity. All animals, particularly threatened species, are monitored by the International Union for Conservation of Nature and Natural Resources (IUCN) and listed in the Red Data books that are produced by this organization.

The status of threatened species is indicated in this book by symbols, the meaning of which is given below. This information is printed with the kind assistance of the IUCN Conservation Monitoring Centre, Cambridge, England.

ENDANGERED (E)
Species in grave danger of extinction which can only survive if effective conservation measures are adopted and the causes of their difficulties cease.

VULNERABLE (V)
Species which may become endangered if their numbers continue to diminish at the present rate.

RARE (R)
Species which are at risk because their population is already small; e.g., those with restricted distribution.

OUT OF DANGER (O)
Species which were once included in one of the above categories but, because of effective conservation measures, are now no longer in danger.

INDETERMINATE (I)
Species which are suspected of being endangered, vulnerable or rare, but about which there is insufficient information to permit classification.

NAMES
Common and scientific names are given for each species. Common names tend to vary greatly, but the most generally accepted version is used. In some instances, where there are two names of equal importance, both are given thus: Common Grey Duiker. Where there is an alternative name that is subordinate to the main choice but still interesting, this is given in brackets: *Mustela nivalis (rixosa)*

RANGE AND HABITAT
The normal range of a species is explained as fully as possible, given the limitations of space. The animal's particular habitat helps to clarify its precise occurrence within a large range. In some instances, where a species has been introduced outside its native range by man, accidentally or deliberately, this is added.

SIZE
Two separate measurements are given: head and body length, and tail length. The vast-range of sizes of the animals has meant that the drawings cannot be to a single scale.

Mammals – the peak of vertebrate adaptability

There are 4,008 species of mammal, which are the most adaptable and diverse group of vertebrates on our planet today. Whales, dolphins and seals are important members of the animal community in the seas, while forms such as otters and beavers are successful in freshwater habitats. On dry land, a huge diversity of mammal types prospers underground, on the land surface and in trees and other vegetation. Mammals have even taken to the air in the form of bats, the night-flying insectivores. In all these niche types, mammals reveal a startling variability in feeding strategies: some feed only on plant material, others on small invertebrates. Many kill and eat other vertebrates, including mammals, while some eat almost anything. To begin to understand the reasons for the adaptability and success of mammalian lines of evolution, it is necessary to look at what a mammal is, how it is constructed, how it operates physically and behaviourally and what its ancestors were like.

A mammal is a warm-blooded, four-limbed, hairy vertebrate (an animal with a backbone). Male mammals inseminate females internally, using a penis, and females, typically, retain their developing foetuses within the uterus, where the bloodstreams of mother and offspring come close together (but do not fuse) in a placenta. The time spent in the womb by the developing foetus is known as the gestation period and varies from group to group. Some mammals, such as rabbits, rodents and many carnivores, are born naked, blind and helpless; while others, such as cattle and deer, are small, but fully formed and capable, versions of the adult. Mothers produce milk for their young from skin-derived mammary glands.

The vast majority of mammals possesses these characteristics, but a few exceptions, real and apparent, must be taken into account. Whales and their relatives and manatees have only forelimbs. There is no doubt, however, from their skeletal structure, that these highly modified aquatic animals are derived from four-legged ancestors. However, the monotremes — the platypus and spiny anteaters — are highly uncharacteristic mammals. They have retained the egg-laying habits of their reptilian ancestors and do not form placentas. The pouched mammals, or marsupials, also have an unorthodox method of reproduction compared with that of the normal placenta-forming mammals. Kangaroos, wallabies and their relatives retain a thin shell around the developing young inside the mother's body, but it breaks down before the offspring emerges to crawl into the pouch and attach itself to a milk-delivering nipple. Some species do have a primitive placenta. The marsupial young is born in a far less advanced state than most mammals and finishes its development in the pouch.

Knowledge of the ancestry of mammals is based largely on the study of fossil remains of parts of skeletons and, to some extent, on surviving primitive mammalian types such as the platypus. Mammals evolved from reptiles, but because the only evidence is in fossil bones, there is no point in the geological record where we can say: Here, reptiles changed into mammals. A more realistic statement is that there is a long evolutionary history of a group of reptiles that possessed mammal-like skeletal characteristics. These reptiles were very successful in the Permian and Triassic periods but were reduced to a mere remnant of their former diversity in the late Triassic and Jurassic times. During the 90 million years between the early Jurassic and the late Cretaceous, however, this mammal-like stock of reptiles gave rise to small forms, which must be regarded as real mammals in the modern sense. From the late Cretaceous period (around 70 million years ago) until now, the mammals have continued their explosive expansion into different types of life and environment to become the dominant group of land-dwelling vertebrates and an important part of aquatic life.

Apart from reproductive sophistication, mammals are remarkable in a number of other ways. They have large, complex brains, and acute and well-integrated sensory systems. They employ a range of vocal, visual and olfactory means of communication with other species and with members of their own species — communication with the latter is important in the organization of family and social groupings. The keratinous hairs that grow out of mammalian skin insulate the body and are part of a complex of temperature-regulation mechanisms with which mammals maintain a constant, high body temperature, irrespective of external climatic conditions. Metabolic heat, especially that produced by brown fat in the body, can be used to offset heat losses and can be transferred around the body via the circulatory system, which is powered by the four-chambered heart. The body can be cooled by the evaporation of sweat secretions at the body surface. All this temperature-control "machinery" is under the control of the hypothalamus in the brain. When temperature control becomes energetically impossible, for example in low temperatures, some mammals are able to hibernate. During the hibernatory sleep, the animal's body temperature drops to close to that of the surroundings, and its heart and respiration slow dramatically so that it uses the minimum of energy. Thus it is able to survive for as long as several months on stored fat.

The astonishing diversity of present-day mammals is illustrated in the following pages, which review each living family and describe representative examples. Briefly, the range of types is as follows. The primitive montremes and marsupials, the majority found in Australia, have already been mentioned. Australia separated from the southern continents before it could receive any eutherian mammals. Thus the mammal fauna was able to radiate into a diverse range of forms, which mirrors the types of placental mammals found in the rest of the world. There are burrowing, tree-dwelling, anteating, herbivorous and predatory marsupials, and marsupial analogues exist or have existed for almost all placental mammals except bats, whales and seals.

The placental mammals break down into four broad groups. The first contains the most primitive creatures: the insectivores such as shrews, hedgehogs and moles, as well as the bats, sloths, anteaters, armadillos and pangolins and the primate group, to which human beings belong. The second group contains two orders of small, largely herbivorous mammals with gnawing teeth: the rodents, and the rabbits and related forms. In the third group is a single, dramatically specialized order: the whales, dolphins and porpoises, which, more than any other animals, have renounced their terrestrial ancestry and become as fully aquatic as any air-breathing animal can be. The fourth and final group contains a complex mixture of orders, which are believed to have some common ancestry, despite the different evolutionary paths they have taken. It includes all the carnivores (animals such as cats, dogs, mustelids and bears), the aardvark, elephants, hyraces, seals, manatees and the dugong, as well as the two major groups of hoofed mammals, the uneven-toed types, such as tapirs, horses and rhinoceroses, and the even-toed (cloven-hoofed) types such as pigs, peccaries, camels, deer, cattle, sheep and goats.

The evolutionary success of the mammals is hard to evaluate. As measured by the dominance of a particular group or its species diversity, success must be the result of an amalgam of intrinsic biological merit and chance. But no other major class of vertebrates has ever conquered such a variety of habitats so completely. The advent of modern man has probably increased the rate of mammalian extinctions in some groups, but others, such as rodents, are evolving into new ecological niches created by man's activities. At the present time, the mammals are an overwhelmingly successful group, and it may be true that its most dominant species — man — holds the future of the planet in his hands.

Spiny Anteaters, Platypus

ORDER MONOTREMATA

Two families with a combined total of only 3 living species make up this order. Although well adapted for their environments, monotremes are considered primitive mammals in that they retain some reptilian characteristics of body structure and they lay eggs. However, they also possess the essential mammalian characteristics of body hair and mammary glands. Monotremes are probably a parallel development, rather than a stage in the evolution of mammals. A problem in understanding the origins of this order is that no fossil monotremes have been found.

TACHYGLOSSIDAE:
Spiny Anteater Family

There are 2 species of spiny anteater, or echidna. Both are covered with coarse hairs, and their backs are set with spines. They have elongated, slender snouts and strong limbs and are powerful diggers. Like all anteaters, they have no teeth and very weak jaws. Termites, ants and other small arthropods are swept into the mouth by a long, sticky tongue, which can reach well beyond the tip of the snout. The insects are then crushed between the tongue and the roof of the mouth.

NAME: **Long-nosed Spiny Anteater,** *Zaglossus bruijni*
RANGE: **New Guinea**
HABITAT: **forest**
SIZE: body: **45–77 cm (17$\frac{3}{4}$–30$\frac{1}{4}$ in)**
tail: vestigial ⓥ

The long-nosed spiny anteater is larger than the short-nosed and has fewer, shorter spines scattered among its coarse hairs. The snout is two-thirds of the head length and curves slightly downward. There are five digits on both hind and forefeet, but on the former, only the three middle toes are equipped with claws. Males have a spur on each of the hind legs. This anteater is primarily a nocturnal animal that forages for its insect food on the forest floor.

The breeding female has a temporary abdominal brood patch, in which her egg is incubated and in which the newborn young remains in safety, feeding and developing. Little is known about the life of this rarely seen animal, but it is believed to have similar habits to those of the short-nosed spiny anteater.

There were once thought to be 3 species in this genus, but now all are believed to be races of this one species. The population of spiny anteaters in New Guinea is declining because of forest clearance and overhunting, and the animal is much in need of protection.

NAME: **Short-nosed Spiny Anteater,** *Tachyglossus aculeatus*
RANGE: **Australia, Tasmania, S.E. New Guinea**
HABITAT: **grassland, forest**
SIZE: body: **30–38 cm (11$\frac{3}{4}$–15 in)**
tail: vestigial

The short-nosed spiny anteater has a compact, round body, closely set with spines. At the end of its naked snout is a small, slitlike mouth, through which its long tongue is extended 15 to 18 cm (6 to 7 in) beyond the snout. The tongue is coated with sticky saliva, so any insect it touches is trapped. Spiny anteaters have no teeth but break up their food between horny ridges in the mouth; termites, ants and other small invertebrates form their main diet.

Spiny anteaters have five digits, all with strong claws, on both hind and forefeet. Males also have spurs on each hind leg which may be used in defence. They are excellent diggers and, if in danger, rapidly dig themselves into the ground. They do not live in burrows but in hollow logs or among roots and rocks. The spiny anteaters' capacity for temperature regulation is poor, and in cool weather they hibernate.

On her abdomen, the breeding female has a temporary patch, or groove, which develops at the start of the breeding season. When she has laid her leathery-shelled egg, she transfers it to the patch, where it incubates for 7 to 10 days. The egg is coated with sticky mucus to help it stay in the groove. When the young hatches, it is only 1.25 cm ($\frac{1}{2}$ in) long and helpless, so it must remain on the mother's abdomen while it develops.

The female has plenty of milk from mammary glands but no nipples, so the baby feeds by sucking special areas of abdominal skin through which the milk flows. Once the spines develop, at about 3 weeks, the young is no longer carried by its mother.

ORNITHORHYNCHIDAE:
Platypus Family

The single species of this family is an extraordinary animal in appearance but perfectly adapted for its way of life. The platypus was discovered 200 years ago, and when the first specimen arrived at London's Natural History Museum, scientists were so puzzled by it that they believed the specimen to be a fake.

Platypuses are now protected by law and are quite common in some areas.

NAME: **Platypus,** *Ornithorhynchus anatinus*
RANGE: **E. Australia, Tasmania**
HABITAT: **lakes, rivers**
SIZE: body: **46 cm (18 in)**
tail: **18 cm (7 in)**

The platypus is a semi-aquatic animal, and many of its physical characteristics are adaptations for its life as a freshwater predator. Its legs are short but powerful, and the feet are webbed, though the digits retain large claws, useful for burrowing. On the forefeet the webs extend beyond the claws and make efficient paddles; on land, the webs can be folded back to free the claws for digging. On each ankle the male platypus has a spur, connected to poison glands in the thighs; these spurs are used against an attacker or against a competing platypus but never against prey. The poison is not fatal to man, but causes intense pain.

The platypus's eye and ear openings lie in furrows, which are closed off by folds of skin when the animal is submerged. Thus, when hunting under water, the platypus relies on the sensitivity of its tactile, leathery bill to find prey. The nostrils are toward the end of the upper bill but can only function when the head is in air. Young platypuses have teeth, but adults have horny, ridged plates on both sides of the jaws for crushing prey.

The platypus feeds mainly at the bottom of the water, making dives lasting a minute or more to probe the mud with its bill for crustaceans, aquatic insects and larvae. It also feeds on frogs and other small animals and on some plants. Platypuses have huge appetites, consuming up to 1 kg (2$\frac{1}{4}$ lb) food each night.

Short burrows, dug in the river bank above the water level, are used by the platypus for refuge or during periods of cool weather. In the breeding season, however, the female digs a burrow 12 m (40 ft) or more in length, at the end of which she lays her 2 or 3 eggs on a nest of dry grass and leaves; the rubbery eggs are cemented together in a raft. She plugs the entrance to the burrow with moist plant matter, and this prevents the eggs from drying out during the 7- to 14-day incubation period. When the young hatch, they are only about 1.25 cm ($\frac{1}{2}$ in) long and helpless. Until they are about 5 months old, they feed on milk, which issues from slits in the mother's abdominal wall. Unlike spiny anteaters, they do not draw up tucks of skin into pseudonipples, but lap and suck the milk off their mother's abdominal fur.

SHORT-NOSED SPINY ANTEATER

LONG-NOSED SPINY ANTEATER

PLATYPUS

Opossums, Colocolo, Rat Opossums

ORDER MARSUPIALIA

There are some 250 species of marsupial mammal in North and South America and in Australasia east of Wallace's line (an imaginary line between Borneo and Sulawesi, and Bali and Lombok). They evolved at about the same time as the true, or placental, mammals, but were replaced by them over much of their range. Only in Australia, which separated from the southern continents of Gondwanaland after it had been populated by some early marsupials but before it received any placental mammals, has the true potential of this order been realized. There, marsupials have adapted to a variety of ecological niches and have exploited all available habitats.

The principal characteristic of marsupial mammals is their method of reproduction. Instead of retaining young inside the uterus until they are well developed, as occurs in placental mammals, gestation is extremely short — as brief as 11 days — and the young finish their development inside a pouch on the mother's belly.

DIDELPHIDAE: Opossum Family

There are over 70 species in the opossum family, found from the southern tip of South America northward to southeast Canada. They are all basically rat-shaped animals, with scaly, almost hairless tails and rather unkempt fur. Some species possess a proper pouch, while others carry their young between two flaps of skin on the belly.

Most opossums are forest dwellers, although one exceptional species has taken to an aquatic way of life. They feed on leaves, shoots, buds and seeds, and insects may also be eaten.

NAME: South American Mouse Opossum, *Marmosa mitis*
RANGE: Belize to N.W. South America; Trinidad, Tobago, Grenada
HABITAT: forest, dense scrubland
SIZE: body: 16.5–18.5 cm (6½–7¼ in)
tail: 26–28 cm (10¼–11 in)

With a long, pointed nose and huge eyes, which indicate its nocturnal way of life, the mouse opossum is more shrewlike than mouselike in appearance. It has no permanent home but constructs temporary daytime nests in tree holes or in old birds' nests. An agile climber, the mouse opossum has a long, prehensile tail, which acts as a fifth limb.

Mouse opossums breed two or three times a year, and litters of up to 10 young are born after a gestation of 17 days. The young must cling to the mother's fur, since mouse opossums do not have proper pouches.

NAME: Virginia Opossum, *Didelphis virginiana*
RANGE: S.E. Canada through USA to Central America: Nicaragua
HABITAT: forest, scrubland
SIZE: body: 32.5–50 cm (12¾–19¾ in)
tail: 25.5–53.5 cm (10–21 in)

The only marsupial found north of Mexico and the largest of the opossum family, the Virginia opossum may weigh up to 5.5 kg (12 lb). It is a successful creature, which has adapted to modern life, scavenging in refuse tips and bins. Should it be threatened by a dog, bobcat, eagle or mink, it may feign death — it "plays possum" — and the predator generally loses interest.

In Canada, opossums breed once a year, in spring, but in the south of the range, two or even three litters of 8 to 18 young may be produced in a year. Usually only about 7 of a litter survive pouch life. In the southern USA, opossums are trapped for their fur and flesh.

NAME: Water Opossum, *Chironectes minimus*
RANGE: Mexico, south through Central and South America to Argentina
HABITAT: freshwater lakes and streams
SIZE: body: 27–32.5 cm (10½–12¾ in)
tail: 36–40 cm (14¼–15¾ in)

The water opossum, or yapok, is the only marsupial to have adapted to aquatic life. It lives in bankside burrows, emerging after dusk to swim and search for fish, crustaceans and other invertebrates. Prey is eaten on the bank. The opossum's long tail aids its movements in water, but it uses the broadly webbed hind feet for propulsion.

In December, water opossums mate, and a litter of about 5 young is born some 2 weeks later. The pouch is made watertight by a strong ring of muscle, and the young remain quite dry, even when the mother is totally immersed in water. It is not known how the young obtain sufficient oxygen in their hermetically sealed environment.

NAME: Short-tailed Opossum, *Monodelphis brevicaudata*
RANGE: Surinam to N. Argentina
HABITAT: forest
SIZE: body: 11–14 cm (4¼–5½ in)
tail: 4.5–6.5 cm (1¾–2½ in)

Although the short-tailed opossum lives in forested country, it is a poor climber and tends to stay on the forest floor. During the day, it shelters in a leafy nest, which it builds in a hollow log or tree trunk. It emerges at night to feed on seeds, shoots and fruit, as well as on insects, carrion and some small rodents, which it kills with a powerful bite to the back of the head.

Litters of up to 14 young are born at any time of year and must cling to their mother's nipples and the surrounding fur, since she has no pouch. When they are older, they ride on her back.

MICROBIOTHERIIDAE: Colocolo Family

There is a single species in this family, which appears to be closely related to the opossums.

NAME: Colocolo, *Dromiciops australis*
RANGE: Chile, W. Argentina
HABITAT: forest
SIZE: body: 11–12.5 cm (4¼–5 in)
tail: 9–10 cm (3½–4 in)

An adaptable animal, the colocolo occurs in both high-altitude and lowland forest. It feeds largely on the leaves and shoots of Chilean bamboo but also eats insects and earthworms. In the colder parts of the range, colocolos hibernate in winter, but in more temperate regions, they remain active all year round.

Colocolos breed in spring, producing litters of up to 5 young. They have no true pouch, and the young must cling to the mother's fur.

CAENOLESTIDAE: Rat Opossum Family

There are only 7 known species of rat opossum, all of which live in inaccessible forest and grassland regions of the High Andes. None is common, and the family is poorly known. Rat opossums are small, shrewlike animals, with thin limbs, a long, pointed snout and slender, hairy tail. Their eyes are small, and they seem to spend much of their lives in underground burrows and on surface runways. It is likely that there are more species yet to be discovered.

NAME: Rat Opossum, *Caenolestes obscurus*
RANGE: Colombia, Venezuela
HABITAT: montane forest
SIZE: body: 9–13 cm (3½–5 in)
tail: 9–12 cm (3½–4¾ in)

The rat opossum lives on the forest floor and shelters in hollow logs or in underground chambers during the day. At dusk, it emerges to forage around in the surface litter for small invertebrate animals and fruit.

Rat opossums may be far more common than is generally thought, but the inhospitableness of their habitat makes their study and collection extremely difficult. Nothing is known of their reproductive habits.

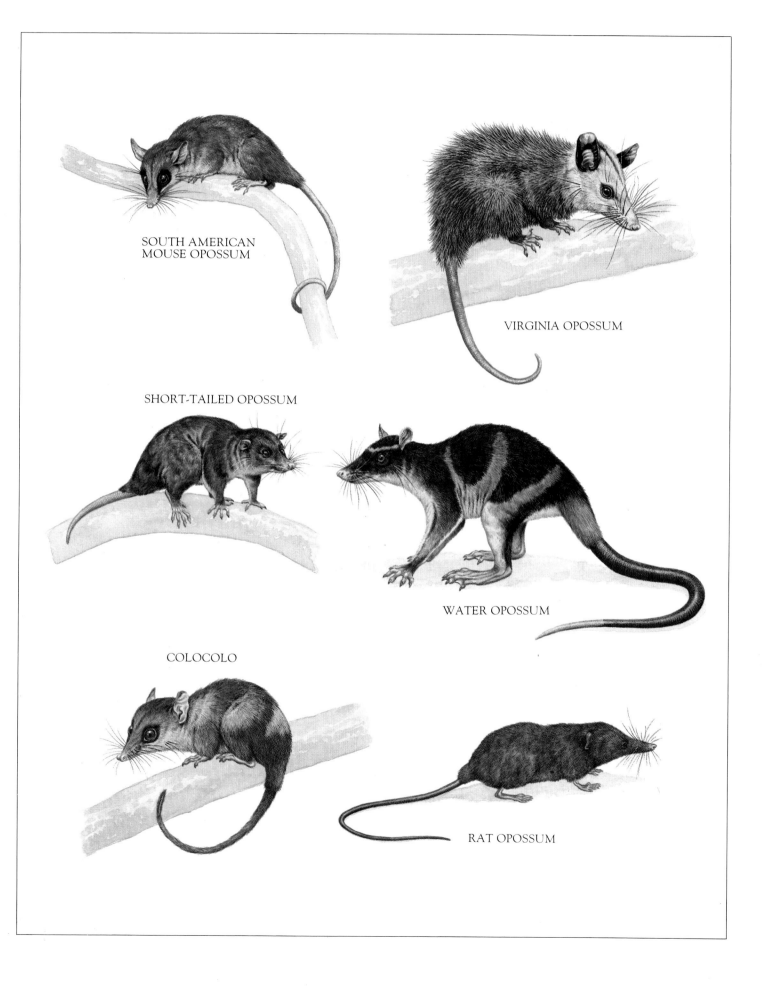

SOUTH AMERICAN
MOUSE OPOSSUM

VIRGINIA OPOSSUM

SHORT-TAILED OPOSSUM

WATER OPOSSUM

COLOCOLO

RAT OPOSSUM

Dasyurid Marsupials

NAME: Pygmy Planigale, *Planigale maculata*
RANGE: N. and N.E. Australia
HABITAT: arid bush and scrub
SIZE: body: 5–5.5 cm (2–2¼ in)
tail: 5.5 cm (2¼ in)

The pygmy planigale shelters in a burrow during the day and emerges at night to search for food. Although it is smaller than a white mouse, it feeds on large insects, such as grasshoppers, which it kills by biting off the head. Small birds are also caught, and in one night, a pygmy planigale may eat its own weight in food.

Little is known of the reproduction and social organization of these animals. They appear to be solitary and to give birth to up to 12 young between December and March.

NAME: Brown Antechinus, *Antechinus stuarti*
RANGE: E. seaboard of Australia
HABITAT: forest
SIZE: body: 9.5–11 cm (3¾–4¼ in)
tail: 10–12 cm (4–4¾ in)

A secretive, nocturnal animal, the brown antechinus is common in the forests surrounding Australia's major cities. It climbs well and probably searches for insect food up in *Eucalyptus* and *Acacia* trees.

Mating, a violent procedure in this species lasting about 5 hours, occurs in August, and a litter of 6 or 7 young is born after a gestation of 30 to 33 days. They cling to the nipples on the mother's belly until they become so large that they impede her movements; they are then left in an underground nest while she hunts. They themselves breed almost a year after birth. As a result of a hormone imbalance, males can mate only once before they die.

NAME: Mulgara, *Dasycercus cristicauda*
RANGE: central Australia
HABITAT: desert, spinifex bush
SIZE: body: 12.5–22 cm (5–8½ in)
tail: 7–13 cm (2¾–5 in)

Perfectly adapted for life in one of the world's most inhospitable environments, the mulgara never comes out of its burrow until the heat of the day has passed, and even then it tends to stay in places that have been in shadow. Its kidneys are highly developed to excrete extremely concentrated urine in order to preserve water, and it never drinks. All the mulgara's nutritional needs are provided by its prey. Insects are its staple diet, but lizards and even newborn snakes are also eaten.

Mulgaras breed from June to September, and the usual litter contains 6 or 7 young. The pouch is little more than two lateral folds of skin.

DASYURIDAE

This family of approximately 49 species contains a wide variety of marsupials, from tiny mouse-sized creatures, which live on the forest floor, to large, aggressive predators. Many zoologists regard it as the least advanced family of Australian marsupials because all members have fully separated digits — fused digits are a characteristic of advanced families of marsupials. Yet its success is undoubted, for representatives are found in all habitats, from desert to tropical rain forest.

Most dasyurids have poorly developed pouches and resort to carrying their young about underneath them, like bunches of grapes. Unlike the American opossums, which carry older young on their backs, dasyurids deposit their young in nests when they become too large to carry around.

NAME: Kowari, *Dasyuroides byrnei*
RANGE: central Australia
HABITAT: desert, grassland
SIZE: body: 16.5–18 cm (6½–7 in)
tail: 13–14 cm (5–5½ in)

A small marsupial, adapted for life in the desert heart of Australia, the kowari lives in underground burrows, singly or in small groups. At night, it emerges to search among the grass tussocks for insects, lizards or even small birds.

Kowaris breed in winter, from May to October, and produce litters of 5 or 6 young after a gestation of 32 days.

NAME: Fat-tailed Dunnart, *Sminthopsis crassicaudata*
RANGE: W. Australia, east to
W. Queensland, W. New South Wales and W. Victoria
HABITAT: woodland, heath, grassland
SIZE: body: 8–9 cm (3–3½ in)
tail: 5.5–8.5 cm (2¼–3¼ in)

The fat-tailed dunnart, like some species of dormouse and sheep, has the ability to store fat in its tail. During the wet season, when insects and spiders are abundant, the dunnart stores excess fat in special cells at the base of its tail. These reserves are used up during periods of drought, and the tail gradually slims down. If the drought persists longer than is usual, the fat-tailed dunnart is able to enter a state of torpor, in which its body temperature falls and its meagre food reserves last longer because its energy needs are less.

Young dunnarts start to breed when they are about 4 months old and produce litters at intervals of about 12 weeks. Their courtship is aggressive, and males indulge in vicious fights for females on heat.

NAME: Quoll, *Dasyurus viverrinus*
RANGE: E. Australia
HABITAT: forest
SIZE: body: 35–45 cm (13¾–17¾ in)
tail: 21–30 cm (8¼–11¾ in)

The quoll is one of 4 species of cat-sized, predatory dasyurid, specialized for life as carnivores. At one time ruthlessly destroyed by poultry keepers, quolls are now known to do as much good as they do harm, by killing rodents, rabbits and invertebrate pests and helping to keep the ecological balance. Quolls live in rock piles or in hollow logs and emerge only at night to search for food.

The breeding season lasts from May to August, and up to 18 young are born after a gestation of about 14 days. The young stay in a well-developed pouch, which they leave before they are weaned at 4½ months. During the latter part of the time for which the young are dependent on their mother, they clamber all over her, clinging to her fur as she feeds. The quoll is one of the few marsupials known in which the litter size at birth is far higher than the number that can be supported — a phenomenon known as superfoetation. Within 48 hours of birth, 10 or more of the young will die, leaving the mother with about 8, which she can support.

NAME: Tasmanian Devil, *Sarcophilus harrisi*
RANGE: Tasmania
HABITAT: dry forest
SIZE: body: 52.5–80 cm (20½–31½ in)
tail: 23–30 cm (9–11¾ in)

This powerfully built marsupial has the reputation of being a vicious and ruthless killer of sheep; in fact, it is more of a scavenger of dead sheep than a killer of live ones. Its massive head and enormous jaws, resembling those of the hyena, allow it to smash through the bones of a carcass. Tasmanian devils live in dens in rock piles and under tree stumps and are normally nocturnal. Sometimes they emerge during the day to lie in the sun. They are more common now than they were a century or more ago, when great numbers were hunted and killed by farmers. Before the thylacine, or Tasmanian wolf, became extinct, the devils lived almost exclusively on the discarded carcasses left by that predator. Hunting may be a new way of life.

Tasmanian devils live for about 8 years and breed in the second year. A litter of about 4 young is born in early winter (May or June), and they quickly crawl into the mother's well-developed pouch, where they remain for 15 weeks. They feed on her milk until they are about 20 weeks old.

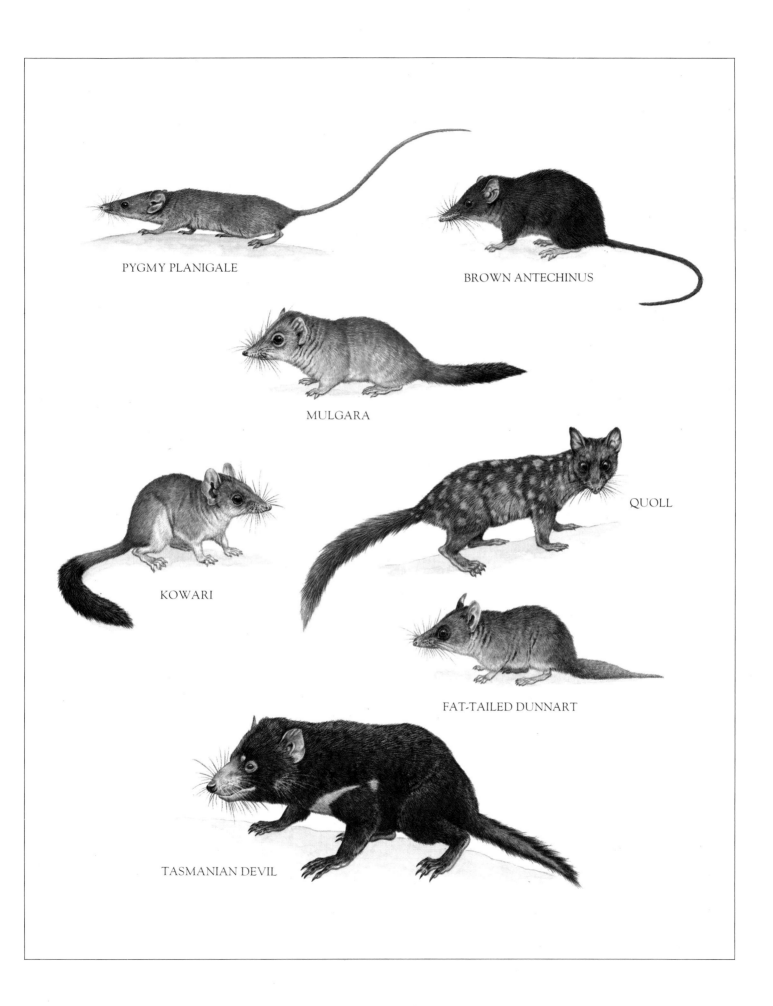

PYGMY PLANIGALE

BROWN ANTECHINUS

MULGARA

KOWARI

QUOLL

FAT-TAILED DUNNART

TASMANIAN DEVIL

Numbat, Koala and other marsupials

MYRMECOBIIDAE:
Numbat Family
The single species in this Australian family is a small marsupial, which is adapted to the niche filled by anteaters in other parts of the world. It was formerly called the banded anteater and feeds in a similar manner to anteaters.

NAME: Numbat, *Myrmecobius fasciatus*
RANGE: S.W. Australia
HABITAT: forest
SIZE: body: 17.5–27.5 cm (6¾–10¾ in)
   tail: 13–17 cm (5–6¾ in) (E)

The numbat feeds principally on termites, although ants and some other small invertebrates are also eaten. Like the anteaters, it has a sticky tongue, about 10 cm (4 in) in length, with which it sweeps insects into its mouth. Its teeth are poorly developed and capable of doing little more than crushing and stilling the wriggling insects. A captive numbat was observed to eat between 10 thousand and 20 thousand termites daily.

Between January and May, the female numbat produces a litter of 4 young. She has no pouch, so the young must cling to her nipples, and she drags them around as she searches for food.

NOTORYCTIDAE:
Marsupial Mole Family
The single species in this Australian marsupial family has a clear resemblance to the placental moles and leads a similar existence.

NAME: Marsupial Mole, *Notoryctes typhlops*
RANGE: S.W. Australia
HABITAT: desert
SIZE: body: 9–18 cm (3½–7 in)
   tail: 1.25–2.5 cm (½–1 in)

The marsupial mole is superbly adapted to a burrowing way of life, having large, shovel-like forepaws, no eyes and silky fur, which helps it move easily through the sand. Unlike true moles, marsupial moles do not dig permanent burrows, for as they travel through the soft sand, the tunnel falls in immediately behind them. They feed on earthworms and other underground invertebrates, such as beetle larvae, and come to the surface quite frequently, although they move awkwardly on land.

Nothing is known of their reproductive habits, but since the pouch contains only two nipples, they presumably bear only 2 young at a time. The testes of the male marsupial mole never descend into a scrotum but remain in the body, close to the kidneys.

PHASCOLARCTIDAE:
Koala Family
The single species in this family is one of Australia's best-known marsupials and also one of the most specialized for life in the trees.

During the first three decades of this century, koalas were hunted for their skins, and in 1924 alone, over 2 million were exported. Today the koala is no longer threatened, and following strict conservation measures, populations are increasing throughout its range.

NAME: Koala, *Phascolarctos cinereus*
RANGE: E. Australia
HABITAT: dry forest
SIZE: body: 60–85 cm (23½–33½ in)
   tail: vestigial

The tree-dwelling koala seldom comes down to the ground and then only to pass from one tree to another. Its diet is limited, consisting of the leaves and shoots of a few species of *Eucalyptus*. An adult will consume just over 1 kg (2¼ lb) leaves a day.

Koalas live singly or in small groups, consisting of a single male with a harem of females. They breed in summer, and the female produces a single young after a gestation of about a month. The tiny koala enters the pouch, which opens backward, and remains there for 5 or 6 months. After this period of pouch life, it rides on its mother's back. After weaning, the mother feeds her young on semi-digested leaves. Many koalas suffer from the infectious fungal disease cryptococcosis, which causes lesions or abscesses in the lungs, joints and brain, and which they can transfer to man, often with fatal results. It is thought that the source of the fungal infection is soil, which koalas regularly eat, apparently as an aid to digestion.

VOMBATIDAE: Wombat Family
The 3 species of wombat all live in Australia, 1 in Tasmania also. They are strong, powerfully built marsupials, which superficially resemble badgers. With their long, bearlike claws, they excavate vast burrow systems and tear up underground roots and tubers for food. They are strictly vegetarian and often raid cultivated fields to feed on the soft, developing ears of corn.

Wombats share with koalas the curious phenomenon of a backward-opening pouch.

NAME: Wombat, *Vombatus ursinus*
RANGE: E. Australia, Tasmania
HABITAT: forest, scrub
SIZE: body: 70 cm–1.2 m (27½ in–4 ft)
   tail: vestigial

The wombat is a common forest animal along Australia's eastern seaboard and is often found at high altitudes in the Snowy Mountains. Its burrows may stretch more than 13 m (42 ft) from the entrance and go down more than 2 m (6½ ft). It is not known whether wombats are gregarious below ground, but the number of burrows occurring together suggests that they are. Above ground, the wombat follows regularly used pathways through the forest.

The female gives birth in late autumn, usually to a single young, which remains in the pouch, where there are two nipples, for about 3 months. Once out of the pouch, it forages with its mother for several months before living independently. It is not unusual for wombats to live for more than 20 years.

TARSIPEDIDAE:
Honey Possum Family
The only member of its family, the honey possum is a zoological enigma because it has no obvious close relatives. In general appearance it resembles the other small possums, but its feet are quite different. The second and third digits on each hind foot are totally fused, but the two tiny claws at the tip of the fused digit are evidence of its original state.

NAME: Honey Possum, *Tarsipes spencerae*
RANGE: S.W. Australia
HABITAT: heathland with bushes and trees
SIZE: body: 7–8.5 cm (2¾–3¼ in)
   tail: 9–10 cm (3½–4 in)

The honey possum feeds on pollen and nectar of the large flowers of *Banksia*, a flowering shrub found in the southwestern corner of Australia. It often hangs upside-down while feeding, using its prehensile tail as a fifth limb. The tongue, with its bristly tip, resembles that of a hummingbird or nectar-feeding bat and can be extended about 2.5 cm (1 in) beyond the tip of the nose. The teeth are poorly developed, but tough ridges on the palate are used to scrape the nectar and pollen off the tongue. Occasionally the honey possum will eat small insects, discarding their tough wings.

In midwinter, the honey possums mate, and females give birth to 2 young after a gestation of about 4 weeks. The young remain in the pouch until they are about 4 months old.

NUMBAT

MARSUPIAL MOLE

KOALA

WOMBAT

HONEY POSSUM

Bandicoots, Phalangers, Pygmy Possums, Ringtails

PERAMELIDAE: Bandicoot Family

There are about 17 species of bandicoot, widely distributed over Australia and New Guinea in a range of habitats, from desert to rain forest. All have long, powerful claws on their forefeet and are good diggers.

NAME: **Gunn's Bandicoot**, *Perameles gunni*
RANGE: **Australia: S. Victoria; Tasmania**
HABITAT: **woodland, heathland**
SIZE: **body: 25–40 cm (9¾–15¾ in)**
    tail: 7.5–18 cm (3–7in)

Like most bandicoots, Gunn's bandicoot is a very aggressive, belligerent creature, which lives alone. Males occupy large territories and consort with females for only as long as is necessary for mating. Primarily a nocturnal animal, it emerges from its nest at dusk to forage for earthworms and other small invertebrates. Probing deep into the soil with its long nose, the bandicoot digs eagerly when a food item is located.

Although the female bandicoot has 8 nipples, she seldom produces more than 4 or 5 young. The gestation period of 11 days is one of the shortest of any mammal and is followed by 8 weeks in the pouch.

NAME: **Brown Bandicoot**, *Isoodon obesulus*
RANGE: **S. Australia, Queensland; Tasmania**
HABITAT: **scrub, forest**
SIZE: **body: 30–35 cm (11¾–13¾ in)**
    tail: 7.5–18 cm (3–7 in)

The brown bandicoot occurs in areas of dense ground cover and can survive in quite dry places, as long as it has somewhere to hide from eagles and foxes. It appears to locate prey, such as earthworms and beetle larvae, by scent and leaves small conical marks with its long nose as it forages about. Ant larvae and subterranean fungi are also eaten, as well as scorpions, which it nips the tails off before consuming them.

Reproduction is closely linked to the local rainfall pattern, and many brown bandicoots breed all year round. A litter of up to 5 young is born after an 11-day gestation and is weaned at 2 months.

THYLACOMYIDAE: Rabbit-bandicoot Family

Only a single species survives of this family, the lesser rabbit-bandicoot, *Macrotis leucura*, having become extinct within the last century. The remaining species is extremely rare, and stringent conservation measures are in force.

NAME: **Rabbit-bandicoot**, *Macrotis lagotis*
RANGE: **central and N.W. Australia**
HABITAT: **woodland, arid scrub**
SIZE: **body: 20–55 cm (7¾–21½ in)**
    tail: 11.5–27.5 cm (4½–10¾ in)   Ⓔ

The rabbit-bandicoot lives alone in a burrow system, which it digs with its powerful forepaws. The burrow descends 2 m (6½ ft) or more, and the bandicoot is thus protected from the heat of the day. After dark, when the air is cool, the rabbit-bandicoot emerges to feed on termites and beetle larvae, which it digs from the roots of wattle trees. It may also eat some of the fungi that grow around the roots.

Courtship is brief and aggressive, as with most bandicoots. The 3 young are born between March and May and spend 8 weeks in the pouch.

PHALANGERIDAE: Phalanger Family

The 12 species in this family occur in Australia, New Guinea and Sulawesi. The totally Australian species are known as possums, those from the islands are called phalangers. All species are about the size of domestic cats and are tree-dwelling and nocturnal. They have a prehensile tail, used as a fifth limb when climbing. Leaves, gum from wattle trees and insects are their main foods, but small birds and lizards may also be eaten if available.

NAME: **Brush-tailed Possum**, *Trichosurus vulpecula*
RANGE: **Australia, Tasmania; introduced into New Zealand**
HABITAT: **forest, woodland**
SIZE: **body: 32–58 cm (12½–22¾ in)**
    tail: 24–35 cm (9½–13¾ in)

The brush-tailed possum is the only native marsupial to have benefited from man's encroachment on virgin land, for it has become adapted to living on man's buildings and to feeding on refuse. In natural conditions, it eats young shoots, flowers, leaves and fruit, with some insects and young birds.

Breeding occurs once or twice a year, normally with 1 young in each litter. The gestation period is 17 days, and the young possum then stays in the pouch for about 5 months.

BURRAMYIDAE: Pygmy Possum Family

There are 7 species of these tiny, mouse-sized possums in Australia and New Guinea. All are arboreal, often living in the tallest trees.

NAME: **Pygmy Glider**, *Acrobates pygmaeus*
RANGE: **E. and S.E. Australia**
HABITAT: **dry forest**
SIZE: **body: 6–8.5 cm (2¼–3¼ in)**
    tail: 6.5–8.5 cm (2½–3¼ in)

The pygmy glider lives at the top of tall forest trees and has evolved a flight membrane, which enables it to glide from one tree to another. As it leaps, the flaps of skin between wrists and heels are stretched out, and the featherlike tail provides directional stability. The tips of its digits are broad and deeply furrowed, to help it cling to surfaces on landing. It feeds on insects, gum, nectar and pollen.

The litter of 2 to 4 young is born in July or August.

PETAURIDAE: Ringtail Family

There are about 22 species of ringtail, in Australia and New Guinea. They are tree-dwellers and make good use of the prehensile, grasping, abilities of their long tails.

NAME: **Sugar Glider**, *Petaurus breviceps*
RANGE: **E. and N. Australia; New Guinea**
HABITAT: **woodland**
SIZE: **body: 11–15 cm (4¼–6 in)**
    tail: 12–18 cm (4¾–7 in)

Its habit of feeding on the sugary sap that oozes from any wound inflicted on the bark of wattle and gum trees is the origin of this animal's common name. It regularly damages the bark of such trees and returns for several days running to lick up the exudates. Sugar gliders are sociable animals, which live in groups of up to 20 in holes in trees. With their gliding membranes outstretched, they can glide for up to 55 m (180 ft), flipping upward at the end of the descent to land, head upward, on a tree trunk.

A litter of 2 or 3 young is born after a 21-day gestation. They leave the pouch when 3 or 4 months old.

NAME: **Greater Glider**, *Schoinobates volans*
RANGE: **E. Australia**
HABITAT: **forest**
SIZE: **body: 30–48 cm (11¾–18¾ in)**
    tail: 45–55 cm (17¾–21½ in)

The largest of Australia's gliding marsupials, the greater glider can weigh up to 1.4 kg (3 lb). It lives in holes, high up in the trees, and feeds exclusively on leaves and shoots. Greater gliders can glide 100 m (330 ft) or more, from tree to tree, moving their long bushy tails to steer themselves.

In midwinter, the female gives birth to 1 young, which then spends 4 months in the pouch.

GUNN'S BANDICOOT

BROWN BANDICOOT

RABBIT-BANDICOOT

BRUSH-TAILED POSSUM

PYGMY GLIDER

SUGAR GLIDER

GREATER GLIDER

Kangaroos I

NAME: Musk Rat-kangaroo,
Hypsiprymnodon moschatus
RANGE: **Australia: N.E. Queensland**
HABITAT: **rain forest**
SIZE: **body: 23.5–33.5 cm (9¼–13¼ in)**
tail: 13–17 cm (5–6¾ in)

The musk rat-kangaroo is a unique member of its family in two ways: first, it has a well-formed first digit, or thumb, on each hind foot, and second, it is the only kangaroo regularly to give birth to twins. It is also unusual in other respects; for example, it often moves around on four legs in the manner of a rabbit. Both male and female produce a pungent musky odour, but the reason for this is not known. Their diet includes a wide range of plant matter, from palm berries to root tubers, and they also eat insects and earthworms.

Little is known of the social habits of the musk rat-kangaroos, but they seem to move singly or in pairs. The young are usually born in the rainy season (February to May), but this can vary.

NAME: Potoroo, *Potorous tridactylus*
RANGE: **E. Australia, S.W. corner of**
W. Australia; Tasmania
HABITAT: **low, thick, damp scrub**
SIZE: **body: 30–40 cm (11¾–15¾ in)**
tail: 15–24 cm (6–9½ in)

Despite its small size and rabbitlike appearance, the potoroo has all the characteristics of reproduction and the gait of the larger kangaroos. Although it may occasionally move on all fours, it usually bounds along on its strong hind legs, covering 30 to 45 cm (11¾ to 17¾ in) with each hop. It is nocturnal, emerging at dusk to forage for plants, roots, fungi and insects.

Potoroos breed at any time of year, and the single young spends 17 weeks in the pouch.

NAME: Rufous Rat-kangaroo,
Aepyprymnus rufescens
RANGE: **E. Australia: Queensland to**
N.E. Victoria
HABITAT: **grassland, woodland**
SIZE: **body: 38–52 cm (15–20½ in)**
tail: 35–40 cm (13¾–15¾ in)

The largest of the rat-kangaroos, the rufous rat-kangaroo builds a grassy nest in which to spend the heat of the day. It has little fear of man and will enter foresters' camps and even feed from the hand. This lack of fear makes the animals vulnerable to attack by introduced dogs and red foxes, and while their populations are in no immediate danger, the future survival of these little kangaroos is a cause for concern. They breed slowly — a maximum of 2 young a year — but nothing more is known of their breeding habits.

MACROPODIDAE:
Kangaroo Family

There are about 57 species of kangaroo, and the family is regarded by most authorities as the most advanced of all the 15 surviving families of marsupials, for both the teeth and feet are greatly modified. The hind feet are extremely large — the origin of the family's scientific name — and the thumb is totally absent; digits 2 and 3 are slender and bound together by skin; digit 4 is massive and armed with a long, tough claw, and digit 5 is only a little smaller. A kangaroo has fewer teeth than other marsupials, and these are high-crowned and deeply folded. They are similar to those of the placental mammals of the family Bovidae (sheep and cattle). Most kangaroos eat only plant matter, but some of the small species also feed on invertebrates.

The main characteristic of kangaroos is their method of moving on two legs. They make a series of great bounds, during which the long hind legs propel the body forward with considerable force. The long, powerful tail acts as a counterbalance, providing the necessary stability when they land. Although hopping may seem an awkward form of locomotion, at speeds of more than 20 km/h (12 mph), it is more efficient in terms of energy use than quadrupedal running. Kangaroos often have to travel long distances to find food, so speed and efficiency are important, and the selective pressures for fast movement are great.

Members of the kangaroo family occur in Australia and New Guinea and also in Tasmania and the Bismarck Islands. Some are forest-dwellers, but others live in extremely hot, arid areas, and during the worst of the day's heat, they seek the shade of a rocky outcrop and regularly salivate over their upper arms to aid evaporative cooling. If intense drought continues for so long that the females can no longer make milk for their young, the sucklings are expelled from the pouch and perish. When the rains arrive and food supplies return, a reserve embryo, which has been held in a suspended state of development, is implanted in the uterus, and a new pregnancy begins without the female having to mate again. In this way, kangaroos cope successfully with their harsh environment. Most kangaroos produce only 1 young at a time. Males and females generally look alike, except for the pouch structure of the female; in some of the larger species, the male is slightly larger than his mate.

NAME: Red-legged Pademelon,
Thylogale stigmatica
RANGE: **Australia: E. Queensland,**
E. New South Wales
HABITAT: **wet forest**
SIZE: **body: 53–62 cm (20¾–24½ in)**
tail: 32–45 cm (12½–17¾ in)

The red-legged is one of 3 species of pademelon, all of which are solidly built forest-dwellers, slightly heavier in the hindquarters than the graceful kangaroos of the open plains. They are adaptable creatures, which can occupy a variety of habitats, provided there is plenty of cover. At dusk, they emerge to forage for leaves, buds, shoots and fruit. Pademelons sometimes occur in herds, but there are also solitary individuals.

A single young is the rule, although twins have been recorded.

NAME: Spectacled Hare-wallaby,
Lagorchestes conspicillatus
RANGE: **N. and central Australia**
HABITAT: **desert grassland**
SIZE: **body: 40–50 cm (15¾–19¾ in)**
tail: 35–45 cm (13¾–17¾ in)

There are 4 species of small hare-wallaby living in the arid and desert grasslands of Australia. They make rough, grassy nests among the tough spinifex vegetation, and if disturbed, behave much like hares, leaping off in a zigzag manner. The spectacled hare-wallaby leads a solitary life, only coming together with another for mating. This isolation is necessary because of the extreme difficulty of eking out an existence in the inhospitable environment.

A single young is produced at any time of year and is itself sexually mature at about a year old.

NAME: Ring-tailed Rock Wallaby,
Petrogale xanthopus
RANGE: **central and E. Australia**
HABITAT: **rocky outcrops, boulder piles**
SIZE: **body: 50–80 cm (19¾–31½ in)**
tail: 40–70 cm (15¾–27½ in) ⓥ

Long exploited for its high-quality fur, the ring-tailed rock wallaby is now found in only a few isolated areas. It is the most handsomely marked of the rock wallabies, all of which live in the most inhospitable regions of the outback. Like all rock wallabies, its feet, with broad, soft pads and strong claws, are adapted for scrambling around on the rocks, and it is remarkably agile. Its long tail does not have a thickened base and is not used as a prop as in true kangaroos. Rock wallabies feed on whatever plant material they can find.

Breeding takes place throughout the year, but if drought conditions persist too long, the rock wallabies sacrifice any young in their pouches.

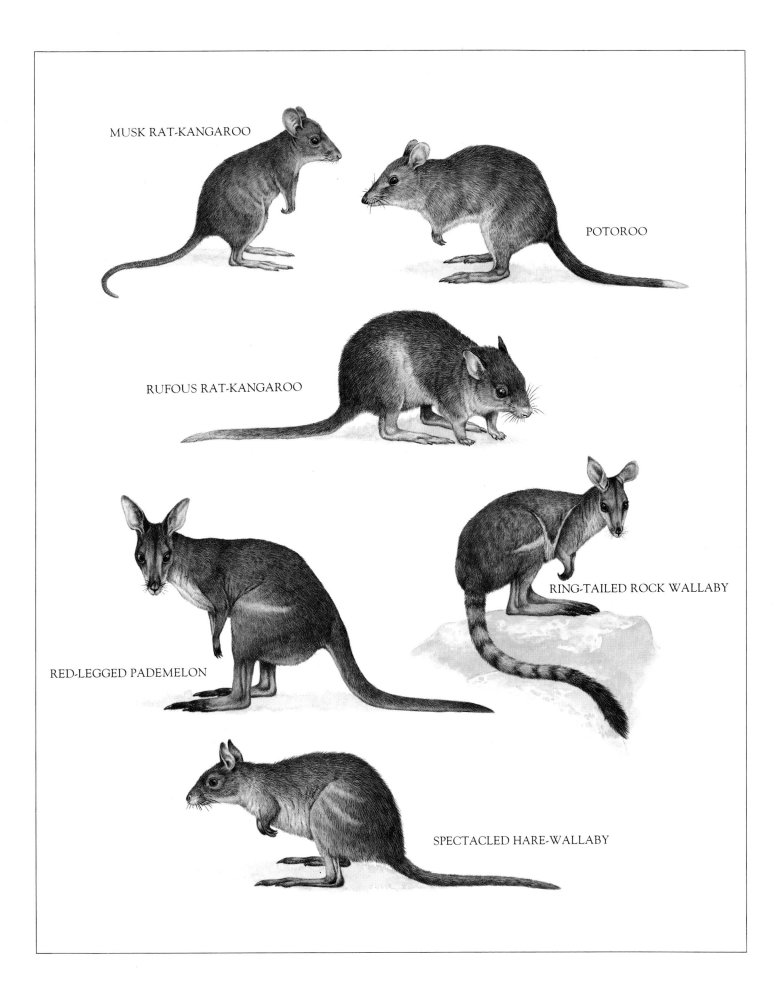

MUSK RAT-KANGAROO

POTOROO

RUFOUS RAT-KANGAROO

RING-TAILED ROCK WALLABY

RED-LEGGED PADEMELON

SPECTACLED HARE-WALLABY

Kangaroos 2

NAME: **Quokka,** *Setonyx brachyurus*
RANGE: **S.W. Australia**
HABITAT: **dense vegetation**
SIZE: **body: 47.5–60 cm (18¾–23½ in)**
    tail: 25–35 cm (9¾–13¾ in)

The quokka was once widespread over the southwest of Australia, but shooting for sport quickly reduced the population to a low level. Today the quokka occurs in only a few swampy valleys in the Darling Range, near Perth, but it is abundant on Rottnest and Bald Islands, just off the coast. Quokkas have a good nose for fresh water, and on Rottnest Island travel as far as 2.5 km (1½ mls) to find it. They have learned to scavenge on refuse tips for food in times of drought and sparse plant supply, and are also able to supplement their protein intake by utilizing urea, a urinary waste product. In this respect, they resemble desert mammals.

The female quokka gives birth to 1 tadpole-sized young after a gestation of 17 days and mates again the following day. The embryo resulting from this second mating is held free in the uterus and will implant and start development only after the earlier offspring has left the pouch. Quokkas first breed when about 2 years old.

It is to be hoped that the development of Rottnest Island as an important recreational site for the people of Perth can be achieved without destroying the habitat of this intelligent little wallaby.

NAME: **Lumholtz's Tree Kangaroo,**
    Dendrolagus lumholtzi
RANGE: **Australia: N.E. Queensland**
HABITAT: **forest**
SIZE: **body: 52–80 cm (20½–31½ in)**
    tail: 42–93 cm (16½–36½ in)

Only 3 species of kangaroo, 2 of which occur in Australia and a third in New Guinea, have taken to life in the trees. Unlike most arboreal mammals, they have few adaptations for this specialized way of life; their hind feet are singularly undeveloped for climbing, but their front feet are better adapted for grasping than those of other kangaroos. They are, however, remarkably agile in trees, and their long tails, although not prehensile, give considerable stability. Grass, leaves and fruit are their main foods, and they frequently come backward down trees to the ground to graze.

Small groups live together and sleep in the same tree. Little is known of their breeding habits other than that they produce 1 young at any time of year. Tree kangaroos are not rare, but their secretive way of life and the denseness of their forest habitat make them hard to observe.

NAME: **Red Kangaroo,** *Macropus rufus*
RANGE: **central Australia**
HABITAT: **grassy arid plains**
SIZE: **body: 1–1.6 m (3¼–5¼ ft)**
    tail: 90 cm–1.1 m (35½ in–3½ ft)

The red kangaroo is the largest living marsupial; an old male can attain a weight of about 70 kg (154 lb). Only the male sports the deep russet-red coat, the female has a bluish-grey coat and is often referred to as the "blue flier". Red kangaroos live on the arid grassland of the desert in small herds, which consist of an adult male and several females. During the heat of the day, they shelter by rocky outcrops or in the shade of stunted trees and emerge during the evening to feed and drink. Although, weight for weight, kangaroos eat as much as sheep, they convert their food more efficiently — a kangaroo is 52 per cent meat, a sheep, 27 per cent.

Breeding occurs throughout the year, and the gestation lasts 30 to 40 days. A few hours before the birth, the mother starts to clean and nuzzle at her pouch. Sitting back, with her tail bent forward between her hind legs, and holding the pouch open with her forelimbs, she licks it inside and out, often continuing until the moment of birth. The newborn offspring weighs only 0.75 gm (1/40 oz) and is 1/30,000 of its mother's weight. Although so tiny, it has well-developed front claws, which it uses to clamber up from the mother's genital opening to the pouch. Once safely in the pouch, it takes one of the nipples into its mouth. The nipple grows with the baby and is about 10 cm (4 in) long at the time of weaning. The quality of the mother's milk changes as the baby grows, becoming richer and more fatty during the later phase of lactation.

The young kangaroo spends about 240 days in the pouch and then accompanies its mother for a further 120 days, occasionally putting its head into her pouch to suck, even though it may contain a younger sibling. During intense drought conditions, most pouch young (known as "joeys") will die, but the loss is made up from a stock of reserve embryos carried in the mother's uterus.

Until recently, red kangaroos were hunted on a massive scale, but this is now subject to strict controls.

NAME: **New Guinea Forest Wallaby,**
    Dorcopsis veterum
RANGE: **New Guinea**
HABITAT: **lowland rain forest**
SIZE: **body: 49–80 cm (19¼–31½ in)**
    tail: 30–55 cm (11¾–21½ in)

In most respects, the New Guinea forest wallaby appears to resemble the Australian pademelons, but very little is known of this retiring species. One difference is that the tip of the tail is armed with a few broad, tough scales, the function of which remains a mystery.

Whenever possible, forest wallabies feed on grass, but they also eat a variety of other plant foods. They are seldom observed by day and may lie up in leafy nests until dusk. Nothing is known of their breeding habits, except that only 1 young is carried at a time.

NAME: **Bridled Nail-tailed Wallaby,**
    Onychogalea fraenata
RANGE: **Australia: central Queensland**
HABITAT: **thick scrub**
SIZE: **body: 45–67 cm (17¾–26¼ in)**
    tail: 33–66 cm (13–26 in) Ⓔ

The 3 species of nail-tailed wallaby derive their name from a scale, like a small finger-nail, hidden in the thick hair at the tip of the long, thin tail. Its function is unknown. In the middle of the nineteenth century, the bridled nail-tailed wallaby was abundant over much of eastern and southeastern Australia, but in this century, it was unrecorded for several decades until a population was discovered in central Queensland, in 1974. Competition by rabbits for food, predation by red foxes and hunting are major causes of its decline. Conservation measures are in hand and will need to be applied for many years if this species is to survive in the wild.

Thick scrub is used for food and cover by nail-tailed wallabies. Nothing is known of their breeding habits except that they produce 1 young at a time.

NAME: **Swamp Wallaby,** *Wallabia bicolor*
RANGE: **E. and S.E. Australia**
HABITAT: **dense thickets, rocky gullies**
SIZE: **body: 45–90 cm (17¾–35½ in)**
    tail: 33–60 cm (13–23½ in)

Swamp wallabies occur in small herds but are often hard to see because of their habit of lying down when danger threatens. Only when the danger is imminent do the wallabies break cover and scatter in different directions, with explosive speed. Their diet is varied, and they readily switch from one plant species to another. This flexibility means that they can become pests of agricultural crops, and some measure of control is often necessary locally.

Breeding takes place throughout the year, and the female produces 1 young, which stays in her pouch for about 300 days. It continues to feed for 60 days after leaving the pouch.

QUOKKA

RED KANGAROO

LUMHOLTZ'S TREE KANGAROO

NEW GUINEA FOREST WALLABY

BRIDLED NAIL-TAILED WALLABY

SWAMP WALLABY

American Anteaters, Sloths

ORDER EDENTATA

This order includes 3 families of mammals, all of which have gone along the evolutionary track of tooth reduction or loss in connection with their specialized diet of insects such as ants and termites. The families are the anteaters, the sloths and the armadillos. The name of the order means, appropriately, "the toothless ones".

MYRMECOPHAGIDAE:
American Anteater Family

There are 4 species of American anteater, found in Mexico and Central and South America as far south as northern Argentina. They normally inhabit tropical forests but also occur in grassland. All forms have extremely elongate snouts and no teeth. Their tongues are long and covered with a sticky salivary secretion, which enables them to trap insects easily. The anteaters break into ant or termite nests by means of their powerful clawed forefeet, each of which has an enlarged third digit. The largest species, the giant anteater, is ground-dwelling, while the other, smaller species are essentially arboreal.

NAME: **Giant Anteater, *Myrmecophaga tridactyla***
RANGE: **Central America, South America to N. Argentina**
HABITAT: **forest, savanna**
SIZE: **body: 1–1.2 m (3¼–4 ft)**
   tail: 65–90 cm (25½–35½ in) Ⓥ

The largest of its family, the remarkable giant anteater has a long snout, a distinctive black stripe across its body and a bushy, long-haired tail. Using its powerful foreclaws, the anteater demolishes ant or termite mounds and feeds on huge quantities of the insects and their eggs and larvae. Its long tongue can be extended as much as 61 cm (24 in) and is covered with sticky saliva, so insects adhere to it. As it wanders in search of food supplies, the anteater walks on its knuckles, thus protecting the sharp foreclaws. Unlike other anteaters, it does not climb trees, although it readily enters water and can swim. Except for females with young, giant anteaters usually live alone. In areas far from habitation, the giant anteater is active in the daytime, but nearer habitation, it is only active at night.

The female produces 1 young after a gestation of about 190 days. The offspring is carried on the mother's back and stays with her until her next pregnancy is well advanced.

NAME: **Tree Anteater, *Tamandua mexicana***
RANGE: **S. Mexico, through Central and South America to Bolivia and Brazil**
HABITAT: **forest**
SIZE: **body: 54–58 cm (21¼–22¾ in)**
   tail: 54.5–55.5 cm (21½–21¾ in)

A tree-dwelling animal, this anteater is smaller than its giant relative and has a prehensile tail, which it uses as a fifth limb; the underside of the tail is naked, which improves its grip. On the ground, the tree anteater moves slowly and clumsily. It is active mainly at night, when it breaks open the nests of tree-living ants and termites and feeds on the insects. Like all anteaters, it has a long, protrusible tongue, which is covered with sticky saliva, enabling it to trap its prey. If attacked, it will strike out at its adversary with its powerful foreclaws.

The female gives birth to 1 young; the length of the gestation is unknown. The youngster is carried on its mother's back but may be set down on a branch while she feeds.

NAME: **Pygmy Anteater, *Cyclopes didactylus***
RANGE: **S. Mexico, through Central and South America to Bolivia and Brazil**
HABITAT: **forest**
SIZE: **body: 15–18 cm (6–7 in)**
   tail: 18–20 cm (7–7¾ in)

An arboreal animal, the pygmy anteater climbs with agility, using its prehensile tail and long feet with special joints, which enable the claws to be turned back under the foot when grasping branches so that they do not become blunted. This anteater rarely comes down to the ground but sleeps in a hollow tree or on a branch during the day and is active at night, searching for ants and termites. Like its relatives, it uses its sharp, powerful foreclaws for breaking into ant and termite nests and its long, sticky tongue for gobbling up the insects. Pygmy anteaters are themselves often attacked by large birds of prey.

Little is known about the pygmy anteaters' breeding habits. The female produces 1 young, which both parents feed on regurgitated insects.

BRADYPODIDAE: Sloth Family

The 5 species of sloth all live in tropical forests of Central and South America. These highly adapted mammals are so specialized for life in the trees that they are unable to walk normally on the ground. However, most of their life is spent among the branches, where they hang upside-down by means of their curved, hooklike claws. They feed on plant material.

NAME: **Three-toed Sloth, *Bradypus tridactylus***
RANGE: **Honduras, south to Brazil, Paraguay and N. Argentina**
HABITAT: **forest**
SIZE: **body: 50–60 cm (19¾–23½ in)**
   tail: 6.5–7 cm (2½–2¾ in)

The three-toed sloth is so well adapted to living upside-down, hanging from branches by its hooklike claws, that its hair grows in the opposite direction from that of other mammals and points downward when the sloth is hanging in its normal position. Each of the outer hairs is grooved, and microscopic green algae grow in these grooves, giving the sloth a greenish tinge, which helps to camouflage it amid the foliage. This sloth's head is short and broad and its neck particularly mobile — it has two more neck vertebrae than normal for mammals — allowing great flexibility of head movement. It climbs slowly, moving one limb at a time, and swims quite well but can only move on the ground by dragging its prone body forward with its hooked limbs. Sloths rarely come to the ground, but the three-toed sloth does descend about once a week to defecate in a hole it makes with its tail. Leaves and tender buds, particularly those of *Cecropia* trees, are the three-toed sloth's main foods. Its sight and hearing are poorly developed, and it depends on smell and touch for finding food.

The female produces 1 young after a gestation of 120 to 180 days; she even gives birth hanging in the trees. The offspring is suckled for about a month and then fed on regurgitated food.

NAME: **Two-toed Sloth, *Choloepus didactylus***
RANGE: **Venezuela, N. Brazil, Guyana, Surinam, French Guiana**
HABITAT: **forest**
SIZE: **body: 60–64 cm (23½–25¼ in)**
   tail: absent or vestigial

The two-toed sloth eats, sleeps, gives birth and even defecates while hanging upside-down in the trees by means of its hooklike claws. On each forefoot are two digits, closely bound together with skin, each of which bears a large, curved claw. Although all its movements are extremely slow, it can strike out quickly to defend itself and can inflict a serious wound with its claws. On the ground, it can only drag itself along, but it swims easily. Mainly active at night, the two-toed sloth sleeps during the day, keeping still to avoid detection by enemies. It feeds on leaves, twigs and fruit.

The female gives birth to a single offspring after a gestation of at least 263 days. The young sloth clings to its mother while she hangs in the trees.

GIANT ANTEATER

TREE ANTEATER

PYGMY ANTEATER

THREE-TOED SLOTH

TWO-TOED SLOTH

Armadillos, Pangolins, Aardvark

DASYPODIDAE: Armadillo Family

The 20 species of armadillo all occur in the New World, from the southern states of the USA through Central and South America to Chile and Argentina.

Armadillos are digging animals, usually active at night. Their skin is dramatically modified to form extremely tough, hornlike, articulated plates, which cover the top of the tail, the back, sides, ears and front of the head and provide the animal with excellent protection. Some armadillo species can curl themselves up into a ball so that their limbs and vulnerable underparts are protected by the armour.

NAME: **Giant Armadillo, *Priodontes maximus***
RANGE: **Venezuela to N. Argentina**
HABITAT: **forest**
SIZE: **body: 75 cm–1 m (29½ in–3¼ ft)**
　　　tail: 50 cm (19¾ in)　　　　　Ⓥ

The largest of its family, the giant armadillo may weigh up to 60 kg (132 lb). Its body is armoured with movable horny plates, and there are only a few hairs on the skin between the plates. It may have as many as 100 small teeth, but these are gradually shed with age. The claws on its forefeet are particularly long, those on the third digits measuring as much as 20 cm (7¾ in). A fairly agile animal, the giant armadillo can support itself on its hind legs and tail, while digging or smashing a termite mound with its powerful forelimbs. It feeds on ants, termites and other insects and also on worms, spiders, snakes and carrion. If attacked, it can only partially roll itself up and is more likely to flee.

The breeding habits of this armadillo are little known. The female produces 1 or 2 young, with tough, leathery skin.

NAME: **Nine-banded Armadillo, *Dasypus novemcinctus***
RANGE: **S. USA, through Central and South America to Uruguay**
HABITAT: **arid grassland, semi-desert**
SIZE: **body: 45–50 cm (17¾–19¾ in)**
　　　tail: 25–40 cm (9¾–15¾ in)

The most common, widespread armadillo, this species has 8 to 11, usually 9, bands of horny plates across its body. It spends the day in a burrow, which may house several individuals, and emerges at night to root about in search of food. It digs with its powerfully clawed forefeet or investigates holes and crevices with its tapering snout to find food such as insects, spiders, small reptiles, amphibians and eggs.

A litter nearly always consists of 4 young, which the mother suckles for about 2 months.

NAME: **Pink Fairy Armadillo, *Chlamyphorus truncatus***
RANGE: **central W. Argentina**
HABITAT: **dry sandy plains**
SIZE: **body: 12.5–15 cm (5–6 in)**
　　　tail: 2.5 cm (1 in)　　　　　Ⓥ

This tiny armadillo, with its pale pink armour, emerges from its burrow at dusk to feed on ants in particular but also on worms, snails and plant material. It digs with its forefeet and supports its rear on its rigid tail, thus freeing its hind limbs for kicking away earth. It has five claws on both fore- and hind feet. These little armadillos have proved difficult to keep in captivity, and their breeding habits are not known.

ORDER PHOLIDOTA
MANIDAE: Pangolin Family

The pangolin family is the only one in its order; it contains 7 species of nocturnal ant- and termite-eating animals, found in Africa and south and Southeast Asia. The typical pangolin has the same general body shape as that of the American giant anteater but is covered with enormous overlapping scales, like the bracts of a pinecone. The scales are movable and sharp edged. The pangolin has no teeth in its elongate head but does have an extremely long, protrusible tongue, with which it catches its prey.

NAME: **Giant Pangolin, *Manis gigantea***
RANGE: **Africa: Senegal, east to Uganda, south to Angola**
HABITAT: **forest, savanna**
SIZE: **body: 75–80 cm (29½–31½ in)**
　　　tail: 50–65 cm (19¾–25½ in)

The largest of its family, the giant pangolin is a ground-dwelling, nocturnal animal. The female is smaller than the male. This pangolin sleeps by day in a burrow and is active mainly between midnight and dawn, when it searches for various species of ants and termites. Using its powerful foreclaws, it can break into nests above or below ground. Its movements are slow and deliberate, and it can walk on its hind limbs, its long tail helping it to balance, or on all fours; it can also swim. If threatened, the giant pangolin can roll itself into a ball, a ploy that protects it from most enemies. If necessary, however, it will lash out with its sharp-scaled tail and spray urine and anal-gland secretions.

There is 1 young, born in an underground nest after a gestation of about 5 months. The newborn pangolin has soft scales, which harden in about 2 days. After a month or so, the youngster accompanies its mother on feeding trips, sitting on the base of her tail, and it is weaned at about 3 months.

NAME: **Tree Pangolin, *Manis tricuspis***
RANGE: **Africa: Senegal to W. Kenya, south to Angola**
HABITAT: **rain forest**
SIZE: **body: 35–45 cm (13¾–17¾ in)**
　　　tail: 49–60 cm (19¼–23½ in)

On its back, this pangolin has distinctive scales, with three pronounced points on their free edge. In older animals the points of the scales become worn. An adept climber, the tree pangolin has a very long prehensile tail, with a naked pad on the underside of the tip that aids grip. During the day, it sleeps on the branch of a tree or in a hole, which it digs in the ground, and emerges at night to feed on tree-dwelling ants and termites, which it detects by smell. It tears open arboreal nests with its powerful forelimbs and sweeps up the insects with deft movements of its long tongue. Like all pangolins, it shows a strong preference for particular species of ant and termite and will reject others. The food is ground down in the pangolin's muscular, horny-surfaced stomach.

The female gives birth to a single young after a gestation of 4 to 5 months. Its scales harden after a couple of days, and at 2 weeks of age, it starts to go on feeding trips with its mother.

ORDER TUBULIDENTATA
ORYCTEROPIDAE:
Aardvark Family

There is one family in this order, containing a single species that lives in Africa. Its relationship to other mammal groups is obscure. The aardvark's teeth are unique and quite unlike those of any other mammal: they have no enamel and consist of dentine columns, interspersed with tubes of pulp.

NAME: **Aardvark, *Orycteropus afer***
RANGE: **Africa, south of the Sahara**
HABITAT: **all regions with termites, from rain forest to dry savanna**
SIZE: **body: 1–1.6 m (3¼–5¼ ft)**
　　　tail: 44.5–60 cm (17½–23½ in)

The aardvark is a solitary, nocturnal, insect-eating animal. Its sight is poor, but its other senses are excellent; it has large ears, which are normally held upright but can be folded and closed, and highly specialized nostrils for sniffing out its prey. Dense hair surrounding the nostrils seals them off when the aardvark digs. It uses its powerful forelimbs to excavate burrows for shelter and to smash down the nests of the ants and termites that are its main foods. It sweeps up the prey with its long, sticky tongue.

The female gives birth to a single young after a gestation of 7 months. The offspring is suckled for 4 months.

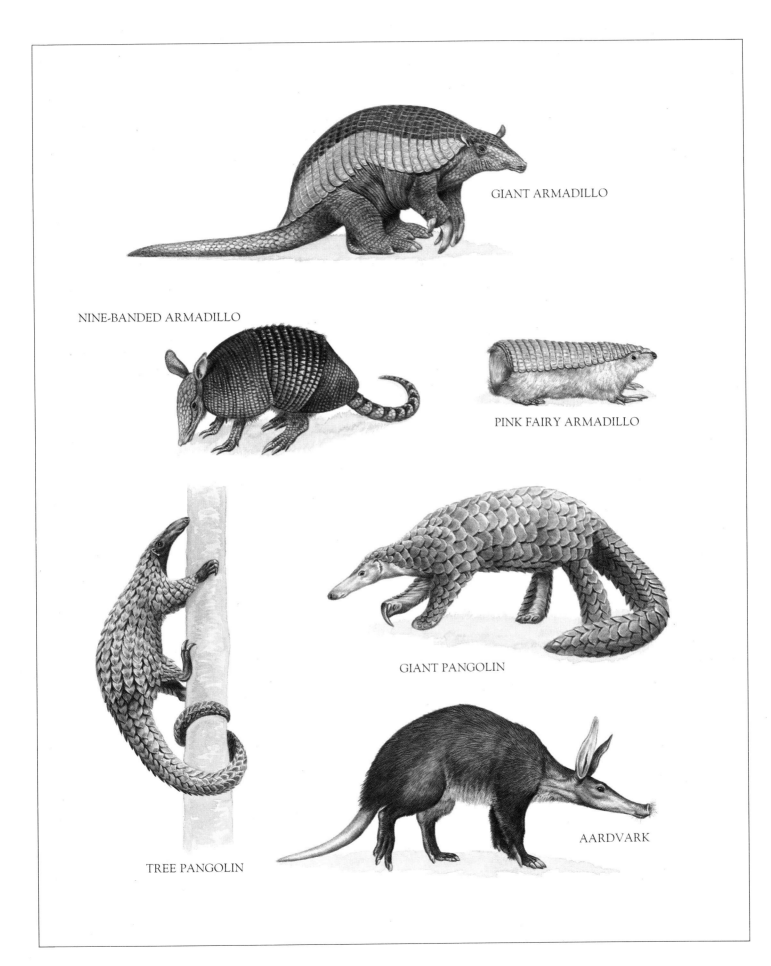

GIANT ARMADILLO

NINE-BANDED ARMADILLO

PINK FAIRY ARMADILLO

GIANT PANGOLIN

TREE PANGOLIN

AARDVARK

ORDER INSECTIVORA

This order includes about 344 species found all over the world, except in Australia and the southern half of South America. Most are ground-dwelling or burrowing animals which feed on insects and invertebrates.

TENRECIDAE: Tenrec Family

This family of insectivores includes the 30 species of shrewlike or hedgehoglike tenrecs, all restricted to Madagascar and the nearby Comoro Islands, and the 3 West African otter shrews. Tenrecs are a most interesting family zoologically because they have become adapted to a number of different ways of life: *Tenrec* resembles the North American opossum; *Setifer*, the hedgehogs; *Microgale*, the shrews; and *Oryzorictes* has several molelike characteristics.

All tenrecs retain some reptilian features, regarded as primitive in mammals, such as the cloaca, where the urogenital and anal canals open into a common pouch.

NAME: **Tail-less Tenrec**, *Tenrec ecaudatus*
RANGE: **Madagascar, Comoro Islands**
HABITAT: **brushland, dry forest clearings, highland plateaux**
SIZE: **body: 27–39 cm (10½–15¼ in)**
 tail: 10–16 mm (½–¾ in)
The tail-less tenrec has a superficial resemblance to a hedgehog, with its sparse coat set with stiff hairs and spines. It is active at night, probing rotting vegetation and leaf litter for insects, worms and roots. Fruit, too, forms part of its diet. In the dry season, tail-less tenrecs hibernate in deep burrows which they plug behind them with soil. Prior to hibernation, the tenrec builds up its brown fat reserves to sustain it through the 6-month sleep.

In early October, immediately after hibernating, the tenrecs mate and produce a litter of up to 25 young in November; usually about 16 survive.

NAME: **Greater Hedgehog Tenrec**, *Setifer setosus*
RANGE: **Madagascar**
HABITAT: **dry forest, highland plateaux**
SIZE: **body: 15–19 cm (6–7½ in)**
 tail: 10–16 mm (½–¾ in)
The greater hedgehog tenrec is set with short sharp spines, which cover its back like a dense, prickly mantle. If disturbed, it rolls itself into a tight ball while emitting a series of squeaks and grunts. The female produces a litter of up to 6 young in January. Like baby hedgehogs, the young have soft spines at birth, but by 2 weeks old the spines have hardened.

NAME: **Streaked Tenrec**, *Hemicentetes semispinosus*
RANGE: **Madagascar**
HABITAT: **scrub, forest edge**
SIZE: **body: 16–19 cm (6¼–7½ in)**
 tail: vestigial
Although less densely spined than the hedgehog tenrec, this species can still protect itself by partially curling up. When alarmed, a small patch of heavy spines in the middle of the back is vibrated rapidly, making a clicking noise. Striped tenrecs do not hibernate, but they do remain inactive during spells of cool weather. Like other tenrecs, they feed on insects and other invertebrates. Females are sexually mature at 8 weeks, and produce a litter of 7 to 11 young between December and March after a gestation of at least 50 days.

NAME: **Shrew Tenrec**, *Microgale longicaudata*
RANGE: **Madagascar**
HABITAT: **forest: sea level to montane**
SIZE: **body: 5–15 cm (2–6 in)**
 tail: 7.5–17 cm (3–6½ in)
The shrew tenrec, as its name implies, occupies the ecological niche filled by shrews in other parts of the world. The coat is short, but dense, and quite lacking in the spines so common in this family. Although it climbs well and the distal third of its long tail is prehensile, the shrew tenrec seems to feed mostly on grubs, worms and small insects on the forest floor. Like shrews, these tenrecs are active at all hours of the day and night, but each individual maintains its own pattern of rest and activity. Little is known of the breeding habits; they are believed to produce litters of 2 to 4 young. They do not appear to hibernate.

NAME: **Rice Tenrec**, *Oryzorictes hova*
RANGE: **Madagascar**
HABITAT: **marshy areas**
SIZE: **body: 8–13 cm (3–5 in)**
 tail: 3–5 cm (1¼–2 in)
Rice tenrecs, so called because they occupy the banks beside rice fields, spend most of their lives underground; their forelimbs are well adapted for digging. They feed on invertebrates, but there is some evidence that they also eat molluscs and crustaceans. Although seen above ground only at night, they may be active underground at all hours. Nothing is known of their breeding habits, but they are sufficiently abundant to achieve pest status in the rice-growing areas of Madagascar.

NAME: **Giant Otter Shrew**, *Potomogale velox*
RANGE: **W. and central equatorial Africa**
HABITAT: **streams: sea level to 1,800 m (6,000 ft)**
SIZE: **body: 29–35 cm (11½–14 in)**
 tail: 24–29 cm (9½–11½ in)
Although geographically separated from the other tenrecs, the 3 otter shrews are believed to be a subfamily of the Tenrecidae. The giant otter shrew is, overall, the largest living insectivore and does bear a strong superficial resemblance to the otter, with its flattened head and heavy tail. Its coat is dense with a glossy over-layer of guard-hairs.

Giant otter shrews live in burrows with entrances below water level. They emerge at dusk to hunt for crabs, fish and frogs, which they pursue through the water with great agility. They live solitary lives, but consort in pairs shortly before mating. Litters of 2 to 3 young are born throughout the year.

SOLENODONTIDAE: Solenodon Family

There are 2 species only of solenodons alive today: *Solenodon cubanus* on Cuba and *S. paradoxus* on the neighbouring island of Haiti and the Dominican Republic. They are rather ungainly, uncoordinated creatures and, although the size of rats, look more like shrews with their probing snouts. Their eyes are small and rheumy and they are far more nocturnal than the wide-eyed rats.

Solenodons grow and breed slowly and this, combined with the predatory attacks of introduced dogs and cats, means that their survival is now threatened. Conservation areas are being established for these animals, but their future is far from secure.

NAME: **Cuban Solenodon**, *Solenodon cubanus*
RANGE: **Cuba**
HABITAT: **montane forest**
SIZE: **body: 28–32 cm (11–12½ in)**
 tail: 17–25 cm (6½–10 in) ®
Solenodons have a varied diet. At night they search the forest floor litter for insects and other invertebrates, fungi and roots. They climb well and feed on fruits, berries and buds but have more predatory habits too. With venom from modified salivary glands in the lower jaw, the solenodon can kill lizards, frogs, small birds or even rodents. Solenodons seem not to be immune to the venom of their own kind, and there are records of cage mates dying after fights. They produce litters of 1 to 3 young.

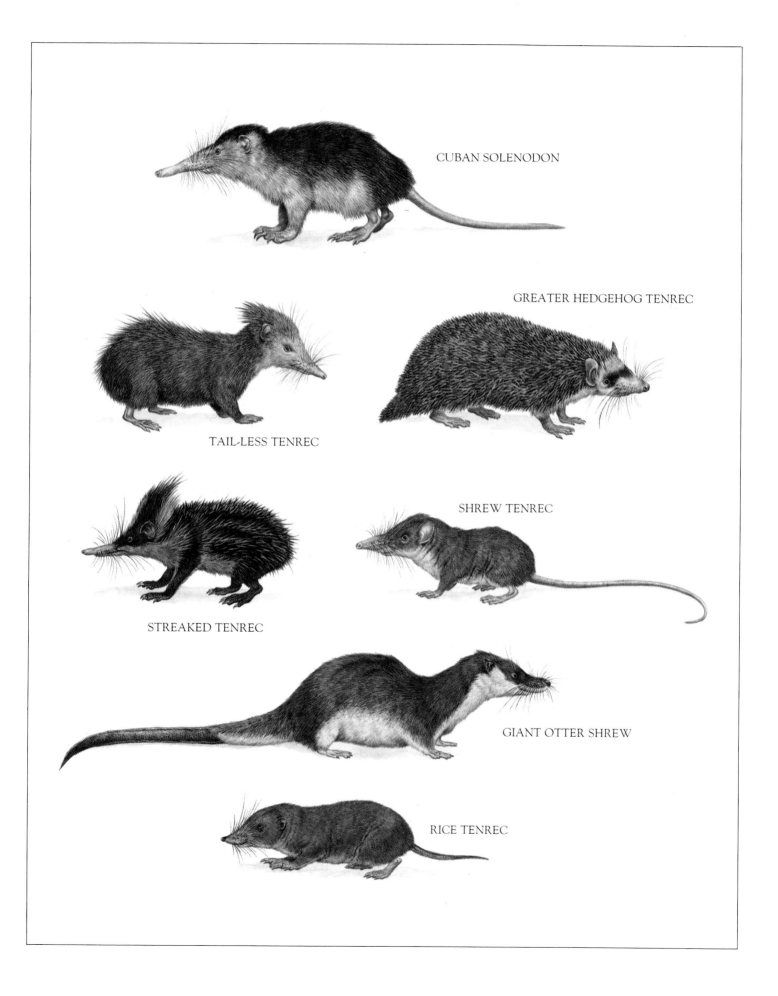

CUBAN SOLENODON

GREATER HEDGEHOG TENREC

TAIL-LESS TENREC

SHREW TENREC

STREAKED TENREC

GIANT OTTER SHREW

RICE TENREC

Golden Moles, Hedgehogs

CHRYSOCHLORIDAE: Golden Mole Family

Golden moles bear a close resemblance to the true moles, but are, in fact, only distantly related. They have cylindrical bodies, short powerful limbs and no visible tail. Their fur is thick and dense, and the metallic lustre it imparts gives the group its name. Golden moles are blind, their eyes being reduced to mere vestiges covered by fused hairy eyelids. The shovel-like paws on their forelimbs are used for digging, and the enlarged flattened claws on the "index" and middle digits act as cutting edges.

The 17 species of golden mole are found only in Africa, south of a line linking Cameroon with Tanzania. They occur in all types of habitat from rugged mountainous zones to sandy plains.

NAME: **Cape Golden Mole, Chrysochloris asiatica**
RANGE: **Zaire, E. Africa, Zimbabwe to Western Cape**
HABITAT: **workable soil up to 2,800 m (9,000 ft)**
SIZE: **body: 9–14 cm (3½–5½ in) tail: absent**

The Cape golden mole is a frequent visitor to gardens and farmland in much of southern Africa. Its presence is usually revealed by raised, cracked tunnel tracks radiating outward from a bush or shed. At night golden moles may travel on the surface, and in damp weather they root about for beetles, worms and grubs.

Once a year, in the rainy season, females produce litters of 2 to 4 young. Shortly before the birth, the mother makes a round grass-lined nest in a special breeding chamber. The young suckle for almost 3 months until their teeth erupt.

NAME: **Hottentot Golden Mole, Amblysomus hottentotus**
RANGE: **W. South Africa**
HABITAT: **sand or peat plains**
SIZE: **body: 8.5–13 cm (3¼–5 in) tail: absent**

The Hottentot golden mole differs from other species in that it has only two claws on each forepaw. When these animals occur in orchards and young plantations, their burrowing may seriously disturb roots and kill the trees, but generally they do more good than harm by eating insects, beetle larvae and other invertebrate pests.

Pairs breed between November and February when rainfall is high; they produce a litter of 2 young.

NAME: **Giant Golden Mole, Chrysospalax trevelyani**
RANGE: **South Africa: E. Cape Province**
HABITAT: **forest**
SIZE: **body: 20–24 cm (8–9½ in) tail: absent**   ®

As its name implies, the giant golden mole is the largest of its family and weighs up to 1.5 kg (3 lb). It is a rare species and on the brink of extinction.

Giant golden moles frequently hunt above ground for beetles, small lizards, slugs and giant earthworms. When disturbed they dart unerringly towards their burrow entrance and safety, but how they are able to locate it is not known. During the winter rainy season, they are believed to produce a litter of 2 young.

ERINACEIDAE: Hedgehog Family

This family contains 17 species of two superficially quite distinct types of animal: the hedgehogs, and the gymnures, or moonrats. Members of the hedgehog subfamily occur right across Europe and Asia to western China, and in Africa as far south as Angola. Moonrats live in Indo-China, Malaysia, Borneo, the Philippines and northern Burma. All feed on a varied diet of worms, insects and molluscs, as well as some berries, frogs, lizards and birds.

In the northern part of their range, hedgehogs hibernate during the winter months, but in warmer areas this is not necessary.

NAME: **Western European Hedgehog, Erinaceus europaeus**
RANGE: **Britain, east to Scandinavia and Romania; introduced in New Zealand**
HABITAT: **scrub, forest, cultivated land**
SIZE: **body: 13.5–27 cm (5½–10½ in) tail: 1–5 cm (½–2 in)**

One of the most familiar small mammals in Europe, the hedgehog gets its name from its piglike habit of rooting around for its invertebrate prey in the hedgerows. It is quite vocal and makes a range of grunting, snuffling noises. The upper part of the head and the back are covered in short, banded spines. If threatened, the hedgehog rolls itself up, and a longitudinal muscle band, running around the edge of the prickly cloak, acts as a drawstring to enclose the creature within its spiked armour. The chest and belly are covered with coarse springy hairs.

Hedgehogs produce 1, sometimes 2, litters of about 5 young each year; the young are weaned at about 5 weeks. In the north of their range, hedgehogs hibernate throughout the winter.

NAME: **Desert Hedgehog, Paraechinus aethiopicus**
RANGE: **N. Africa, Middle East to Iraq**
HABITAT: **arid scrub, desert**
SIZE: **body: 14–23 cm (5½–9 in) tail: 1–4 cm (½–1½ in)**

The desert hedgehog resembles its slightly larger European cousin, but its coloration is more variable. Generally the spines are sandy-buff with darker tips, but dark and white forms are not uncommon. Desert hedgehogs dig short, simple burrows in which they pass the day. At night, when the air is cool, they emerge to search for invertebrates and the eggs of ground-nesting birds. Scorpions are a preferred food; the hedgehogs nip off the stings before eating them. In common with most desert mammals, these hedgehogs probably have highly adapted kidneys, enabling them to exist for long periods without water.

In July or August desert hedgehogs breed, producing a litter of about 5 young.

NAME: **Moonrat, Echinosorex gymnurus**
RANGE: **Cambodia, east to Burma**
HABITAT: **forest, mangrove swamps**
SIZE: **body 26–44 cm (10–17¼ in) tail: 20–21 cm (8–8¼ in)**

One of the largest insectivores, the moonrat has a long snout, an unkempt appearance and an almost naked, scaly tail. It can defend itself by producing a foul, fetid odour from a pair of anal glands which repels all but the most persistant predators. Moonrats live in crevices between tree roots, or in hollow logs; they emerge around dusk to forage for molluscs, insects and worms. Some fruit, and fish and crabs too, may be eaten. Little is known of the moonrat's breeding habits, but they seem to breed throughout the year, producing 2 young at a time.

NAME: **Mindanao Moonrat, Podogymnura truei**
RANGE: **Philippines: Mindanao**
HABITAT: **upland forest and forest edge, from 1,600 to 2,300 m (5,250–7,500 ft)**
SIZE: **body: 13–15 cm (5–6 in) tail: 4–7 cm (1½–2¾ in)**   ⓥ

This curious creature is restricted to a small natural range and has never been common. Now, because of logging operations and slash-and-burn agriculture, much of its habitat is being destroyed and its survival is seriously threatened. It has long soft fur and a tail with more hairs than that of *Echinosorex*. It feeds on insects, worms and even carrion, which it finds in grasses and among stands of moss. Nothing is known of the breeding habits of Mindanao moonrats.

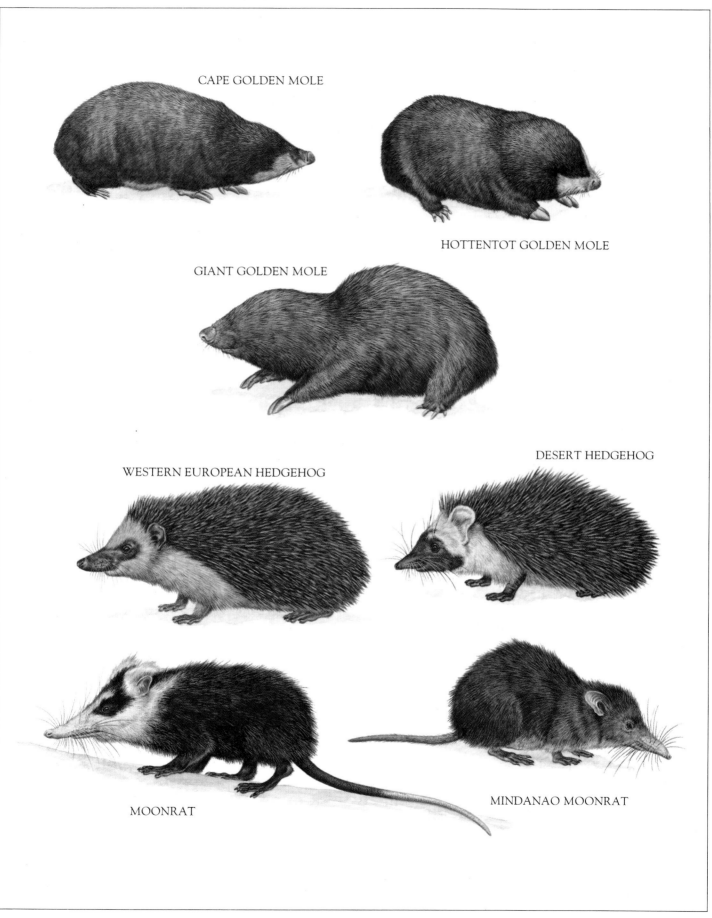

CAPE GOLDEN MOLE

HOTTENTOT GOLDEN MOLE

GIANT GOLDEN MOLE

DESERT HEDGEHOG

WESTERN EUROPEAN HEDGEHOG

MOONRAT

MINDANAO MOONRAT

Shrews

NAME: **Masked Shrew,** *Sorex cinereus*
RANGE: **northern N. America to New Mexico**
HABITAT: **moist forest**
SIZE: **body: 4.5–9.5 cm ($1\frac{3}{4}$–$3\frac{3}{4}$ in)
    tail: 2.5–8 cm (1–3 in)**

The masked shrew is a common species throughout North America, though less abundant in the arid regions to the south. It inhabits the surface layer of forest, living in and around burrows made by itself and other woodland animals. It is active day and night with about seven periods of feeding activity in each 24 hours. Earthworms and snails are preferred foods, but masked shrews eat a wide range of invertebrate prey.

Generally solitary animals, male and female pair only for mating. Several litters of up to 10 young are born during the late spring and summer. The young are weaned after about a month, but the family may remain together for another month, the only time during which these shrews are sociable.

NAME: **Short-tailed Shrew,** *Blarina brevicauda*
RANGE: **E. USA**
HABITAT: **almost all terrestrial habitats**
SIZE: **body: 7.5–10.5 cm (3–4 in)
    tail: 1.5–3 cm ($\frac{1}{2}$–$1\frac{1}{4}$ in)**

This abundant and widespread species is unusual in two ways: first, it seems to be partly gregarious since, in captivity, it seeks out the company of other shrews; second, it often climbs trees for food — most shrews climb only rarely.

In much of the USA, the short-tailed shrew is an important controlling influence on larch sawflies and other destructive forest pests on which it feeds. It builds a grassy nest under a stump or log and produces three or four litters a year of up to 9 young each. The gestation period is 17 to 21 days.

NAME: **Giant Mexican Shrew,** *Megasorex gigas*
RANGE: **W. coastal strip of Mexico**
HABITAT: **rocky or semi-desert, dry forest**
SIZE: **body: 8–9 cm (3–$3\frac{1}{2}$ in)
    tail: 4–5 cm ($1\frac{1}{2}$–2 in)**

The giant Mexican shrew has large prominent ears. In common with most desert mammals, it is active only at night, when it emerges from crevices between boulders or under rocks to search for insects and worms. Although it eats up to three-quarters of its own weight every 24 hours, it seems only to feed at night, unlike other shrews which feed throughout the period. Its breeding habits are not known.

SORICIDAE: Shrew Family

There are more than 200 species of shrew, distributed throughout most of the world, except Australia and New Zealand, the West Indies and most of South America. Shrews are insectivores and most lead inoffensive lives among the debris of the forest floor or on pastureland, consuming many types of invertebrate. Water shrews are known to overpower frogs and small fish which they kill with venomous bites — the saliva of many shrews contains strong toxins. Carrion may also be included in the shrew's diet.

Shrews are active creatures with high metabolisms. Their hearts may beat more than 1,200 times every minute, and, relative to their body size, they have enormous appetites. Even in cold northern regions, they do not hibernate in winter; it would be impossible for them to build up sufficient reserves. Although shrews are heavily preyed upon by owls and hawks, acrid-smelling secretions from well-developed flank glands seem to deter most mammalian predators.

Some species of shrew are reported to eat their own faeces and perhaps those of other creatures. By doing so, they boost their intake of vitamins B and K and some other nutrients. This habit may be related to the shrews' hyperactive life and enhanced metabolism. Shrews rely heavily on their senses of smell and hearing when hunting; their eyes are tiny and probably of little use.

NAME: **Ceylon Long-tailed Shrew,** *Crocidura miya*
RANGE: **Sri Lanka**
HABITAT: **damp and dry forest, savanna**
SIZE: **body: 5–6.5 cm (2–$2\frac{1}{2}$ in)
    tail: 4–4.5 cm ($1\frac{1}{2}$–$1\frac{3}{4}$ in)**

This shrew, with its long, lightly haired tail, spends much of its life among the debris of the forest floor where damp, cool conditions encourage a rich invertebrate population on which it feeds. It also eats small lizards and young birds on occasion. Strong-smelling scent glands seem to protect the shrew itself from much predation.

The breeding season of the long-tailed shrew lasts from March until November; a female produces about five litters in this time, each of about 6 young. After 8 days, the young leave the nest for the first time, each gripping the tail of the one in front in its mouth as the caravan, led by the mother, goes in search of food. This habit seems to be restricted to this genus of shrews.

NAME: **Pygmy White-toothed Shrew,** *Suncus etruscus*
RANGE: **S. Europe, S. Asia, Africa**
HABITAT: **semi-arid grassland, scrub, rocky hillsides**
SIZE: **body: 3.5–5 cm ($1\frac{1}{4}$–2 in)
    tail: 2.5–3 cm (1–$1\frac{1}{4}$ in)**

Usually regarded as the world's smallest terrestrial mammal, a fully grown pygmy shrew weighs about 2 g (1/14 oz). How such a tiny mammal can survive is not fully understood, but it must have a constant and reliable source of food, and that is one reason why it is restricted to the warmer parts of the Old World. Its coat is dense to prevent undue heat loss from its tiny body.

Pygmy shrews feed on spiders and insects almost as large as themselves, including grasshoppers and cockroaches. Nothing is known of their breeding habits, but they remain quite abundant, and it may be that their small size protects them from heavy predation, since they share their habitat with larger, more tempting species.

NAME: **Mouse Shrew,** *Myosorex varius*
RANGE: **South Africa, north to the Limpopo**
HABITAT: **moist areas, forest, scrub, river banks**
SIZE: **body: 6–11 ($2\frac{1}{4}$–$4\frac{1}{4}$ in)
    tail: 3–5.5 cm ($1\frac{1}{4}$–$2\frac{1}{4}$ in)**

Perhaps the most primitive of existing shrews, the mouse shrew has two more teeth in its lower jaw than normal and thus resembles the extinct early mammals, although otherwise it is like most shrews. Mouse shrews do not appear to make or use burrows but seek holes and hollows for daytime shelter. Nests of shredded grass are made for sleeping and for use as nurseries. Females produce up to six litters a year of 2 to 4 young each.

NAME: **Armoured Shrew,** *Scutisorex somereni*
RANGE: **Africa: Uganda, near Kampala**
HABITAT: **forest**
SIZE: **body: 12–15 cm ($4\frac{3}{4}$–6 in)
    tail: 7–9.5 cm ($2\frac{3}{4}$–$3\frac{3}{4}$ in)**

One of the most puzzling members of the shrew family, the armoured shrew has a spine which is fortified and strengthened by a mesh of interlocking bony flanges and rods. Despite this, it moves much like other shrews, although its predatory behaviour is characterized by rather ponderous and apparently well thought-out movements. There are reports that an armoured shrew can support the weight of a grown man without being crushed.

Armoured shrews appear to eat plant food as well as invertebrates. They are believed to breed throughout the year.

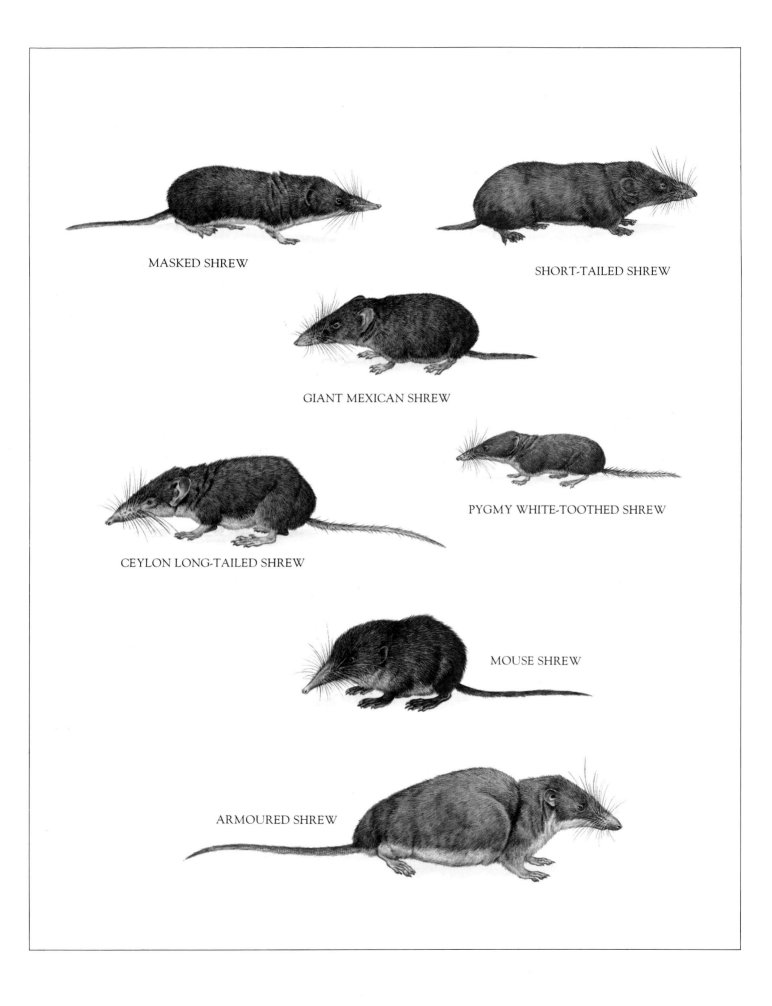

MASKED SHREW

SHORT-TAILED SHREW

GIANT MEXICAN SHREW

PYGMY WHITE-TOOTHED SHREW

CEYLON LONG-TAILED SHREW

MOUSE SHREW

ARMOURED SHREW

Moles, Elephant Shrews, Flying Lemurs

TALPIDAE: Mole Family

The majority of the 20 species of mole lead an underground life, but 2 species of desman and the star-nosed mole are adapted for an aquatic life. Moles are widespread throughout Europe and Asia, south to the Himalayas, and from southern Canada to northern Mexico. They need habitats with soft soil so that they are able to dig their extensive burrow systems.

All moles have highly modified hands and forearms, which act as pickaxe and shovel combined. Moles seldom come above ground and their eyes are tiny and covered with hairy skin. Their tactile sense is highly developed, however, and their facial bristles respond to the tiniest vibrations. Moles can move backwards or forwards with equal ease — when reversing, the stumpy tail is held erect and the sensory hairs on it provide warning of any approaching danger.

NAME: **Pacific Mole, *Scapanus orarius***
RANGE: **N. America: British Columbia to Baja California**
HABITAT: **well-drained deciduous forest**
SIZE: **body: 11–18.5 cm (4½–7¼ in)**
tail: 2–5.5 cm (¾–2¼ in)

The Pacific, or coast, mole and its close relatives, the broad-footed mole and Townsend's mole, all have nostrils which open upward. Their eyes are much more visible than those of other species, but this does not necessarily mean that their sight is better. Like other moles, they live underground and rarely venture up to the surface.

Coast moles feed on earthworms and soil-dwelling larvae and do much good by devouring the larvae of insect pests. Between 2 and 5 young are born in early spring after a 4-week gestation.

NAME: **European Mole, *Talpa europaea***
RANGE: **Europe, E. Asia**
HABITAT: **pasture, forest, scrub**
SIZE: **body: 9–16.5 cm (3½–6½ in)**
tail: 3–4 cm (1¼–1½ in)

The extensive burrow systems in which European moles live are excavated rapidly: a single individual can dig up to 20 m (66 ft) in one day. The moles feed primarily on earthworms but also eat a wide range of other invertebrates, as well as snakes, lizards, mice and small birds.

In early summer the female produces a litter of up to 7 young, born in a leaf-lined underground nest and weaned at about 3 weeks. Occasionally there is a second litter.

NAME: **Hairy-tailed Mole, *Parascalops breweri***
RANGE: **S.E. Canada, N.E. USA**
HABITAT: **well-drained soil in forest or open land**
SIZE: **body: 11.5–14 cm (4½–5½ in)**
tail: 2–3.5 cm (¾–1½ in)

As its name implies, this species is characterized by its almost bushy tail. Otherwise it is similar to other moles in habits and appearance. Hairy-tailed moles dig extensive tunnels at two levels: an upper level just beneath the surface, used in warm weather, and a lower tunnel, used as a winter retreat. The moles mate in early April and litters of 4 or 5 young are born in mid-May.

The eastern mole, *Scalopus aquaticus*, also a North American species, closely resembles the hairy-tailed, but has a nearly naked tail.

NAME: **Star-nosed Mole, *Condylura cristata***
RANGE: **S.E. Canada, N.E. USA**
HABITAT: **any area with damp soil**
SIZE: **body: 10–12.5 cm (4–5 in)**
tail: 5.5–8 cm (2¼–3¼ in)

This species of mole has a fringe of 22 fingerlike tentacles, surrounding the nostrils, which it uses to search for food on the bottoms of ponds and streams. Although star-nosed moles dig and use tunnel systems, they seldom feed within them. They are excellent swimmers and divers and feed largely on aquatic crustaceans, small fish, water insects and other pond life. Their fur is heavy and quite waterproof. The female gives birth to a litter of 2 to 7 young in spring. The young are born with well-developed nostril tentacles.

NAME: **Russian Desman, *Desmana moschata***
RANGE: **E. Europe to central W. Asia**
HABITAT: **pools and streams in densely vegetated areas**
SIZE: **body: 18–21.5 cm (7–8½ in)**
tail: 17–21.5 cm (6¾–8½ in) Ⓥ

The largest member of the mole family, the desman has forsaken the underground for the aquatic life, although it does excavate short burrows as bankside residences. When swimming, the desman's flattened tail acts as a rudder and propulsive force; its webbed feet, supplied with fringing hairs, make most effective paddles. Desmans feed on a variety of aquatic life from fishes and amphibians to insects, crustaceans and molluscs.

Now much reduced in numbers, this species is the subject of intensive conservation measures and there are projects to reintroduce desmans into parts of their former range where they have been eliminated.

MACROSCELIDIDAE: Elephant Shrew Family

The 16 species of African elephant shrews have extraordinary trunklike noses. They remain a zoological enigma and are classified among the Insectivora on account of their dentition.

NAME: **Elephant Shrew, *Macroscelides proboscideus***
RANGE: **Africa: Namibia, South Africa: Cape Province**
HABITAT: **plains, rocky outcrops**
SIZE: **body: 9.5–12.5 cm (3¾–5 in)**
tail: 9.5–14 cm (3¾–5½ in)

The abundant elephant shrews are active by day and night. They catch and eat termites and sometimes dig burrows into termite mounds. Seeds, fruit and berries, too, are part of their diet, and to reach these the elephant shrews hop and jump from twig to branch on their powerful hind legs, using their long tails as counterbalances.

A litter of 1 or 2 well-developed young is born during the rainy season. The young can walk and jump almost as soon as they are born and appear to suckle for only a few days.

ORDER DERMOPTERA
CYNOCEPHALIDAE: Flying Lemur Family

There are only 2 species of flying lemur, or colugo, and, like elephant shrews, they are a puzzle, having characteristics of lemurs and of insectivores. For this reason they have their own order.

NAME: **Philippine Flying Lemur/Colugo, *Cynocephalus volans***
RANGE: **Philippines**
HABITAT: **forest**
SIZE: **body: 38–42 cm (15–16½ in)**
tail: 22–27 cm (8¾–10½ in)

The flying lemur "flies" with the aid of its gliding membrane, or patagium, which stretches from the neck to the wrists and ankles and to the tip of the tail. With limbs and membrane outstretched, the flying lemur can glide through the trees for up to 135 m (450 ft). Almost helpless on the ground, it is an agile climber and feeds on shoots, buds, fruit and flowers from a great range of forest trees. Its great enemy is the Philippine eagle which feeds almost exclusively on flying lemurs.

In February flying lemurs mate, and after a 2-month gestation, the female gives birth to a single young, occasionally to twins. She carries her young with her until it grows too large and heavy.

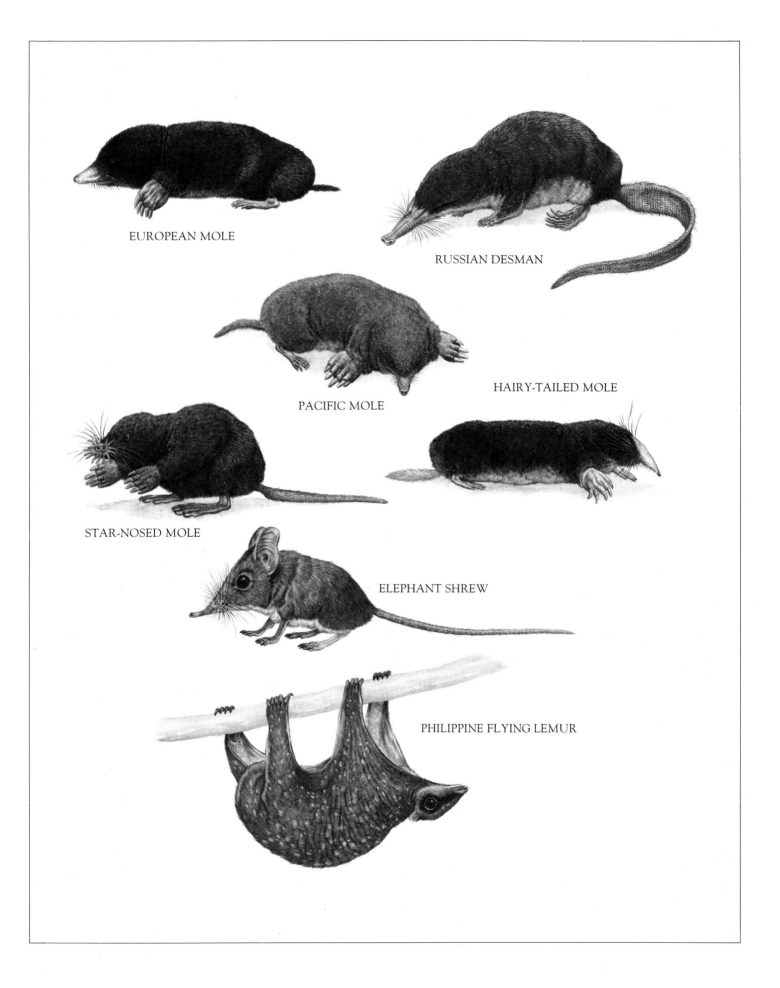

EUROPEAN MOLE

RUSSIAN DESMAN

PACIFIC MOLE

HAIRY-TAILED MOLE

STAR-NOSED MOLE

ELEPHANT SHREW

PHILIPPINE FLYING LEMUR

Fruit Bats

NAME: Greater Fruit Bat, *Pteropus giganteus*
RANGE: S. and S.E. Asia
HABITAT: forest, scrub
SIZE: body: 35–40 cm (13¾–15¾ in)
  wingspan: 1.5 m (5 ft)
  tail: absent

The wingspan of the greater fruit bat is the largest of any bat. It is a highly sociable creature and roosts by day in large trees in flocks of several thousand. At dusk the flocks take to the air and disperse in search of food. The bat crushes fruit between its peglike teeth to obtain the juice and spits out the seeds and flesh. Soft flesh, such as banana, is swallowed.

There is no general breeding season for this species but, in each part of its huge range, births are more or less synchronized. One young is born after a gestation of about 6 months and is carried about by its mother until it is 8 weeks old.

NAME: Hammerheaded Bat, *Hypsignathus monstrosus*
RANGE: Africa: Gambia to Uganda and Angola
HABITAT: mangrove and other swamps
SIZE: body: 25–30 cm (10–12 in)
  wingspan: 70–95 cm (27–37 in)
  tail: absent

The hammerheaded bat derives its name from a curious nasal swelling, the function of which remains a mystery. It is one of the noisiest bats; males gather in special trees and chorus to one another for hours.

Hammerheaded bats roost in small numbers. They feed on the juices of mangoes and soursops and may also have carnivorous tendencies. Females produce a single young after a gestation of 5½ months.

NAME: Dog Bat, *Rousettus aegyptiacus*
RANGE: Africa, east to India and Malaysia
HABITAT: forest; caves, tombs, temples
SIZE: body: 11–13 cm (4¼–5 in)
  wingspan: 30–45 cm (12–18 in)
  tail: 1.5 cm (½ in)

Dog bats roost deep in caves or tombs and seem to rely on echolocation for flight navigation in these places. They are the only fruit bats to use this phenomenon, so important to the insectivorous bats. They feed on fruit juices and flower nectar and play a useful role as pollinators.

The breeding season of dog bats is from December to March. Females produce 1 young after a gestation of 15 weeks. The young bat clings to its mother and is transported everywhere until it can itself fly. It begins to feed on fruit at the age of 3 months.

ORDER CHIROPTERA

One species in every four mammals is a bat, yet remarkably little is known of this order. Bats are the only mammals capable of sustained flight, as opposed to gliding, which they achieve by means of their well-designed wings. The four elongated fingers of each hand support the flight membrane, which is also attached to the ankles and sometimes incorporates the tail.

PTEROPODIDAE:
Fruit Bat Family

There are about 130 species of these large, fruit-eating bats, sometimes known as flying foxes, found in the tropical and subtropical regions of the Old World. Their eyes are large, their ears simple in structure like those of rodents, and they have a keen sense of smell. On each hand there is a sturdy thumb, equipped with a robust claw; some species have an extra claw at the tip of the second digit where it protrudes from the wing membrane. Using these claws and their hind feet, fruit bats make their way along the branches of their feeding trees in search of fruit; some also feed on pollen and nectar. Males and females look alike.

NAME: Franquet's Fruit Bat, *Epomops franqueti*
RANGE: Africa: Nigeria to Angola, east to Zimbabwe and Tanzania
HABITAT: forest, open country
SIZE: body: 13.5–18 cm (5–7 in)
  wingspan: 23–25 cm (9–10 in)
  tail: absent

Franquet's fruit bat is also known as the epauleted bat on account of the distinct patches of white fur on each shoulder. Rather than crushing fruit to release the juice, this bat sucks it out. It encircles the fruit with its lips, pierces the flesh with its teeth and, while pushing the tongue up against the fruit, it sucks, using the action of its pharyngeal pump.

Franquet's bats breed throughout the year, producing a single young at a time after a 3½-month gestation.

NAME: Harpy Fruit Bat, *Harpyionycteris whiteheadi*
RANGE: Philippines, Sulawesi
HABITAT: forest, up to 1,700 m (5,575 ft)
SIZE: body: 14–15 cm (5½–6 in)
  wingspan: 23–30 cm (9–12 in)
  tail: absent

The prominent incisor teeth of the harpy fruit bat appear to operate almost like the blades of a pair of scissors, and the bat seems to use them to snip off figs and other fruit from the trees.

NAME: Tube-nosed Fruit Bat, *Nyctimene major*
RANGE: Sulawesi to Timor, New Guinea, N. Australia, Solomon Islands
HABITAT: forest
SIZE: body: 7–12 cm (3–4¾ in)
  wingspan: 20–28 cm (8–11 in)
  tail: 1.5–2.5 cm (½–1 in)

This bat has a pair of nasal scrolls which stand out on each side of the head like snorkel tubes. Their exact function is not clear, but they may help the bat locate ripe fruit by bestowing a "stereo" effect on the nose. Guavas, figs and even the pulp of young coconuts make up the diet of the tube-nosed bat. Pieces of fruit are torn out with the teeth and then chewed and kneaded against the chest and belly. Only the juice is consumed, the rest is dropped to the ground.

Tube-nosed bats seem less social than other fruit bats and usually roost alone. They cling to the trunks of trees and are afforded some camouflage by their spotted wings. They breed in September and October and produce 1 young.

NAME: Long-tongued Fruit Bat, *Macroglossus minimus*
RANGE: Burma, east to Malaysia and Bali
HABITAT: forest, plantations
SIZE: body: 6–7 cm (2¼–2¾ in)
  wingspan: 14–17 cm (5½–6¾ in)
  tail: vestigial

One of the smallest fruit bats, the long-tongued has adopted a solitary way of life, thus making itself less obvious to predators. By day it roosts in rolled-up banana or hemp leaves, emerging at dusk to feed on pollen and nectar from plants. It feeds on fruit, too, and is considered a plantation pest in some parts of its range. These bats are quite vocal and make a shrieking noise at night. They breed in August and September, producing a single young after a gestation of 12 to 15 weeks.

NAME: Queensland Blossom Bat, *Syconycteris australis*
RANGE: S. New Guinea, south to Australia: New South Wales
HABITAT: wet and dry forest
SIZE: body: 5–6 cm (2–2½ in)
  wingspan: 12–15 cm (4¾–6 in)
  tail: vestigial

The smallest fruit bat, the Queensland blossom bat inhabits eucalyptus and acacia forests and seems to feed almost exclusively on the pollen and nectar of the many species of these trees. The bat feeds by inserting its long brushlike tongue deep into the long-necked blooms. Females give birth to a tiny infant in summer, that is, in November or December.

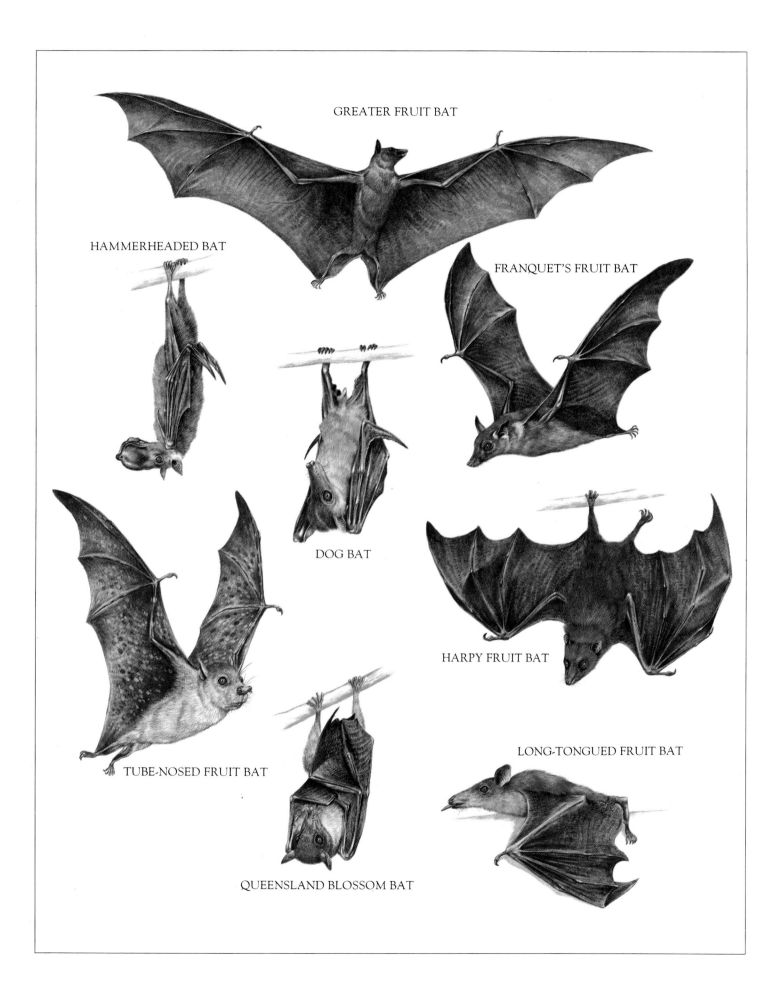

GREATER FRUIT BAT

HAMMERHEADED BAT

FRANQUET'S FRUIT BAT

DOG BAT

HARPY FRUIT BAT

TUBE-NOSED FRUIT BAT

LONG-TONGUED FRUIT BAT

QUEENSLAND BLOSSOM BAT

Mouse-tailed Bats, Sheath-tailed Bats

RHINOPOMATIDAE: Mouse-tailed Bat Family

Mouse-tailed bats derive their name from their long, naked tails which equal the head and body in length. There are 4 species, found from the Middle East through India to Thailand and Sumatra. These bats have occupied certain pyramids in Egypt for 3 millenniums or more.

NAME: **Greater Mouse-tailed Bat,** *Rhinopoma microphyllum*
RANGE: **Middle and Near East**
HABITAT: **treeless arid land**
SIZE: **body: 6–8 cm (2¼–3 in)**
　　wingspan: 17–25 cm (6¾–10 in)
　　tail: 6–8 cm (2¼–3 in)

Colonies of thousands of mouse-tailed bats occupy roosts in large ruined buildings, often palaces and temples. They feed exclusively on insects and, in those areas where a cool season temporarily depletes the food supply, the bats may enter a deep sleep resembling torpor. Prior to this, they lay down thick layers of fat which may weigh as much as the bats themselves, and with this they survive for many weeks with neither food nor water. As they sleep, the accumulated fat is used up and by the time the cold season is passed, nothing of it remains.

Mouse-tailed bats mate at the beginning of spring and the female produces a single offspring after a gestation of about 4 months. The young bat is weaned at 8 weeks but does not attain sexual maturity until its second year.

EMBALLONURIDAE: Sheath-tailed Bat Family

There are 40 species of bat in this family and all have a characteristic membrane which spans the hind legs; the tail originates beneath the membrane and penetrates it through a small hole. The advantages of this arrangement are not clear, but the bat is able to adjust this versatile "tailplane" during flight by movements of its legs only, and it may improve its aerial abilities.

Sheath-tailed bats range over tropical and subtropical regions of the world in a variety of habitats, but never far from trees. They are primarily insect-eaters, but may supplement their diet with fruit.

Scent-producing glandular sacs on the wings, in the crooks of the elbows, are another characteristic of these bats. The thick, pungent secretions of these glands are more profuse in males than females and may assist females in their search for mates.

NAME: **Proboscis Bat,** *Rhynchonycteris naso*
RANGE: **S. Mexico to central Brazil**
HABITAT: **forest, scrub; near water**
SIZE: **3.5–4.5 cm (1¼–1¾ in)**
　　wingspan: 12–16 cm (4¾–6¼ in)
　　tail: 1–2 cm (½ in)

Characteristic features of the proboscis bat are its long snout and the tufts of grey hair on its forearms. Proboscis bats are slow fliers, ill suited for hunting up in the tree-tops where they would be easy prey for birds, so they have developed the habit of feeding on insects which live just above the surface of ponds, lakes and rivers. They sometimes roost in rocky crevices, but often just cling to rocks or to concrete, where their colour pattern gives them a resemblance to patches of lichen. Individuals roost a considerable distance from one another, perhaps to enhance the effect of the camouflage.

Between April and July, females give birth to 1 young. Until the young bat is about 2 months old and able to fend for itself, its mother chooses a dark safe roost inside a log, or deep within a pile of stones, in which to leave it.

NAME: **Two-lined Bat,** *Saccopteryx leptura*
RANGE: **Mexico to Bolivia and Brazil**
HABITAT: **lowland forest**
SIZE: **4–5 cm (1½–2 in)**
　　wingspan: 18–22 cm (7–8¾ in)
　　tail: 1 cm (½ in)

One of the most strikingly marked of all bats, the two-lined bat has a pair of wavy white lines running down its back from the nape of the neck to the rump. The lines serve to break up the body shape of the bat and also provide some camouflage. The male has a pair of well-developed saclike glands in the wing membrane, just above the crooks of the arms.

In Mexico, two-lined bats have been observed roosting under concrete bridges and on the walls of buildings. Each individual seems faithful to its roosting place, returning each morning to the precise spot that it left the evening before. Females give birth to young in the rainy season when their food supply — beetles and moths — is most plentiful. The young are weaned before they are 2 months old.

NAME: **Old World Sheath-tailed Bat,** *Emballonura monticola*
RANGE: **Thailand to Malaysia, Java, Sumatra, Borneo and Sulawesi**
HABITAT: **rain forest**
SIZE: **body: 4–6 cm (1½–2¼ in)**
　　wingspan: 16–18 cm (6¼–7 in)
　　tail: 1 cm (½ in)

Old World sheath-tailed bats generally roost in rock fissures or caves in the forest in groups of a dozen or so. At dusk they all depart simultaneously to feed and then return together at dawn. They appear to feed in the top layer of the tallest trees and to supplement their insect diet with fruit and even flowers. Although little is known of the breeding habits of these bats, they are likely to breed throughout the year, producing 1 young at a time.

NAME: **Tomb Bat,** *Taphozous longimanus*
RANGE: **Madagascar, east to S. Asia**
HABITAT: **coconut groves, scrub, ruined tombs and palaces**
SIZE: **body: 7–9 cm (2¾–3½ in)**
　　wingspan: 25–33 cm (9¾–13 in)
　　tail: 2–3.5 cm (¾–1½ in)

Tomb bats are fond of roosting in man-made structures and are often found in tombs. Tomb bats appear in Chinese paintings of 2,000 years ago — almost the oldest recorded artistic impression of any bat. Neat little creatures with short shiny coats, tomb bats fly up as high as 100 m (330 ft) at dusk in search of their insect prey. As the night progresses, they gradually descend. While hunting, tomb bats emit loud cries, easily audible to human ears.

NAME: **White Bat,** *Diclidurus virgo*
RANGE: **S. Mexico, Central America**
HABITAT: **forest, open land**
SIZE: **body: 5–8 cm (2–3¼ in)**
　　wingspan: 18–30 cm (7–11¾ in)
　　tail: 1.5–2.5 cm (½–1 in)

The white bat, with its white fur and wing membrane, is a truly spectral creature. Since almost all other bats are dark, it is a mystery as to why this species should be white. However, it is just as successful at hunting insect prey as its dark relatives, which suggests that there is no evolutionary disadvantage in the light coloration. Another curious feature of this species is the presence of saclike glands in the tail membrane; there are no wing glands. The white bat roosts in caves or crevices, usually alone, occasionally in pairs. It appears to breed throughout the year.

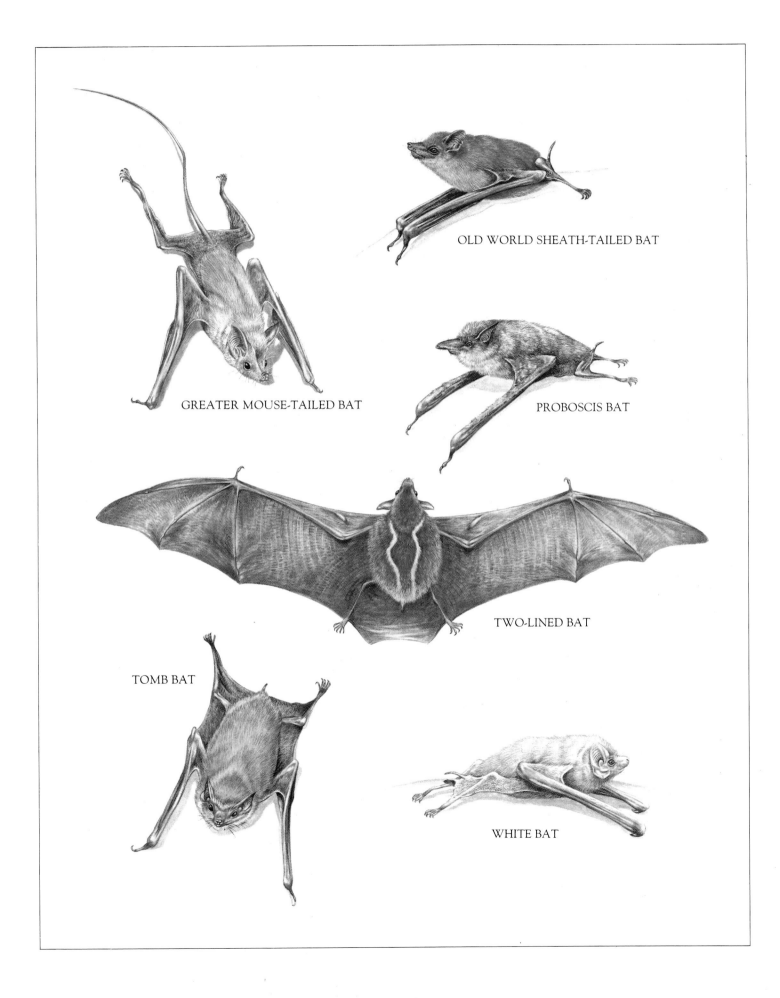

GREATER MOUSE-TAILED BAT

OLD WORLD SHEATH-TAILED BAT

PROBOSCIS BAT

TWO-LINED BAT

TOMB BAT

WHITE BAT

Hog-nosed Bat, Slit-faced Bats, False Vampires

CRASEONYCTERIDAE:
Hog-nosed Bat Family
The single species in this family was first discovered in 1973 in the karst region of western Thailand. It is the smallest bat yet recorded and may be the world's smallest mammal. Its distribution may be far wider than at present known.

NAME: **Hog-nosed Bat,** *Craseonycteris thonglongyai*
RANGE: **Thailand**
HABITAT: **limestone caves**
SIZE: **body: 3–3.5 cm (1¼ in)**
wingspan: 11–12.5 cm (4¼–5 in)
tail: absent

A fully grown, adult hog-nosed bat weighs no more than 2 g (1/14 oz). Its piglike nose is thought to be an adaptation for gleaning small insects and other invertebrates off the surface of leaves. At dusk, hog-nosed bats emerge from their roosting caves and forage around the tops of bamboo clumps and in the dense foliage of teak trees, which their tiny bodies can easily penetrate. Nothing is known of their breeding habits.

NYCTERIDAE:
Slit-faced Bat Family
There are 10 species in this family. All are distinguished by a pair of slits in the sides of the face which extend laterally from the nostrils to just above the eyes. Like most complex facial features in bats, these slits are thought to play a part in beaming ultrasonic "radar" signals. Another unusual feature of the slit-faced bat is the tail, which extends to the end of the tail membrane and then terminates in a "T" shaped bone, unique among mammals.

NAME: **Egyptian Slit-faced Bat,** *Nycteris thebaica*
RANGE: **Middle East, Africa, south of the Sahara; Madagascar**
HABITAT: **dry plains, forest**
SIZE: **body: 4.5–7.5 cm (1¾–3 in)**
wingspan: 16–28 cm (6¼–11 in)
tail: 4–7.5 cm (1½–3 in)

Slit-faced bats feed on a variety of invertebrate animals which they catch in trees or even on the ground; scorpions seem to be a particularly favoured food. The bats generally give birth to a single offspring in January or February and are thought to produce a second later in the year.

MEGADERMATIDAE:
False Vampire Family
In the struggle for survival among the many bat species, one family has evolved as specialist predators on other bats. The false vampire swoops silently down on a smaller bat and seems to chew it for a while before devouring the meat — a habit that led to the belief that these bats sucked blood. This is now known not to be the case and, in fact, not all false vampires are even carnivorous; those that are use their predatory skills only to supplement an insectivorous diet. There are 5 species of false vampire, all with long ears and fluted nose leaves.

NAME: **Greater False Vampire,** *Megaderma lyra*
RANGE: **India to Burma, S. China, Malaysia**
HABITAT: **forest, open land**
SIZE: **body: 6.5–8.5 cm (2½–3¼ in)**
wingspan: 23–30 cm (9–11¾ in)
tail: absent

The greater false vampire regularly supplements its diet of insects, spiders and other invertebrates with prey such as bats, rodents, frogs and even fish. Groups of 3 to 50 false vampires roost together and are usually the sole inhabitants of their caves. Presumably their predatory tendencies deter other bats from sharing their abode.

False vampires mate in November and 1 young is born after a gestation of 20 weeks. Shortly before females are due to give birth, males leave the roost and return 3 or 4 months later, to resume communal roosting.

NAME: **Heart-nosed Bat,** *Cardioderma cor*
RANGE: **E. central Africa: Ethiopia to Tanzania**
HABITAT: **forest, scrub**
SIZE: **body: 7–9 cm (2¾–3½ in)**
wingspan: 26–35 cm (10–13½ in)
tail: absent

The heart-nosed bat closely resembles the greater false vampire in appearance and way of life, but has a wider, heart-shaped nose leaf. Like its relatives, it feeds on vertebrate animals as well as insects. Before the heat of the day has passed, the heart-nosed bat emerges from its roost and swoops to catch lizards; it will even enter houses in pursuit of rodents or wall lizards. It is a strong flier and can take off from the ground carrying a load as big as itself. Heart-nosed bats attack smaller bats in flight, "boxing" them with their powerful wings to upset their directional stability.

NAME: **Australian False Vampire,** *Macroderma gigas*
RANGE: **N. and W. tropical Australia**
HABITAT: **forest with rock caves**
SIZE: **body: 11.5–14 cm (4½–5½ in)**
wingspan: 46–60 cm (18–23½ in)
tail: absent Ⓔ

On account of its pale coloration, the Australian false vampire is popularly known as the ghost bat. It is one of the most carnivorous of bats and feeds almost exclusively on mice, birds, geckos and other bats. Its method of attack is to flop down on the unsuspecting prey and enmesh it in its strong wings, then to deliver a single killing bite to the back of the prey's neck. It can rise from the ground carrying a dead rodent and fly to a feeding perch in cave or tree. Like all false vampires, this species has long ears which are joined by a membrane extending about half-way up their length.

Males forsake the communal roost in September or October, just before the young are due to be born. By January the young bats are as big as their mothers and accompany them on hunting trips. The males move back into the roosts by April. This handsome species is now rare and active conservation measures are urgently required to ensure its survival.

NAME: **Yellow-winged Bat,** *Lavia frons*
RANGE: **Africa: Senegal to Kenya**
HABITAT: **swamps, lakes in forest and open country**
SIZE: **body: 6.5–8 cm (2½–3 in)**
wingspan: 24–30 cm (9½–11¾ in)
tail: absent

The fur colour of the yellow-winged bat is variable, but the ears and wings are always yellowish-red. It roosts in trees and bushes, and only the flickering of its long ears gives away its presence. These bats often fly in the daytime but seem to feed only at night. In contrast to the other false vampires, yellow-winged bats appear to restrict themselves to insect food. Their method of hunting is rather like that of the flycatcher birds: they wait until an insect flies near them before swooping down from the perch to snap it up.

Yellow-winged bats breed throughout the year; males do not leave the communal roost for the birth season.

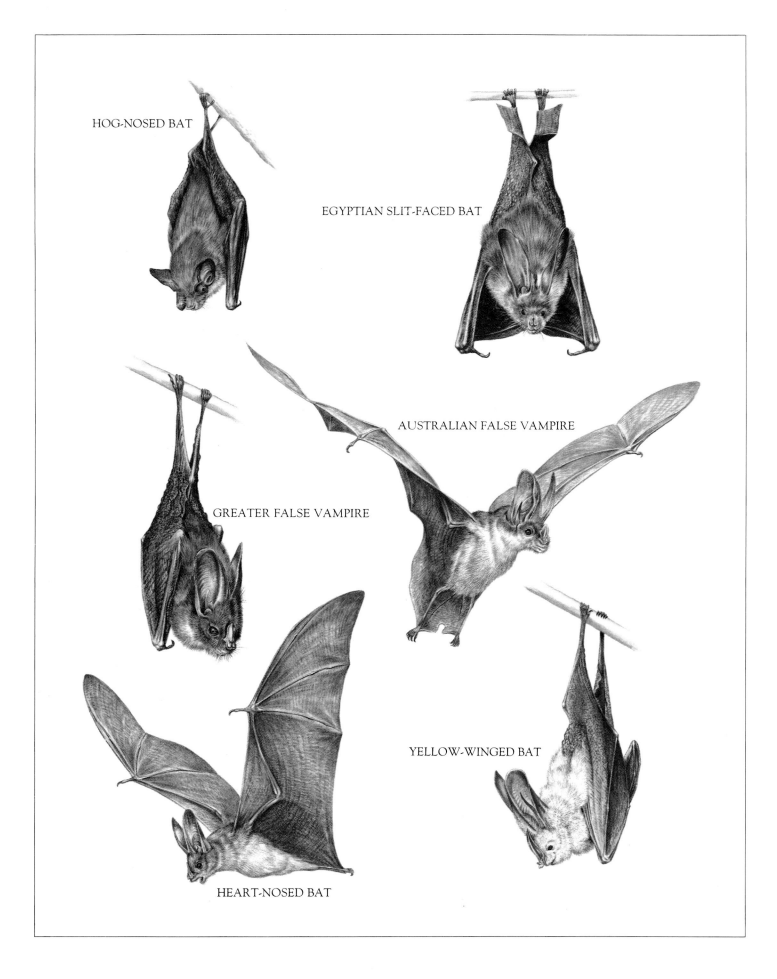

HOG-NOSED BAT

EGYPTIAN SLIT-FACED BAT

AUSTRALIAN FALSE VAMPIRE

GREATER FALSE VAMPIRE

YELLOW-WINGED BAT

HEART-NOSED BAT

Horseshoe Bats, Old World Leaf-nosed Bats

RHINOLOPHIDAE: Horseshoe Bat Family

The shape of the fleshy structure surrounding the nose distinguishes the horseshoe bat from other insectivorous bats. While most small bats emit their ultrasonic cries through open mouths, horseshoe bats "shout" through their nostrils. The nose leaf acts as an adjustable megaphone, enabling the bat to direct its "radar" beam wherever it wishes. Two other structures on the face, the lancet above the nostrils and the sella, which partially separates them, are immensely muscular and can vibrate at the same frequency as the sound pulse. Horseshoe bats are also distinctive in that, instead of folding their wings to their sides when roosting, they wrap them around their bodies, which makes them appear like the cocoons of enormous insects.

There are 51 species of horseshoe bat, found in temperate and tropical parts of the Old World as far east as Japan and Australia. Male and female generally look alike.

NAME: **Greater Horseshoe Bat,** *Rhinolophus ferrumequinum*
RANGE: **Europe, Asia, N. Africa**
HABITAT: **forest; open and cultivated land**
SIZE: **body: 11–12.5 cm (4½–5 in)**
wingspan: 33–35 cm (13–14 in)
tail: 2.5–4 cm (1–1½ in)

The greater horseshoe bat is rather slow and fluttering in flight and not adept at catching insects in the air. It feeds largely on the ground, swooping down on beetles with unerring accuracy. These bats hibernate from October to March and often choose winter quarters deep within caves, crevices or potholes. Thousands hibernate together and make long migrations to reach these quarters. The female gives birth to 1 young in April and carries it about until it is 3 months old.

NAME: **Lesser Horseshoe Bat,** *Rhinolophus hipposideros*
RANGE: **Europe, Asia, N. Africa**
HABITAT: **open country with caves**
SIZE: **body: 7–10 cm (2¾–4 in)**
wingspan: 22.5–25 cm (9–10 in)
tail: 1.5–2.5 cm (½–1 in)

The lesser horseshoe bat is similar to the greater and equally fluttering in flight, but it is more manoeuvrable and hunts far more in the air. In summer it roosts in trees, hollow logs and houses, making incessant chattering noises. In winter the bats make short migrations to winter hibernation quarters in caves, which are frost-free but not necessarily dry.

NAME: **Philippine Horseshoe Bat,** *Rhinolophus philippinensis*
RANGE: **Philippine Islands**
HABITAT: **primary forest, broken land**
SIZE: **body: 7–9 cm (2¾–3½ in)**
wingspan: 23–26 cm (9–10¼ in)
tail: 1.5–2.5 cm (½–1 in)

Within the rich bat fauna of its native islands, the Philippine horseshoe bat occupies its own special niche. It feeds on large slow-flying insects and on heavily armoured ground beetles and, using its sharp teeth, can slice through the thick wing cases and wings before devouring the insects.

Hibernation is not necessary in the Philippines climate and the bats remain active throughout the year. Breeding also occurs throughout the year. Young mature in their second year, and males are smaller than females.

HIPPOSIDERIDAE: Old World Leaf-nosed Bat Family

The 40 species of Old World leaf-nosed bats resemble the horseshoe bats and are sometimes classified with them as a subfamily. They have a horseshoe-shaped nose leaf and the area above the nostrils is often highly modified into prongs and forks, creating curious facial structures. The family is distributed in tropical and subtropical regions of Africa and southern Asia, east to Australia and the Solomon Islands. These bats are extremely numerous and of incalculable benefit to mankind, since they feed entirely on insects and destroy many insect pests.

NAME: **Persian Trident Bat,** *Triaenops persicus*
RANGE: **Egypt, east to Iran, south to the Gulf of Eilat**
HABITAT: **arid land, semi-desert**
SIZE: **body: 3.5–5.5 cm (1½–2¼ in)**
wingspan: 15–19 cm (6–7½ in)
tail: 1.5 cm (½ in)

The Persian trident bat and the other 2 species in its genus are distinguished from the rest of their family by the structures above the nose disc. These bats roost in underground cracks and tunnels. They emerge while it is still light to fly fast and low over the ground until they reach their feeding areas where, high in the foliage, small insects are hunted down.

Trident bats breed some time between December and May, the actual timing of the births coinciding with the rains. A single young is born and is left in the roost while its mother hunts.

NAME: **Flower-faced Bat,** *Anthops ornatus*
RANGE: **Solomon Islands**
HABITAT: **forest, mixed agricultural land**
SIZE: **body: 4.5–5 cm (1¾–2 in)**
wingspan: 14–16.5 cm (5½–6½ in)
tail: 0.5–1.5 cm (¼–½ in)

The flower-faced bat is known only from a handful of specimens collected early this century. It derives its name from the flowerlike appearance of the nose disc. This crenellated structure has several layers, superimposed one upon the other like the whorls of petals in a flower. Why such a bizarre device is required by this species is not known.

NAME: **Trident Leaf-nosed Bat,** *Asellia tridens*
RANGE: **N. Africa, east to India**
HABITAT: **arid scrub**
SIZE: **body: 5–6 cm (2–2¼ in)**
wingspan: 21–22 cm (8¼–8¾ in)
tail: 2–2.5 cm (¾–1 in)

The trident bat has a three-pronged nasal lancet. Hundreds of these bats roost together in underground tunnels and cracks. They emerge in early evening to skim over the surface of the land toward palm groves where, in the shade and moisture, insects abound. Beetles and moths are favoured foods.

NAME: **Large Malay Leaf-nosed Bat,** *Hipposideros diadema*
RANGE: **Thailand, east to Malaysia**
HABITAT: **forest, often near human habitation**
SIZE: **body: 7–10 cm (2¾–4 in)**
wingspan: 22.5–25 cm (9–10 in)
tail: 2.5–3 cm (1–1¼ in)

This sociable bat roosts in groups of many hundreds in caves or old buildings. At dusk the bats depart to hunt; they forage around flowers, snapping up insects, and occasionally tear open figs to dig out insect larvae, at the same time consuming fig pulp and seeds. The single young is usually born in November or December.

NAME: **Tail-less Leaf-nosed Bat,** *Coelops frithi*
RANGE: **Bangladesh through Indo-China to Java**
HABITAT: **forest**
SIZE: **body: 3–4.5 cm (1–1¾ in)**
wingspan: 11–13 cm (4¼–5¼ in)
tail: absent

The tail-less leaf-nosed bat has a much less complex nose disc than its relatives. Its ears are shorter and more rounded than is usual in this family. Small groups of a dozen or more bats shelter during the day in hollow trees or in human dwellings. In Java, it has been observed that births occur towards the end of February.

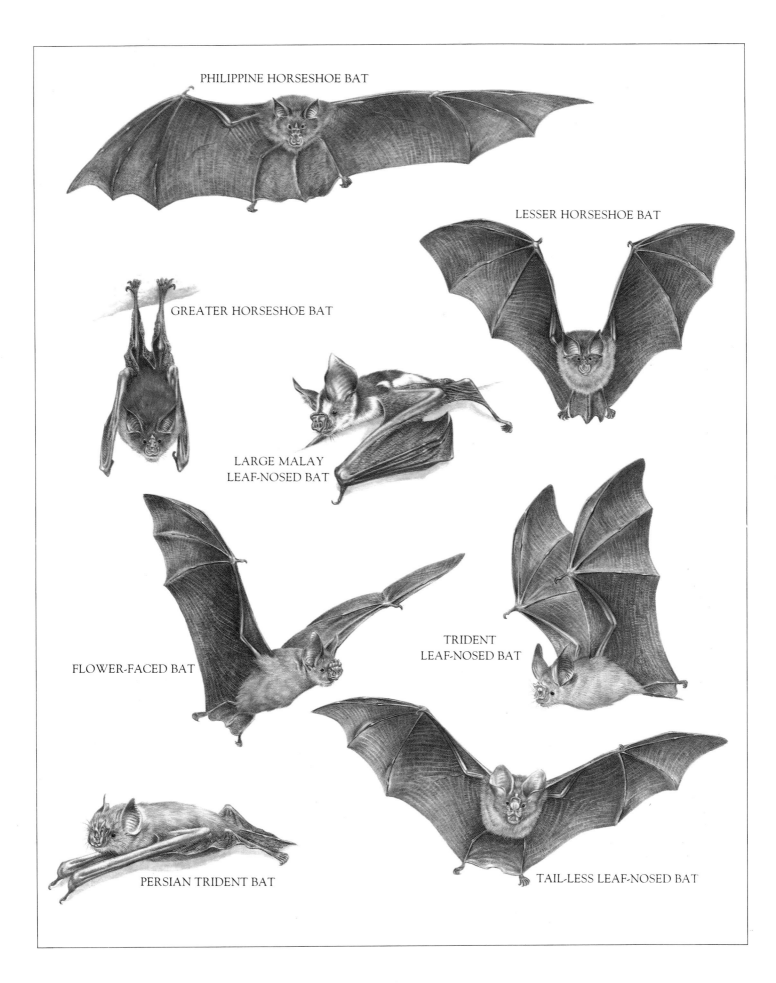

PHILIPPINE HORSESHOE BAT

LESSER HORSESHOE BAT

GREATER HORSESHOE BAT

LARGE MALAY
LEAF-NOSED BAT

FLOWER-FACED BAT

TRIDENT
LEAF-NOSED BAT

PERSIAN TRIDENT BAT

TAIL-LESS LEAF-NOSED BAT

Fisherman Bats, Moustached Bats, Vampire Bats, Free-tailed Bats

NOCTILIONIDAE:
Fisherman Bat Family
There are 2 species in this family, both found in Central America and northern South America. They inhabit swampy forests and mangroves and, as their name implies, feed on fish.

NAME: **Fisherman Bat,** *Noctilio leporinus*
RANGE: **Mexico, south to Argentina; Antilles, Trinidad**
HABITAT: **forest, mangrove swamps**
SIZE: **body: 10–13 cm (4–5¼ in)**
   wingspan: 28–30 cm (11–12 in)
   tail: 1–2.5 cm (½–1 in)

With powerful, stiff-winged flight, the fisherman bat swoops to within an inch of the water, dipping briefly to impale a small fish with its long sharp claws and scoop it into its mouth. Fisherman bats also use their feet to catch insects, both in the water and the air, and these make up a substantial part of their diet in some regions. Just how they locate their prey is unknown, but, since they emit floods of ultrasound while hunting, it is thought that they can detect tiny ripples made by a surfacing fish.

In November or December fisherman bats mate, and a single young is born after a gestation of about 16 weeks.

MORMOOPIDAE:
Moustached Bat Family
There are 8 species in this family which ranges from southern Arizona to Brazil. The bats derive their name from the fringe of hairs surrounding the mouth. Some species have wing membranes which meet and fuse in the middle of the back, which gives the back a naked appearance.

NAME: **Moustached Bat,** *Pteronotus parnelli*
RANGE: **N. Mexico to Brazil; West Indies**
HABITAT: **lowland tropical forest**
SIZE: **body: 4–7.5 cm (1½–3 in)**
   wingspan: 20–33 cm (8–13 in)
   tail: 1.5–3 cm (½–1¼ in)

In addition to the fringe of hairs surrounding its mouth, this bat has a plate-like growth on its lower lip and small fleshy papillae projecting down from the upper lip. This mouth structure aids the bat in the collection of its insect food.

Moustached bats are gregarious and roost in large groups in caves. They lie horizontally and do not hang in the usual manner of bats. Young are usually born in May when the bats' food supply is most abundant. Most females have just 1 young a year.

NAME: **Leaf-chinned Bat,** *Mormoops megalophylla*
RANGE: **Arizona to northern S. America; Trinidad**
HABITAT: **forest, scrub, near water**
SIZE: **body: 6–6.5 cm (2–2½ in)**
   wingspan: 25–28 cm (10–11 in)
   tail: 2.5 cm (1 in)

A pair of fleshy flaps on the chin, and peglike projections on the lower jaw distinguish this species. Leaf-chinned bats shelter in caves, tunnels and rock fissures. They hunt somewhat later than most bats, emerging after dark to look for insects. They fly close to the ground and often feed near pools and swamps.

DESMODONTIDAE:
Vampire Bat Family
Highly specialized mammals, vampire bats are totally adapted to a diet of blood. They are the only mammals to qualify for the title of parasite. There are 3 species.

NAME: **Vampire Bat,** *Desmodus rotundus*
RANGE: **N. Mexico to central Chile, Argentina and Uruguay**
HABITAT: **forest**
SIZE: **body: 7.5–9 cm (3–3½ in)**
   wingspan: 16–18 cm (6¼–7 in)
   tail: absent

The vampire, like most bats, hunts at night. It alights on the ground a few feet away from its victim and walks over on all fours. With its four razor-sharp canine teeth, the vampire makes a small incision on a hairless or featherless part of the animal. The edges of the bat's long, protruded tongue are bent downward and form a tube, through which saliva is pumped out to inhibit clotting and blood is sucked in.

About half an hour's feeding each night is sufficient for a vampire bat, and although the host loses only a small amount of blood, the bat may transmit various diseases in its saliva, including rabies.

MOLOSSIDAE:
Free-tailed Bat Family
The 80 species of free-tailed bat occur in the warmer parts of the Old and New World. The typical free-tailed bat has a rodentlike tail, which extends well beyond the free edge of the tail membrane, and rather narrow wings, which beat more rapidly than those of other insectivorous bats. Free-tailed bats feed entirely on insects, favouring hard-shelled species. They roost in vast hordes, and their droppings create guano used in the fertilizer industry.

NAME: **Egyptian Free-tailed Bat,** *Tadarida aegyptiaca*
RANGE: **N. Africa, Middle East**
HABITAT: **arid scrub land**
SIZE: **body: 5.5–7 cm (2¼–2¾ in)**
   wingspan: 17–19 cm (6¾–7½ in)
   tail: 3–5 cm (1¼–2 in)

One of the commonest mammals in the Middle East today, the Egyptian free-tailed bat roosts in groups of many thousands. Almost any sizeable crevice suffices, even if it is already occupied by a bird or another mammal. The bats mate in late winter and the female gives birth to a single young after a gestation of 77 to 84 days. A second pregnancy may follow immediately.

NAME: **Wroughton's Free-tailed Bat,** *Otomops wroughtoni*
RANGE: **S. India, Sri Lanka**
HABITAT: **open, partially forested land**
SIZE: **body: 9.5–11 cm (3¾–4¼ in)**
   wingspan: 28–30 cm (11–12 in)
   tail: 3–5 cm (1¼–2 in)

This species roosts in small groups or even alone and, although few specimens have been collected, it is probably not as rare as it seems. Females give birth to a single young in December.

NAME: **Velvety Free-tailed Bat,** *Molossus ater*
RANGE: **S. Mexico, Central America, Trinidad**
HABITAT: **scrub, savanna, forest**
SIZE: **body: 6–9 cm (2½–3½ in)**
   wingspan: 26–28 cm (10–11 in)
   tail: 3–4 cm (1¼–1½ in)

This species has dense, short-piled fur of velvety texture. It hunts insects and catches as many as possible at a time, cramming them into capacious cheek pouches. On returning to its roost, the bat devours the contents of the pouches. This hunting behaviour may have evolved to reduce the amount of time the bats are out and at risk from the attentions of predators. One or two litters are born in summer.

NAME: **Mastiff Bat,** *Eumops perotis*
RANGE: **S.W. USA, Mexico, N. South America**
HABITAT: **forest, near human habitation**
SIZE: **body: 7–10 cm (2¾–4 in)**
   wingspan: 28–30 cm (11–12 in)
   tail: 4–6 cm (1½–2¼ in)

Mastiff bats emerge from their roost after dark and hunt small insects, usually ants, bees and wasps, wherever they may be found. In the mating season, the throat glands of males swell enormously and the odorous secretions of these glands may attract females. Young are born in late summer, and twins are not uncommon.

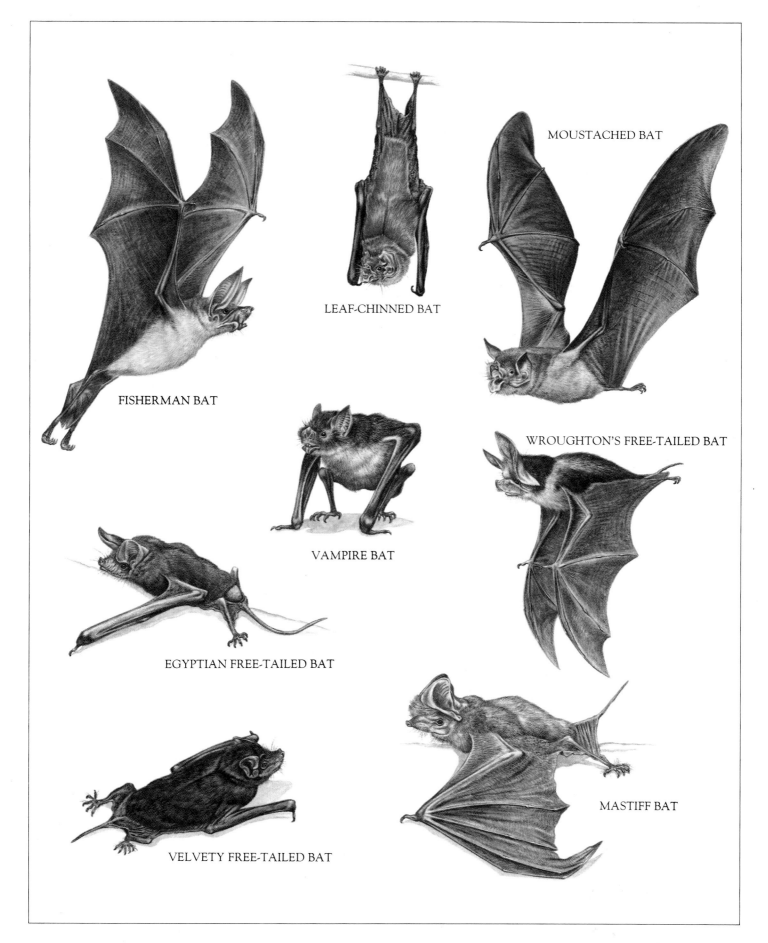

MOUSTACHED BAT

LEAF-CHINNED BAT

FISHERMAN BAT

WROUGHTON'S FREE-TAILED BAT

VAMPIRE BAT

EGYPTIAN FREE-TAILED BAT

MASTIFF BAT

VELVETY FREE-TAILED BAT

New World Leaf-nosed Bats

NAME: **Long-tongued Bat,** *Glossophaga soricina*
RANGE: **N. Mexico to Brazil, Paraguay and Argentina; West Indies**
HABITAT: **woodland, often arid**
SIZE: **body: 5–6.5 cm (2–2½ in)**
wingspan: 20–24 cm (7¾–9½ in)
tail: 0.5 cm (¼ in)

The mammalian equivalent of the hummingbird, this bat hovers in front of flowers and scoops up pollen and nectar from deep within them with the aid of its long tongue. Tiny bristles on its surface help the pollen to stick fast.

Female bats form maternity colonies in summer, rejoining their original roosts when the young are born. Twins do occur, but a single young is more common.

NAME: **American False Vampire,** *Vampyrum spectrum*
RANGE: **S. Mexico to Peru and Brazil; Trinidad**
HABITAT: **forest, often near human habitation**
SIZE: **body: 12.5–13.5 cm (5–5¼ in)**
wingspan: 80 cm–1 m (31½ in–3¼ ft)
tail: absent

This species, formerly believed to be a blood-sucker, is now known to kill and eat rodents, birds and other bats. It is surprisingly agile on all fours and stalks mice stealthily; it then kills by an accurate pounce, which breaks the prey's neck or shatters its skull. The bats breed in June and the females are most solicitous parents, licking their young incessantly and feeding them pieces of chewed mouse flesh as they approach weaning.

NAME: **Tent-building Bat,** *Uroderma bilobatum*
RANGE: **S. Mexico to Peru and Brazil; Trinidad**
HABITAT: **forest, plantations**
SIZE: **body: 5.5–7.5 cm (2¼–3 in)**
wingspan: 20–24 cm (7¾–9½ in)
tail: absent

The tent-building bat is alert and active by day and has developed a simple way of creating a shady refuge. Fan-shaped palm leaves are partially bitten through in a semicircular line, a third of the way along a frond. The end of the frond collapses, making a tent-shaped refuge, from which the bats fly out in search of ripe fruit. A colony of 20 or 30 bats may share a tent. Pregnant females establish maternity tents and young are born between February and April. The young remain in the maternity tent until they are able to fly. Males live alone or in small groups during the breeding season, returning to the females when the young are weaned.

PHYLLOSTOMATIDAE:
New World Leaf-nosed Bat Family

As night descends over the Central and South American jungles, leaf-nosed bats emerge from their roosts to feed on the pollen and nectar of flowers, as well as on the insects attracted to the blooms. There is such a variety of plants that about 140 species of leaf-nosed bat have evolved to exploit this particular way of life. Generally the bats are small, but the American false vampire has a wingspan which may exceed 1 m (3¼ ft), making it the largest New World bat.

Leaf-nosed bats may be tail-less or have tiny tails embedded in the tail membrane. The characteristic linking most members of the family is the presence of a nose leaf — a flap of tissue above the nostrils. Unlike the leaves of the Old World leaf-nosed bats (Hipposideridae), these are little more than simple flaps of skin.

These abundant little bats do much good by transferring pollen from one flower to another, and a number of plants have become specially adapted to pollination by bats by flowering at night and producing a heavy, musky odour, which attracts them.

NAME: **Spear-nosed Bat,** *Phyllostomus hastatus*
RANGE: **Belize to Peru, Bolivia and Brazil; Trinidad**
HABITAT: **forest, broken country**
SIZE: **body: 10–13 cm (4–5¼ in)**
wingspan: 44–47 cm (17¼–18½ in)
tail: 2.5 cm (1 in)

One of the larger American bats, the spear-nosed has virtually abandoned an insectivorous diet for a carnivorous one. It feeds on mice, birds and small bats and occasionally on insects and fruit. Huge flocks of these bats shelter in caves and abandoned buildings and at dusk depart together for their feeding grounds. Young are born in May and June in the communal roost.

NAME: **Yellow-shouldered Bat,** *Sturnira lilium*
RANGE: **N. Mexico to Paraguay and Argentina; Jamaica**
HABITAT: **lowland forest**
SIZE: **body: 6–7 cm (2¼–2¾ in)**
wingspan: 24–27 cm (9½–10½ in)
tail: absent

The yellow-shouldered bat feeds on ripe fruit and roosts alone or in small groups in old buildings, hollow trees or in the crowns of palm trees. In the north of their range, these bats breed throughout the year, but in the south, the young are born between May and July. Females generally produce 1 young at a time.

NAME: **Little Big-eared Bat,** *Micronycteris megalotis*
RANGE: **S.W. USA to Peru, Brazil; Trinidad, Tobago and Grenada**
HABITAT: **dry scrub to tropical rain forest**
SIZE: **body: 4–6.5 cm (1½–2½ in)**
wingspan: 16–20 cm (6¼–7¾ in)
tail: 0.5–1 cm (¼–½ in)

Little big-eared bats roost in small groups in a great variety of shelter holes. They emerge as dusk is falling and swoop around fruit trees, chasing heavy, slow-flying insects, such as cockchafers, or plucking cockroaches from the ground. They appear to supplement this diet with the pulp of fruit such as guava and bananas. Young are born between April and June, most females having a single offspring.

NAME: **Short-tailed Leaf-nosed Bat,** *Carollia perspicillata*
RANGE: **S. Mexico to S. Brazil**
HABITAT: **forest, plantations**
SIZE: **body: 5–6.5 cm (2–2½ in)**
wingspan: 21–25 cm (8¼–9¾ in)
tail: 0.5–1.5 cm (¼–½ in)

This bat feeds almost entirely on ripe fruit, such as bananas, figs, guavas and plantains, which it locates by smell. There is no defined breeding season for this species, and females produce their single youngster at any time of the year.

NAME: **Jamaican Fruit-eating Bat,** *Artibeus jamaicensis*
RANGE: **N. Mexico to Brazil and N. Argentina; West Indies**
HABITAT: **scrub, forest**
SIZE: **body: 7.5–9 cm (3–3½ in)**
wingspan: 23–26 cm (9–10¼ in)
tail: absent

This bat is one of the most efficient mammalian food processors: it feeds on fruit, which passes through its gut in as little as 15 minutes. There is no time for bacterial action to destroy material, so the bat is an important disseminator of seeds. The breeding season lasts from February to July, and females give birth to 1, sometimes 2, young.

NAME: **Cuban Flower Bat,** *Phyllonycteris poeyi*
RANGE: **Cuba**
HABITAT: **primary forest, cultivated land**
SIZE: **body: 7.5–8 cm (3–3¼ in)**
wingspan: 21–23 cm (8¼–9 in)
tail: 1–1.5 cm (½ in)

The Cuban flower bat has little or no nose leaf. It is a gregarious bat and roosts in thousands in caves and rock fissures. With its long, slender tongue, it sucks and laps up nectar and pollen from many types of flowers, and it also feeds on fruit. These bats breed throughout the year; young are left in the roost until they are able to fly.

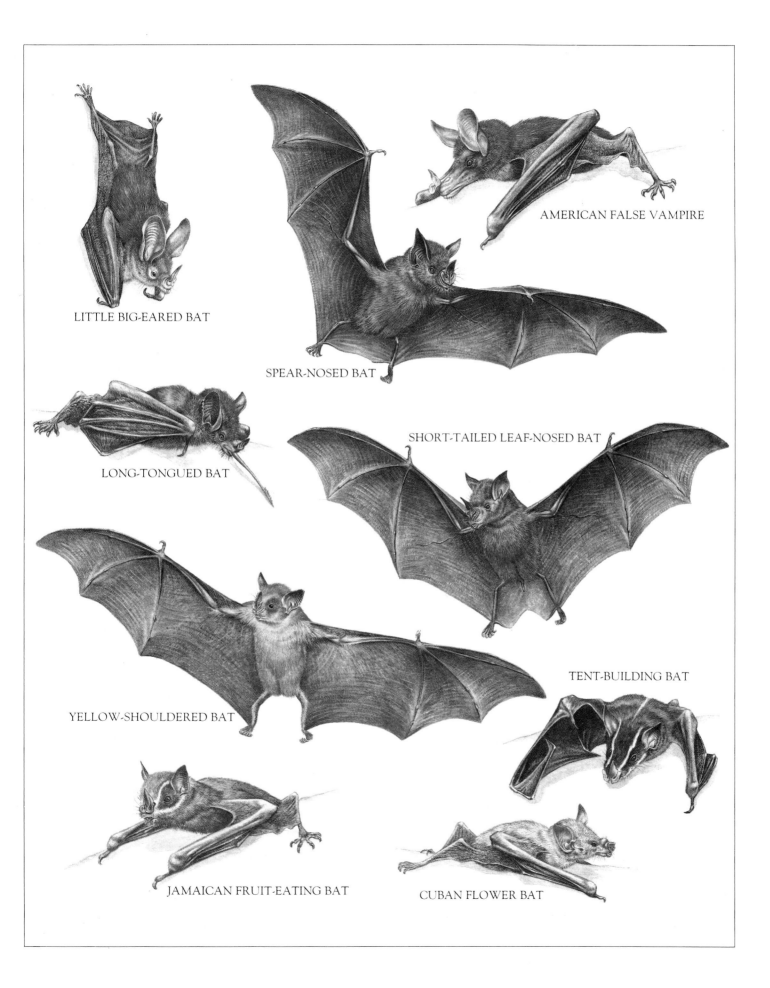

LITTLE BIG-EARED BAT

SPEAR-NOSED BAT

AMERICAN FALSE VAMPIRE

LONG-TONGUED BAT

SHORT-TAILED LEAF-NOSED BAT

YELLOW-SHOULDERED BAT

TENT-BUILDING BAT

JAMAICAN FRUIT-EATING BAT

CUBAN FLOWER BAT

Evening Bats

NAME: **Little Brown Bat,** *Myotis lucifugus*
RANGE: **N. America: from 62°N, south to Mexico**
HABITAT: **forest, built-up areas**
SIZE: **body: 4 cm (1½ in)**
　　wingspan: 14–18 cm (5½–7 in)
　　tail: 2.5 cm (1 in)

A common species in North America, the little brown bat adapts equally well to cold and hot climates. It feeds on whatever insects are abundant locally. In summer, the females segregate themselves and roost in maternity sites. A single young, sometimes twins, is born in May or June after a gestation of 50 to 60 days. The young are mature at 1 year.

In the warmer parts of their range, little brown bats do not hibernate, but northern populations may migrate hundreds of miles to suitable hibernation sites.

NAME: **Fish-eating Bat,** *Pizonyx vivesi*
RANGE: **coasts of Baja California and W. Mexico**
HABITAT: **caves, coastal rock piles**
SIZE: **body: 7–8.5 cm (2¾–3¼ in)**
　　wingspan: 25–32 cm (9¾–12½ in)
　　tail: 5–6.5 cm (2–2½ in)

A bat specialized for feeding on fish, this species has feet with long rakelike toes, ending in razor-sharp claws. Late in the evening, the bat flies low over the sea and takes small fish and crustaceans from the surface by impaling them on its claws. It is not known how it locates its prey, but its "radar" system might be capable of noting irregularities in the water caused by a fish moving close to the surface.

The female bat bears a single young in May or June and carries her infant with her until it is half-grown. It is then left in a secure roost with other young, while the mother hunts.

NAME: **Common Long-eared Bat,** *Plecotus auritus*
RANGE: **N. Europe, east to N.E. China and Japan**
HABITAT: **sheltered, lightly wooded areas**
SIZE: **body: 4–5 cm (1½–2 in)**
　　wingspan: 23–28 cm (9–11 in)
　　tail: 3–4.5 cm (1¼–1¾ in)

The distinguishing feature of this bat is its long ears, which are three-quarters the length of its head and body. In summer, these bats roost in buildings and trees and hunt primarily for the night-flying noctuid moths. They also feed on midges, mosquitoes and other flies, often picking them off vegetation in dive-bombing, swooping flights.

The female gives birth to a single young in June, and females and young form nursery roosts. Long-eared bats are mature at about a year old.

VESPERTILIONIDAE:
Evening Bat Family

There are some 275 species in this family, found around the world from the tropics to as far as about 68° North. Many species hibernate for 5 or 6 months to survive the winter in harsh northern latitudes. Nearly all species are insectivorous, although one or two feed on fish which they scoop from the water. Insects are usually caught in the air, the bat tossing the insect into its tail membrane with its wing. All these bats make use of echolocation for finding prey and for plotting their flight course.

Evening bats are extremely numerous in the cool, northern parts of the world. Without their massive consumption of blackflies, midges and mosquitoes, life for humans during the short northern summers would be distinctly more uncomfortable.

NAME: **Common Pipistrelle,** *Pipistrellus pipistrellus*
RANGE: **Europe, east to Kashmir**
HABITAT: **open land**
SIZE: **body: 3–4.5 cm (1¼–1¾ in)**
　　wingspan: 19–25 cm (7½–9¾ in)
　　tail: 2.5–3 cm (1–1¼ in)

Perhaps the commonest European bats, pipistrelles roost in groups of up to a thousand or more in lofts, church spires, farm buildings and the like. In winter, these bats migrate to a suitable dry cave to hibernate in colonies of 100,000 or more. They feed on insects, eating small prey in flight but taking larger catches to a perch to eat. Births occur in mid-June and twins have been recorded. The bats actually mate in September, prior to hibernation, and the sperm is stored in the female for 7 or 8 months before fertilization occurs and gestation starts.

NAME: **Big Brown Bat,** *Eptesicus fuscus*
RANGE: **N. America: Alaska to Central America; West Indies**
HABITAT: **varied, often close to human habitation**
SIZE: **body: 5–7.5 cm (2–3 in)**
　　wingspan: 26–37 cm (10¼–14½ in)
　　tail: 4.5–5.5 cm (1¾–2¼ in)

Big brown bats eat almost all insects except, it seems, moths, and also manage to catch water beetles. They have been recorded flying at a speed of 25 km/h (15½ mph). They will enter houses to hibernate, although huge numbers migrate to caves in Missouri and other southern states for this purpose. Young are born from April to July; a single young is the rule west of the Rockies, but twins are common in the east.

NAME: **Red Bat,** *Lasiurus borealis*
RANGE: **N. America, West Indies, Galápagos Islands**
HABITAT: **forested land with open space**
SIZE: **body: 6–8 cm (2¼–3¼ in)**
　　wingspan: 36–42 cm (14¼–16½ in)
　　tail: 4.5–5 cm (1¾–2 in)

The red bat's fur varies in shade from brick-red to rust, suffused with white. Males are more brightly coloured than females. The species is unique among bats in the size of its litters, regularly giving birth to 3 or 4 young in June or early July. The female carries her young with her at first, even though their combined weights may exceed her own body weight. North American red bats migrate southward in autumn and north again in spring.

NAME: **Noctule,** *Nyctalus noctula*
RANGE: **Europe, east to Japan**
HABITAT: **forest, open land**
SIZE: **body: 7–8 cm (2¾–3¼ in)**
　　wingspan: 32–35.5 cm (12½–14 in)
　　tail: 3–5.5 cm (1¼–2¼ in)

One of the largest evening bats, the noctule feeds mainly on maybugs, crickets and dorbeetles, and there are reports of its killing house mice. In winter, noctules hibernate in trees or lofts, usually in small groups, but roosts of a few hundred do occur. They breed in June, producing 1, or sometimes 2 or even 3, young.

NAME: **Barbastelle,** *Barbastella barbastellus*
RANGE: **Europe, east to central USSR**
HABITAT: **open land, often near water**
SIZE: **body: 4–5 cm (1½–2 in)**
　　wingspan: 24.5–28 cm (9½–11 in)
　　tail: 4–4.5 cm (1½–1¾ in)

In early evening, often before sunset, barbastelles emerge to hunt for insects, flying low over water or bushes. Males and females segregate for the summer and females form maternity colonies. From late September, barbastelles congregate in limestone regions to hibernate in deep, dry caves.

NAME: **Painted Bat,** *Kerivoula argentata*
RANGE: **Africa: S. Kenya, Namibia, Natal**
HABITAT: **arid woodland**
SIZE: **body: 3–5.5 cm (1¼–2¼ in)**
　　wingspan: 18–30 cm (7–11¾ in)
　　tail: 3–5.5 cm (1¼–2¼ in)

Small groups of painted bats are often found in the most unlikely roosts, such as in the suspended nests of weaver finches and sunbirds or under the eaves of African huts. Doubtless the bright and broken coloration of these bats is a form of camouflage to protect them while they roost in vulnerable sites. Nothing is known of their breeding biology.

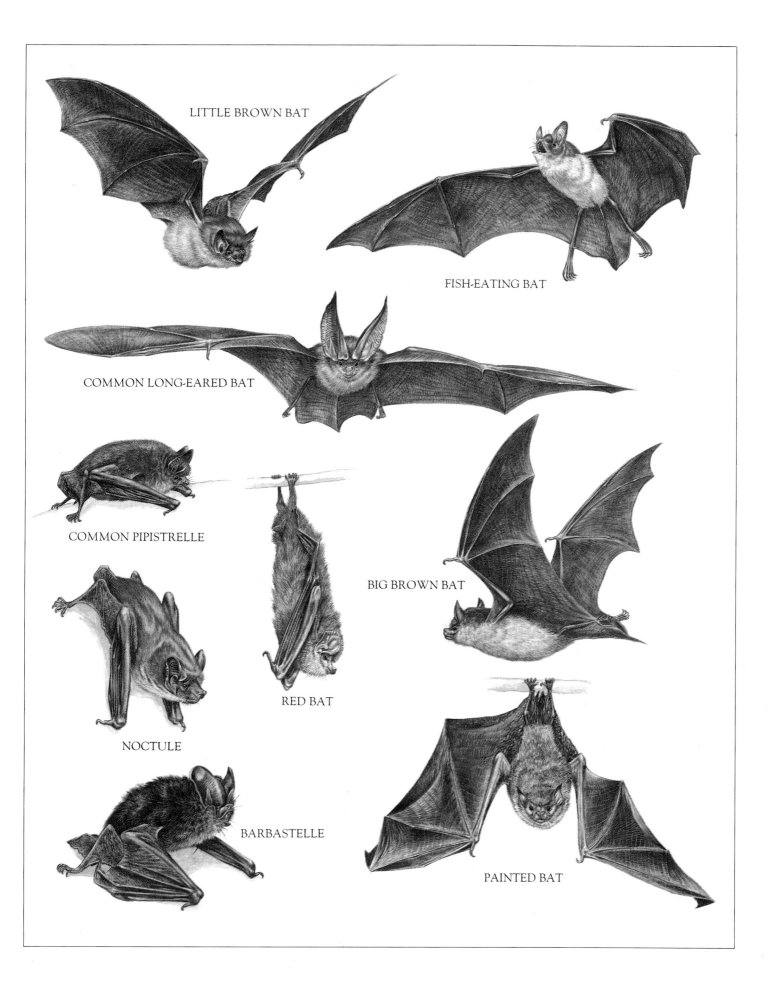

LITTLE BROWN BAT

FISH-EATING BAT

COMMON LONG-EARED BAT

COMMON PIPISTRELLE

BIG BROWN BAT

NOCTULE

RED BAT

BARBASTELLE

PAINTED BAT

Funnel-eared Bats and relatives

NATALIDAE: Funnel-eared Bat Family

The 3 species of funnel-eared bat occur in cavernous country in tropical South America. The family name comes from the large funnel-shaped ears, the outer surfaces of which bear glandular projections. Males and females look alike, except that on their muzzles males have thick glandular, or sensory, pads whose function is not known.

NAME: **Mexican Funnel-eared Bat, *Natalus stramineus***
RANGE: **N. Mexico to Brazil; Lesser Antilles**
HABITAT: **tropical lowlands**
SIZE: **body: 3–5.5 cm (1¼–2¼ in)**
   wingspan: 18–24 cm (7–9½ in)
   tail: 5–6 cm (2–2¼ in)

Funnel-eared bats roost in caves and mines, sometimes in huge numbers but more often in groups of about a dozen. The bats emerge at dusk to search for slow-flying insects, and they themselves fly with fluttering, rather mothlike movements. Breeding takes place at any time of the year, and the sexes segregate just before the young are born.

FURIPTERIDAE: Smoky Bat Family

There are 2 species in this family, both characterized by their smoky coloration, by being thumbless and by having extremely long tail membranes. The family occurs in tropical South America and in Trinidad.

NAME: **Smoky Bat, *Furipterus horrens***
RANGE: **Costa Rica, northern South America: E. Peru, Guianas, Brazil; Trinidad**
HABITAT: **forest**
SIZE: **body: 4–6 cm (1½–2¼ in)**
   wingspan: 22–30 cm (8½–11¾ in)
   tail: 4–6 cm (1½–2¼ in)

Unlike other bats, the smoky bat has no clawed thumbs, so that when it alights on the wall of its cave or tunnel roost, it must perform an aerial somersault to grasp the surface with its hind feet. Another curious feature of this bat and its fellow species is the high forehead: the snout and the brow join almost at right angles.

Practically nothing is known of the biology of either species of smoky bat because few zoologists have studied them. They certainly feed on insects, but how they avoid competing with the many other species of insect-eating bat within their range is not understood.

THYROPTERIDAE: Disc-winged Bat Family

The 2 species in this family occur in Central America and in tropical South America, as far as Peru and southern Brazil. They derive their name from suction pads at the base of each thumb and on the ankles. These discs are attached to the bat by short stalks and enable it to climb rapidly up smooth leaves and bare tree trunks. A curious characteristic of disc-winged bats is that they usually hang head upward when roosting — a feat made possible by the powerful suction discs.

NAME: **Honduran Disc-winged Bat, *Thyroptera discifera***
RANGE: **Belize, south to Ecuador and Peru**
HABITAT: **forest**
SIZE: **body: 3–5 cm (1¼–2 in)**
   wingspan: 18–25 cm (7–9¾ in)
   tail: 2.5–3 cm (1–1¼ in)

This small insect-eating bat has a similar high forehead to that of the smoky bat, but again the reasons for this are not known. It can climb up smooth surfaces, even glass, by means of its sucker discs; the suction pressure from a single disc is quite sufficient to support the entire weight of the bat. Suction is applied or released by specialized muscles in the forearms.

Disc-winged bats have unusual roosting habits: they roost in the young, curled leaves of plants such as bananas. These leaves, before they unfurl, form long tubes, and several bats may roost in a single leaf, anchored by their suckers to the smooth surface. The bats must find a new leaf practically every day as the previous roost unfurls.

Breeding takes place throughout the year, and the sexes do not appear to segregate before the females give birth. A single young is the normal litter and the mother carries it about with her until it weighs more than half her body weight. Her flying ability is not impaired by this extra load. Subsequently the young bat is left in the roost while she hunts.

MYZOPODIDAE: Sucker-footed Bat Family

The single species of sucker-footed bat is restricted to the island continent of Madagascar. The Myzopodidae is the only family of bats found solely on that island and represents a relict species, cut off from Africa when Madagascar broke free from that land mass.

NAME: **Sucker-footed Bat, *Myzopoda aurita***
RANGE: **Madagascar**
HABITAT: **forest**
SIZE: **body: 5–6 cm (2–2¼ in)**
   wingspan: 22–28 cm (8½–11 in)
   tail: 4.5–5 cm (1¾–2 in)

Superficially the sucker-footed bat resembles the disc-winged bats of the New World, for there are large suction discs at the base of each thumb and on each ankle. But the discs of this bat are quite immobile, in sharp contrast to those of the disc-winged bat which are connected to the body by mobile stalks, and they appear to be less efficient. The bat roosts inside curled leaves or hollow plant stems, and sometimes the smooth trunk of a tree may also serve it as a temporary nest.

Nothing is known of the breeding or feeding habits of these bats and they seem to be rather rare. Logging and the destruction of large areas of their forest habitat may be posing a threat to their survival.

MYSTACINIDAE: New Zealand Short-tailed Bat Family

The sole member of its family, the short-tailed bat is one of New Zealand's two native species of mammal (the other is a Vespertilionid bat called *Chalinolobus tuberculatus*).

NAME: **New Zealand Short-tailed Bat, *Mystacina tuberculata***
RANGE: **New Zealand**
HABITAT: **forest**
SIZE: **body: 5–6 cm (2–2¼ in)**
   wingspan: 22–28 cm (8½–11 in)
   tail: 1.5–2 cm (½–¾ in)

The short-tailed bat has a thick moustache fringing its small mouth; each bristle of the moustache has a spoon-shaped tip. Its thumbs bear not only the usual heavy claws but also have tiny secondary talons at their base, and its hind feet are equipped with razor-sharp claws.

An agile bat on all fours on the ground, the short-tailed bat runs rapidly, even up steeply sloping objects. To assist it in this, the wing membranes are furled in such a way that the forearms can be used as walking limbs. Beetles and other ground-dwelling insects seem to form the bulk of the diet of this species. They do not hibernate (the other New Zealand bat does) and roost in small groups in hollow trees. A single young is born in October.

MEXICAN FUNNEL-EARED BAT

SMOKY BAT

HONDURAN DISC-WINGED BAT

SUCKER-FOOTED BAT

NEW ZEALAND SHORT-TAILED BAT

Tree Shrews

NAME: **Common Tree Shrew,** *Tupaia glis*
RANGE: **S. and S.E. Asia: India to Vietnam and Malaysia, S. China, Indonesia**
HABITAT: **rain forest, woodland, bamboo scrub**
SIZE: **body: 14–23 cm (5½–9 in)**
tail: 12–21 cm (4¾–8¼ in)

A squirrel-like creature, with a long, bushy tail, the common tree shrew is active and lively, climbing with great agility in the trees but also spending much of its time on the ground, feeding. Its diet is varied and includes insects (particularly ants), spiders, seeds, buds and probably also small birds and mice. It normally lives alone or with a mate.

Breeding seems to occur at any time of year, and a rough nest is made in a hole in a fallen tree or among tree roots. In Malaysia, where breeding of this species has been most closely observed, females produce litters of 1 to 3 young after a gestation of 46 to 50 days. The newborn young are naked, with closed eyes, but are ready to leave the nest about 33 days after birth.

NAME: **Mountain Tree Shrew,** *Tupaia montana*
RANGE: **Borneo**
HABITAT: **montane forest**
SIZE: **body: 11–15 cm (4¼–6 in)**
tail: 10–15 cm (4–6 in)

The mountain tree shrew has a long, bushy tail and a slender, pointed snout. Although agile in trees, it spends much of its time on the ground, searching for food. Insects, fruit, seeds and leaves are all included in its diet, and it will sit back on its haunches to eat, holding the food in both its forepaws. This species is thought to be slightly more social than others of its genus and may live in small groups.

Breeding takes place at any time of year, and a litter, normally of 2 young, is born after a 49- to 51-day gestation.

NAME: **Madras Tree Shrew,** *Anathana elliotti*
RANGE: **central and S. India**
HABITAT: **rain forest, thorny woodland**
SIZE: **body: 16–18 cm (6¼–7 in)**
tail: 16–19 cm (6¼–7½ in)

The Madras tree shrew is squirrel-like and similar in most aspects to the tree shrews of the *Tupaia* genus. It is identified, however, by its larger ears, heavier snout and the pale stripe on each shoulder. Active during the day, it moves in trees and on the ground, searching for insects and probably fruit to eat. Little is known about its breeding habits, but they are probably similar to those of the *Tupaia* tree shrews.

ORDER SCANDENTIA
TUPAIIDAE: Tree Shrew Family

There is a single family in the order Scandentia, containing approximately 16 species of tree shrew, which live in the forests of eastern Asia, including Borneo and the Philippines. These biologically interesting, but visually undistinguished, mammals have affinities with the order Insectivora, because of their shrewlike appearance, and with the order Primates because of their complex, convoluted brains; they have been included in both of these orders. Most modern zoologists agree, however, that tree shrews should be placed in a distinct order so that their uniqueness is emphasized, rather than hidden in a large, diverse order. Until much more is known of their biology, tree shrews will remain an enigma.

Tree shrews resemble slim, long-nosed squirrels in general appearance, and their ears are squirrel-like in shape and relative size. Their feet are modified for an arboreal existence, having naked soles equipped with knobbly pads, which provide tree shrews with a superb ability to cling to branches. This ability is enhanced by the presence of long, flexible digits, with sharp, curved claws. The 16 species range from 10 to 22 cm (4 to 8½ in) in body length, with tails of 9 to 22.5 cm (3½ to 8¾ in). They run rapidly through the forest canopy, and most are active in the daytime, searching for insects and fruit to eat. They drink frequently and are also fond of bathing.

Despite their name, tree shrews are not exclusively arboreal, and many species spend a good deal of time on the ground. Their senses of smell, sight and hearing are good. Tree shrews usually live in pairs, and males, particularly, are aggressive toward one another. The borders of a pair's territory are marked with urine and glandular secretions.

There are 1 to 4 young in a litter, usually 1 or 2, born in a separate nest from the normal sleeping quarters. The female visits her young only once a day or even every other day, but they are able to take sufficient milk in a short period to sustain them during her long absences. Males and females look similar, but males are usually larger.

Tree shrews bear a close resemblance to fossils of the earliest mammals, so it may be assumed that the first true mammals looked, and possibly behaved, like these fascinating little animals.

NAME: **Bornean Smooth-tailed Tree Shrew,** *Dendrogale melanura*
RANGE: **Borneo**
HABITAT: **montane forest above 900 m (3,000 ft)**
SIZE: **body: 11–15 cm (4¼–6 in)**
tail: 9–14 cm (3½–5½ in)

The smallest of the tree shrews, this species is distinguished by its smooth, short-haired tail, which ends in a point. Its body fur, too, is short and close, dark reddish-brown on the back and lighter orange-buff on the underparts. More arboreal than the other members of its family, it finds much of its insect food on the lower branches of trees. Its breeding habits are not known.

NAME: **Philippine Tree Shrew,** *Urogale everetti*
RANGE: **Philippines: Mindanao**
HABITAT: **rain forest, montane forest**
SIZE: **body: 17–20 cm (6¾–7¾ in)**
tail: 11–17 cm (4¼–6¾ in)

A particularly elongate snout and a rounded, even-haired tail are characteristics of the Philippine tree shrew. Its fur is brownish, but with orange or yellow underparts. Active during the day, it climbs well and runs fast on the ground. Its diet is varied and includes insects, lizards, young birds and birds' eggs and fruit.

In the wild, Philippine tree shrews are thought to nest on the ground or on cliffs. Their breeding habits have been observed in captivity, where females have produced 1 or 2 young after a gestation of 54 to 56 days.

NAME: **Feather-tailed Tree Shrew,** *Ptilocercus lowi*
RANGE: **Malaysia, Sumatra, Borneo**
HABITAT: **rain forest**
SIZE: **body: 12–14 cm (4¾–5½ in)**
tail: 16–18 cm (6¼–7 in)

This tree shrew is easily identified by its unusual tail, which is naked for much of its length but has tufts of hair on each side of the terminal portion, making it resemble a feather. Other features characteristic of this species are its ears, which are large and membranous and stand away from the head, and its hands and feet, which are larger, relative to body size, than those of other tree shrews. Said to be nocturnal, the feather-tailed tree shrew spends much of its life in trees and is a good climber, using its tail for balance and support and spreading its toes and fingers wide for grip. Insects, fruit and some lizards are its main foods.

Feather-tailed tree shrews nest in holes in trees or branches, well off the ground, but their breeding habits are not known. They generally live in pairs.

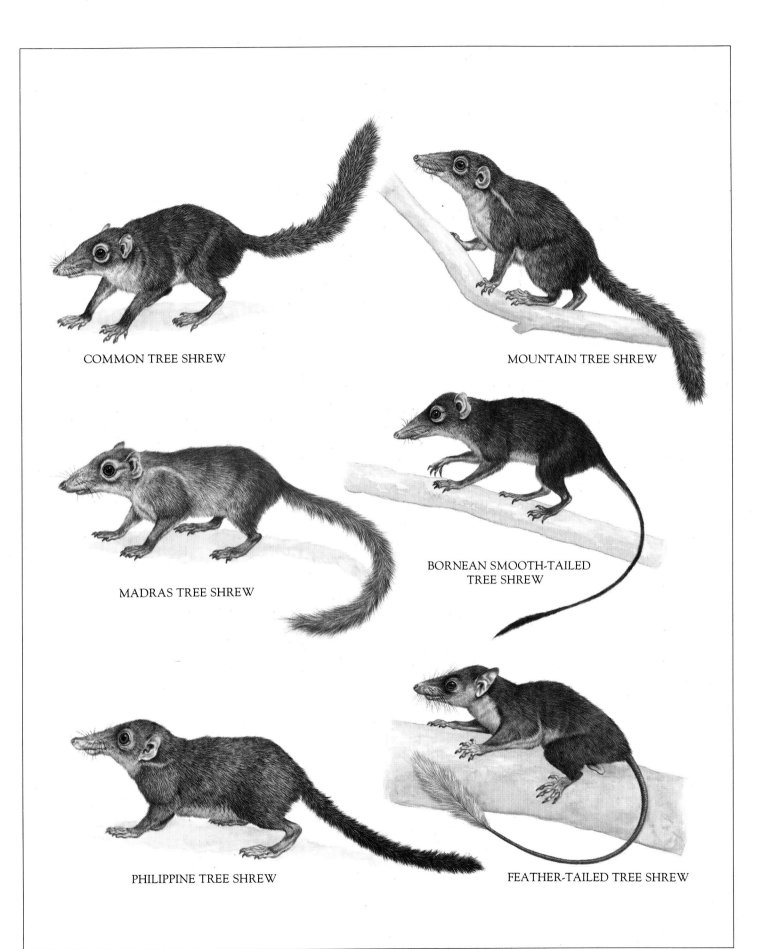

COMMON TREE SHREW

MOUNTAIN TREE SHREW

MADRAS TREE SHREW

BORNEAN SMOOTH-TAILED
TREE SHREW

PHILIPPINE TREE SHREW

FEATHER-TAILED TREE SHREW

Mouse-lemurs, Lemurs, Aye-aye

ORDER PRIMATES

There are about 179 species of primate, divided into two main groups: first, the prosimians, or primitive primates, which include lemurs, the aye-aye, lorises and tarsiers; and second, the higher primates such as marmosets, monkeys, apes and man.

CHEIROGALEIDAE:
Mouse-lemur Family

The 7 species of mouse-lemur occur in forest areas throughout Madagascar and are among the smallest primates.

NAME: **Russet Mouse-lemur**, *Microcebus rufus*
RANGE: **E. Madagascar**
HABITAT: **forest**
SIZE: **body: 12.5–15 cm (5–6 in)**
tail: 12.5–15 cm (5–6 in)

Mainly nocturnal in its habits, this tiny primate moves swiftly and nimbly on fine branches amid dense foliage, using its tail as a balancing organ, and will leap across gaps between trees. It also comes down to the ground to forage in leaf litter for beetles. Insects and small vertebrates are its main foods, supplemented by fruit and buds. Though they normally move and hunt alone, russet mouse-lemurs often sleep in small groups in nests made in hollow trees or constructed from leaves.

LEMURIDAE: Lemur Family

There are about 16 species of lemur, all found in Madagascar and the nearby Comoro Islands. Most live in wooded areas and are agile tree-climbers.

NAME: **Ring-tailed Lemur**, *Lemur catta*
RANGE: **S. Madagascar**
HABITAT: **dry, rocky country with some trees**
SIZE: **body: 45 cm (17¾ in)**
tail: 55 cm (21½ in)

The ring-tailed lemur has a pointed muzzle, large eyes and triangular ears. Its fur is thick and soft, and its bushy, distinctively ringed tail accounts for more than half its total length. Both sexes have special scent glands on the lower forelimbs; males have larger glands than females, with a horny spur near each. Males also have scent glands on the upper arms, under the chin and by the penis, while females have scent glands in the genital region. The lemurs use the secretions in these glands to mark territory boundaries, but will also mark whenever excited or disturbed. Troops of up to 20, sometimes 40, animals occupy a territory, with females and young

forming the core, while males move from troop to troop. Females are dominant over all males. Ring-tailed lemurs are active throughout the day, sometimes climbing up trees but also spending much time on the ground on all fours, with tail erect. They feed on fruit, leaves, bark, grass and resin, which they chisel from the trees with their lower incisors.

Females usually produce 1 young at a time, sometimes 2 or 3, after a gestation of about 136 days. The young is born well haired and with eyes open and is independent at 6 months.

NAME: **Ruffed Lemur**, *Varecia variegata*
RANGE: **N.E. and E. Madagascar**
HABITAT: **rain forest**
SIZE: **body: 60 cm (23½ in)**
tail: 60 cm (23½ in)

Distinguished by its long ruff, this lemur has white and black, brown or rufous fur. It is a nimble climber, most active at dusk and during the first part of the night, when it forages for fruit, leaves and bark; it rarely descends to the ground.

In November, the female produces 1 to 3 young in a nest in a hole in a tree or on a forked branch, which she lines with her own fur. The gestation period is 99 to 102 days.

INDRIIDAE:
Leaping Lemur Family

The 4 species of leaping lemur occur in the scrub country and forests of Madagascar. In all of these lemurs the snout is shortened and bare of fur, giving them a resemblance to monkeys.

NAME: **Verreaux's Sifaka**, *Propithecus verreauxi*
RANGE: **W. and S. Madagascar**
HABITAT: **dry and rain forest**
SIZE: **body: 45 cm (17¾ in)**
tail: 55 cm (21½ in) Ⓔ

This large, long-limbed sifaka has a naked black face, large eyes, and ears nearly concealed by fur. Coloration is highly variable, ranging from yellowish-white to black or a reddish-brown. Troops of up to 10 animals occupy a well-defined territory, the boundaries of which are marked with urine or with secretions of the male's throat gland. They feed in the morning and afternoon on leaves, buds and fruit and are relaxed in their movements, spending much of the day resting and sun-bathing. Sifakas are primarily arboreal but sometimes come down to the ground.

Young are born from the end of June to August after a gestation of about 130 days. Females usually produce 1 young, which is suckled for about 6 months.

NAME: **Woolly Lemur**, *Avahi laniger*
RANGE: **E. and N.W. Madagascar**
HABITAT: **forest**
SIZE: **body: 30–45 cm (11¾–17¾ in)**
tail: 33–40 cm (13–15¾ in) Ⓥ

A nocturnal animal, the long-limbed woolly lemur sleeps up in the trees by day and is active at night, when it searches for fruit. leaves and buds to eat. It is an agile climber and only occasionally descends to the ground, where it moves in an upright position.

Young are usually born in late August or September after a gestation of about 150 days. The female normally produces a single young, which she suckles for about 6 months.

NAME: **Indri**, *Indri indri*
RANGE: **N.E. Madagascar**
HABITAT: **forest up to 1,800 m (6,000 ft)**
SIZE: **body: 61–71 cm (24–28 in)**
tail: 3–6 cm (1¼–2¼ in) Ⓔ

The largest of the lemurs, the indri is immediately identified by its stumpy tail. It has a longer muzzle than the sifakas and a naked black face. Indris live in family groups and are active during the day at all levels of the forest, searching for leaves, shoots and fruit.

Mating takes place in January or February, and females give birth to 1 young after a gestation of 4½ to 5½ months.

DAUBENTONIIDAE:
Aye-aye Family

The single species of aye-aye is a nocturnal, arboreal animal, found in the dense forests of Madagascar.

NAME: **Aye-aye**, *Daubentonia madagascariensis*
RANGE: **formerly N.E. Madagascar, now only in nature reserves**
HABITAT: **rain forest**
SIZE: **body: 36–44 cm (14¼–17¼ in)**
tail: 50–60 cm (19¾–23½ in) Ⓔ

Well adapted for life in the trees, the extraordinary aye-aye has specialized ears, teeth and hands. It emerges from its nest at night to search for food, mainly insect larvae, plant shoots, fruit and eggs. All its digits are long and slender, but the third is particularly long, and using this, it taps on a tree trunk to locate wood-boring insects. It listens for movement with its large, sensitive ears, and when a grub is located, it probes with its long finger to winkle out the prey. Sometimes it will first tear open the wood with its powerful teeth. It also uses its teeth to open eggs or coconuts.

Every 2 or 3 years, females produce a single young, which they suckle for over a year.

RUSSET MOUSE-LEMUR

RUFFED LEMUR

RING-TAILED LEMUR

WOOLLY LEMUR

VERREAUX'S SIFAKA

INDRI

AYE-AYE

Lorises, Tarsiers

LORISIDAE: Loris Family

The 12 species that make up this family of primates occur in Africa, southern India, Southeast Asia and neighbouring islands. Most specialists divide the family into two groups: first, the lorises, potto and angwantibo, which have short tails or no tails at all and are slow and deliberate in their movements; and second, the bushbabies and galagos, which have long limbs and tails and are expert leapers.

NAME: **Slender Loris**, *Loris tardigradus*
RANGE: **Sri Lanka, S. India**
HABITAT: **rain forest, open woodland, swamp forest**
SIZE: **body: 18–26 cm (7–10¼ in)**
tail: absent or vestigial

The slender loris spends most of its life in trees, where it moves slowly and deliberately on its long, thin limbs. It has a strong grip with its efficient grasping hands, and its thumbs and great toes are opposable. A nocturnal animal, it spends the day sleeping up in the trees, its body rolled up in a ball. Toward evening, it becomes active and hunts for insects — particularly grasshoppers — lizards, small birds and their eggs, as well as some shoots and leaves. It approaches prey stealthily, with its usual deliberate movements, and then quickly grabs it with both hands.

In India, the slender loris is known to breed twice a year, births occurring most often in May and December. Usually 1 young, sometimes 2, is born, which makes its own way to the mother's teats, clinging to her fur.

NAME: **Slow Loris**, *Nycticebus coucang*
RANGE: **S. and S.E. Asia: E. India to Malaysia; Sumatra, Java, Borneo, Philippines**
HABITAT: **dense rain forest**
SIZE: **body: 26–38 cm (10¼–15 in)**
tail: vestigial

A plumper, shorter-limbed animal than its relative the slender loris, the slow loris is, however, similar in its habits. It spends the day sleeping up in a tree, its body rolled into a tight ball. At night, it feeds in the trees on insects, birds' eggs, small birds and shoots and fruit, seldom coming down to the ground. A slow, but accomplished, climber, its hands and feet are strong and capable of grasping so tightly that it can hang by its feet. The thumb and great toe are opposable to the digits.

Breeding takes place at any time of year, and 1 young, sometimes 2, is born after a gestation of 193 days. Slow lorises are thought to live in family groups.

NAME: **Potto**, *Perodicticus potto*
RANGE: **W., central and E. Africa**
HABITAT: **forest, forest edge**
SIZE: **body: 30–40 cm (11¾–15¾ in)**
tail: 5–10 cm (2–4 in)

A thickset animal with dense fur, the potto has strong limbs and grasping feet and hands; the great toe and thumb are opposable. At the back of its neck are four horny spines, projections of the vertebrae that pierce the thin skin. The potto sleeps up in the trees during the day and hunts at night. Although slow and careful in its movements, the potto can quickly seize insect prey with its hands; snails, fruit and leaves are also eaten.

The female gives birth to 1 young a year, after a gestation of 6 to 6½ months. Although weaned at 2 or 3 months, the young may stay with its mother for up to a year.

NAME: **Angwantibo**, *Arctocebus calabarensis*
RANGE: **W. Africa: Nigeria, Cameroon, south to Congo River**
HABITAT: **dense rain forest**
SIZE: **body: 25–40 cm (9¾–15¾ in)**
tail: about 1.25 cm (½ in)

Also known as the golden potto because of the sheen to its fur, the arboreal angwantibo has strong hands and feet, adapted for grasping branches. Its first finger is a mere stump, and the second toe is much reduced. A skilled, but slow, climber, the angwantibo is active at night, when it feeds mostly on insects, especially caterpillars, and also on snails, lizards and fruit. Outside the breeding season, it is usually solitary.

The female gives birth to 1 young after a gestation of 131 to 136 days. Her offspring is weaned at 4 months and fully grown at 7.

NAME: **Greater Bushbaby**, *Otolemur crassicaudatus*
RANGE: **Africa: Somalia, Kenya, south to South Africa: Natal**
HABITAT: **forest, wooded savanna, bushveld, plantations**
SIZE: **body: 27–47 cm (10½–18½ in)**
tail: 33–52 cm (13–20½ in)

The greater bushbaby is a strongly built animal, with a pointed muzzle and large eyes. Its hands and feet are adapted for grasping, with opposable thumbs and great toes. Much of its life is spent in the trees, where it is active at night and feeds on insects, reptiles, birds and birds' eggs and plant material. It makes a rapid pounce to seize prey and kills it with a bite. It has a call like the cry of a child (hence the name bushbaby) made most frequently in the breeding season.

The female gives birth to a litter of 1 to 3 young between May and October, after a gestation of 126 to 136 days.

NAME: **Lesser Bushbaby**, *Galago senegalensis*
RANGE: **Africa: south of the Sahara to the Vaal River (not Guinea or central African rain forest)**
HABITAT: **savanna, bush, woodland**
SIZE: **body: 14–21 cm (5½–8¼ in)**
tail: 20–30 cm (7¾–11¾ in)

More active and lively than its relative the greater bushbaby, this species moves with great agility in the trees and hops and leaps with ease. Like all its family, it sleeps up in the trees during the day and hunts for food at night. Spiders, scorpions, insects, young birds, lizards, fruit, seeds and nectar are all included in its diet. It lives in a family group, the members of which sleep together but disperse on waking.

Breeding habits vary slightly in different areas of the range. In regions where there are two rainy seasons, females have two litters a year, each usually of only 1 offspring; where there is only one rainy season, females produce one litter, often, but not always, of twins. The gestation period varies between 128 and 146 days, and the young are fully grown at about 4 months.

TARSIIDAE: Tarsier Family

The 3 species in this family occur in the Philippines and Indonesia. These little prosimian primates are primarily nocturnal and arboreal and climb and jump with great agility. They are better adapted to leaping than any other primates, with their elongate hind legs and feet. Their long tails are naked.

NAME: **Western Tarsier**, *Tarsius bancanus*
RANGE: **Sumatra, Borneo**
HABITAT: **secondary forest, scrub**
SIZE: **body: 8.5–16 cm (3¼–6¼ in)**
tail: 13.5–27 cm (5¼–10½ in)

The western tarsier, like its two relatives, is identified by its long, naked tail and extremely large, round eyes. Its forelimbs are short and its hind limbs long because of the adaptation of the tarsus, or ankle bones; this enables the tarsier to leap. The specialized tarsus is, of course, the origin of both the common and scientific names. A nocturnal, mainly arboreal animal, the tarsier sleeps during the day, clinging to a branch with its tail. At dusk, it wakes to prey on insects, its main food, which it catches by making a swift pounce and seizing the insect in its hands.

Breeding occurs at any time of year, and females give birth to 1 young after a gestation of about 6 months. The young tarsier is born well furred, with its eyes open, and is capable of climbing and hopping almost immediately.

SLENDER LORIS

SLOW LORIS

POTTO

ANGWANTIBO

GREATER BUSHBABY

LESSER BUSHBABY

WESTERN TARSIER

Marmosets and Tamarins

NAME: **Goeldi's Marmoset,** *Callimico goeldii*
RANGE: **upper Amazon basin**
HABITAT: **scrub, forest**
SIZE: **body: 18–21.5 cm (7–8½ in)**
　tail: 25–32 cm (9¾–12½ in)　　Ⓘ

Few Goeldi's marmosets have been seen or captured, and details of their habits are not well known. This marmoset is identified by the long mane around its head and shoulders and by some long hairs on its rump. It forages in all levels of the trees and bushes, searching for plant matter, such as berries, and for insects and small vertebrates. An agile animal, it walks and runs well and leaps expertly from branch to branch. It often goes down to the ground and will seek refuge on the ground when alarmed by a predator such as a bird of prey.

Strong, long-lasting pair bonds exist between males and females, and Goeldi's marmosets usually live in family groups of parents and offspring. The female bears a single young as a rule, after a gestation of 150 days. These marmosets are highly vocal and communicate with a variety of trills and whistles.

The exact state of the population of this species of marmoset is uncertain, but it is known to be rare and to have a patchy, localized distribution. In recent years it has suffered badly from the destruction of large areas of forest and from illegal trapping. More information is needed on these marmosets in order to set up suitable reserves and to ensure the survival of the species.

NAME: **Pygmy Marmoset,** *Cebuella pygmaea*
RANGE: **upper Amazon basin**
HABITAT: **tropical forest**
SIZE: **body: 14–16 cm (5½–6¼ in)**
　tail: 15–20 cm (6–7¾ in)

The smallest marmoset and one of the smallest primates, the pygmy marmoset is active in the daytime but may rest at noon. Particularly vulnerable to attack by large birds of prey because of its size, it tries to keep out of sight to avoid danger and moves either extremely slowly or in short dashes, punctuated by moments of frozen immobility. Its cryptic coloration also helps to hide it from predators. It is primarily arboreal and sleeps in holes in trees but does come down to the ground occasionally to feed or to move from one tree to another. It eats fruit, insects, small birds and birds' eggs and is also thought to feed on sap from trees, which it obtains by gnawing a hole in the bark.

Pygmy marmosets live in troops of 5 to 10 individuals in which the females are dominant.

CALLITRICHIDAE:
Marmoset and Tamarin Family

Marmosets and tamarins make up 1 of the 2 families of primates to occur in the New World. Together with the monkeys, family Cebidae, they are known as the flat-nosed, or platyrrhine, monkeys. There are about 17 species of marmoset and tamarin, but more may sometimes be listed, depending on whether some variations are regarded as subspecies or species in their own right. Apart from the mouse-lemurs, marmosets are the smallest primates, varying from mouse- to squirrel-size. Their fur is soft, often silky, and many have tufts or ruffs of fur on their heads. Their tails are furry and not prehensile.

Active in the daytime, marmosets and tamarins are primarily tree-dwelling but do not have grasping hands or opposable thumbs like most primates. Neither do they swing from branch to branch, but they are rapid and agile in their movements and bound swiftly through the trees in a similar manner to squirrels. Their diet is varied, including both plant and animal material. At night, marmosets sleep curled up in holes in trees.

They are social animals and live in small family groups. The usual number of young is 2, and the male assists his mate by carrying one of the twins on his back. Often extremely vocal, marmosets make a variety of high-pitched cries.

NAME: **Silvery Marmoset,** *Callithrix argentata*
RANGE: **Brazil, Bolivia**
HABITAT: **forest, tall grass**
SIZE: **body: 15–30 cm (6–11¾ in)**
　tail: 18–40 cm (7–15¾ in)　　Ⓥ

Recognized by its silky, silvery white body fur, this marmoset has no hair on its face and ears, which are reddish in colour; there is often some grey on its back, and its tail is black. With quick, jerky movements, the silvery marmoset runs and hops in trees and bushes, looking for food such as fruit, leaves, insects, spiders, small birds and birds' eggs. It usually moves in groups of 2 to 5. Like most marmosets, it has a range of expressions, including facial grimaces and raising of the eyebrows, which are used to threaten rivals or enemies.

The female gives birth to 1 or 2 young, occasionally 3, after a gestation of 140 to 150 days. The male assists at the birth and is largely responsible for the care of the young.

NAME: **Golden Lion Tamarin,** *Leontopithecus rosalia*
RANGE: **S.E. Brazil**
HABITAT: **coastal forest**
SIZE: **body: 19–22 cm (7½–8½ in)**
　tail: 26–34 cm (10¼–13¼ in)　　Ⓔ

A beautiful animal, this tamarin has a silky golden mane covering its head and shoulders and concealing its ears. In common with the other members of its family, it leaps from branch to branch with great agility as it searches for fruit, insects, lizards, small birds and birds' eggs to eat. It lives in small groups and is highly vocal.

The female gives birth to 1 or 2 young, rarely 3, after a gestation of 132 to 134 days. The father assists in the care of the young, giving them to the mother at feeding time and later preparing their first solid food by squashing and softening it in his fingers.

NAME: **Emperor Tamarin,** *Saguinus imperator*
RANGE: **W. Brazil, E. Peru, N. Bolivia**
HABITAT: **lowland forest**
SIZE: **body: 18–21 cm (7–8¼ in)**
　tail: 25–32 cm (9¾–12½ in)　　Ⓘ

Easily identified by its flowing white moustache, the emperor tamarin is one of 11 species included in this genus. It is an active, agile animal, moving with quick, jerky movements in the shrubs and trees as it searches for fruit, tender vegetation, insects, spiders, small vertebrate animals and birds' eggs. It lives in small groups and makes a great variety of shrill sounds.

The female produces twin offspring after a gestation of about 5 months. The father assists at the birth and cleans the newly born young. Like many other marmosets, he also helps at feeding time by handing the baby to the mother.

NAME: **Black and Red Tamarin,** *Saguinus nigricollis*
RANGE: **S. Colombia to adjacent areas of Ecuador and Brazil**
HABITAT: **primary and secondary forest**
SIZE: **body: 15–28 cm (6–11 in)**
　tail: 27–42 cm (10½–16½ in)

This tamarin is typical of its genus, with its unspecialized, short, broad hands, equipped with claws, and a small body. The hairs around its mouth are white, but the skin under the moustache is pigmented, as are the genitalia. Family groups, consisting of a male, a female and 1 or 2 young, live in a defined territory — the female marks branches on the boundaries of the territory with secretions of her anal glands and urine. Insects, leaves and fruit are the main foods of these tamarins.

The female gives birth to 2 young after a gestation of 140 to 150 days.

SILVERY MARMOSET

PYGMY MARMOSET

GOELDI'S MARMOSET

EMPEROR TAMARIN

GOLDEN LION
TAMARIN

BLACK AND RED TAMARIN

New World Monkeys I

NAME: **Douroucouli/Night Monkey,**
 Aotus trivirgatus
RANGE: **Panama to Paraguay (patchy
 distribution)**
HABITAT: **forest**
SIZE: **body: 24–37 cm (9½–14½ in)**
 tail: 31–40 cm (12¼–15¾ in)

The douroucouli has a heavily furred
body and tail and is characterized by its
large eyes and round head. It is usually
greyish in colour, with dark markings
on the head, and has an inflatable sac
under the chin that amplifies its calls. A
nocturnal monkey, it can see very well
at night and moves with agility in the
trees, leaping and jumping with ease. It
rarely descends to the ground. Fruit,
leaves, insects and spiders are its main
foods, but it also takes some small mammals and birds. Douroucoulis usually
live in pairs, accompanied by their offspring. Several families may group
together during the day to sleep in a
hollow tree or in a nest among foliage.

The female gives birth to a single
young, which clings to its mother for the
first few weeks of life before it starts to
climb alone.

NAME: **Monk Saki,** *Pithecia monachus*
RANGE: **upper Amazon basin**
HABITAT: **forest**
SIZE: **body: 35–48 cm (13¾–18¾ in)**
 tail: 31–51 cm (12¼–20 in)

The monk saki has long, shaggy hair
framing its face and on its neck and a
thick, bushy tail. A shy, wary animal, it
is totally arboreal, living high in the
trees and sometimes descending to
lower levels but not to the ground. It
generally moves on all fours but may
sometimes walk upright on a large
branch and will leap across gaps. During
the day, it moves in pairs or small family
groups, feeding on fruit, berries, honey,
some leaves, small mammals, such as
mice and bats, and birds.

The female gives birth to 1 young.

NAME: **Black-bearded Saki,** *Chiropotes
 satanas*
RANGE: **N. South America to Brazil**
HABITAT: **forest**
SIZE: **body: 36–52 cm (14¼–20½ in)**
 tail: 36–50 cm (14¼–19¾ in)

The black-bearded saki is identified by
its prominent beard and the long black
hair on its head; the rest of the coat is
reddish-chestnut or blackish-brown.
The tail is thick and heavily furred.
Little is known of the habits of this
monkey in the wild other than that it
lives in large trees and feeds mainly on
fruit. It requires many square kilometres
of undisturbed habitat for successful
breeding, however, and the widespread
felling of primary forest poses a threat to
the long-term survival of this species.

CEBIDAE: New World Monkeys

Most of the New World monkeys of
the family Cebidae are much larger in
size than any of the marmosets and
tamarins — the other group of flat-
nosed (platyrrhine) American monkeys.
There are about 32 species, including
capuchins, howler monkeys, woolly
monkeys, sakis and uakaris. Typically
these monkeys have long, hairy tails,
which in some species are prehensile
and of great importance in arboreal
locomotion.

With a few exceptions, cebid monkeys conform to the flat-nosed appearance so characteristic of New World
monkeys. The nostrils are wide apart
and open to the sides; they are a major
distinguishing feature between New
and Old World monkeys which have
nostrils placed close together and opening forward. The long, thin fingers of the
hands are useful manipulative organs
and bear strong nails, but the thumbs
are not opposable. The big toe, however, is large and can be opposed against
the other toes for gripping branches.
Thus equipped, cebid monkeys are excellent runners and leapers in wooded
habitats in Central and South America,
from Mexico in the north to Argentina
in the south. They are gregarious and
live in family-based groups with much
vocal and visual communication. Diet is
largely vegetarian.

NAME: **White-fronted Capuchin,** *Cebus
 albifrons*
RANGE: **parts of Colombia, Venezuela,
 upper Amazon area; Trinidad**
HABITAT: **forest**
SIZE: **body: 30–38 cm (11¾–15 in)**
 tail: 38–50 cm (15–19¾ in)

A lively, intelligent monkey like all the
capuchins, this species is slender and
long-limbed, with a partially prehensile
tail. There is considerable variation in
colour over the range, but these
capuchins are usually different shades
of brown. Alert and fast-moving, they
are inquiring by nature, have great
manual dexterity and investigate all
sorts of plants and fruit in the hope that
they may be edible. Shoots, fruit, insects, young birds and birds' eggs are all
part of their diet. Primarily arboreal,
these capuchins do sometimes descend
to the ground and may venture across
open country. They are gregarious and
live in territories in groups of 20 or 30.

The female usually gives birth to 1
young, although twins have been
known. The offspring is suckled for
several months and is carried around
by both parents.

NAME: **Dusky Titi,** *Callicebus moloch*
RANGE: **Colombia to Bolivia**
HABITAT: **forest, thickets**
SIZE: **body: 28–39 cm (11–15¼ in)**
 tail: 33–49 cm (13–19¼ in)

An inhabitant of densely vegetated
areas, the dusky titi often occurs in
damp, waterlogged forest. It can move
quite fast if necessary but rarely does so
and generally stays within a fairly small
area, feeding on fruit, insects, spiders,
small birds and birds' eggs. Active in the
daytime, it moves in pairs or family
groups, which communicate by means
of a wide repertoire of sounds. Dusky
titis have rounded heads and thick, soft
coats and frequently adopt a characteristic posture, with the body hunched,
limbs close together and tail hanging
down.

The female gives birth to 1 young.

NAME: **Squirrel Monkey,** *Saimiri sciureus*
RANGE: **Colombia to Amazon basin**
HABITAT: **forest, cultivated land**
SIZE: **body: 26–36 cm (10¼–14¼ in)**
 tail: 35–42 cm (13¾–16½ in)

A slender monkey, with a long, mobile
tail, the squirrel monkey has a short,
brightly coloured coat. It is highly active
and lively, feeding during the day on
fruit, nuts, insects, spiders, young birds
and eggs, and it occasionally comes to
the ground to feed. Squirrel monkeys
sometimes raid plantations to eat the
fruit. They are gregarious and live in
bands of 12 to 30 or more.

The female gives birth to 1 young
after a gestation of 24 to 26 weeks. The
newborn infant is able to climb soon
after birth and receives little attention
from its parents.

NAME: **Bald Uakari,** *Cacajao calvus*
RANGE: **W. Brazil**
HABITAT: **forest**
SIZE: **body: 51–57 cm (20–22½ in)**
 tail: 15–16 cm (6–6¼ in) Ⓥ

A distinctive monkey, the bald uakari
has a naked face, long, shaggy hair and
a beard. It is normally white but looks
reddish in the sunlight. Its tail is fairly
short — the 3 species of uakari are the
only New World monkeys to have
short tails. Extremely agile on all fours,
this uakari rarely leaps, since it does not
have the long, counterbalancing tail of
other types of cebid monkey. It fre-
quents the tree-tops, feeding largely on
fruit but also on leaves, insects, small
mammals and birds, and seldom
descends to the ground. Bald uakaris
live in small troops, consisting of
several adult males and females and
their young of different ages, and are
active in the daytime.

WHITE-FRONTED CAPUCHIN

DOUROUCOULI

DUSKY TITI

SQUIRREL MONKEY

MONK SAKI

BLACK-BEARDED SAKI

BALD UAKARI

New World Monkeys 2

NAME: **Red Howler,** *Alouatta seniculus*
RANGE: **Colombia to mouth of Amazon River, south to Bolivia**
HABITAT: **forest, mangroves**
SIZE: **body: 80–90 cm (31½–35½ in)**
tail: 80–90 cm (31½–35½ in)

One of the largest New World monkeys, the red howler has reddish-brown fur and a sturdy body and legs. Its tail is prehensile, with an extremely sensitive naked area on the underside near the tip. All male howlers are renowned for the incredibly loud calls produced by their specialized larynxes, and this apparatus is most developed in the red howler. The chief adaptation of the larynx is the greatly expanded hyoid bone surrounding it that makes a resonating chamber for the sound. The jaw is expanded and deepened to accommodate the bulbous larynx and sports a thick beard. The male red howler occupies a territory and leads a troop of, usually, 6 to 8 animals. In order to defend his territory, he shouts for long periods at rival groups to signal his possession. Most shouting is done in the early morning and late afternoon, but red howlers may be heard at any time of day — from over 3 km (1¾ mls) away.

Red howlers live and move adeptly in the trees, although they frequent large branches because of their sturdy build. Their digits are organized in such a way as to facilitate the grasping of branches, with the first two fingers separated and opposable to the other three; this arrangement does, however, make delicate manipulation of food items difficult. Red howlers leap well and use their prehensile tails for support. They sometimes go down to the ground and will even cross open land; they are good swimmers. Leaves and some fruit are the red howlers' staple diet. At night, these monkeys sleep in the trees on branches.

Breeding appears to occur at any time of year, and the female gives birth to 1 young after an average gestation of 20 weeks. The young howler clings to its mother's fur at first and later rides on her back; it is suckled for 18 months to 2 years.

NAME: **Black Howler,** *Alouatta caraya*
RANGE: **S. Brazil to N. Argentina**
HABITAT: **forest**
SIZE: **body: 80–90 cm (31½–35½ in)**
tail: 80–90 cm (31½–35½ in)

Only male black howlers are actually black; females have brown coats. These howlers live in troops, probably containing more than 1 male, and occupy territories, which they defend by their powerful shouts in the same way that red howlers do. The bulbous larynx, concealed under the bushy beard,

amplifies the shouts. Black howlers tend to have quieter voices and smaller territories than red howlers.

A strongly built monkey, with powerful limbs and a prehensile tail, the black howler lives in the trees, feeding on leaves and fruit. The female gives birth to 1 young after a gestation of about 20 weeks, and her offspring stays with her for up to 2 years.

NAME: **Black Spider Monkey,** *Ateles paniscus*
RANGE: **N. South America to Brazil and Bolivia**
HABITAT: **forest**
SIZE: **body: 40–60 cm (15¾–23½ in)**
tail: 60–80 cm (23½–31½ in) Ⓥ

Only surpassed by the gibbons for grace and agility in the trees, the black spider monkey, with its extremely long limbs and tail, is the most adept and acrobatic of all New World monkeys. Light in build, with a small head, the spider monkey has the most highly developed prehensile tail of all mammals and uses it as a fifth limb to grasp branches or food items as it moves through the trees. The monkey's whole weight can be supported by the tail, and when hanging by the tail with the long arms outstretched it has an amazing reach. Part of the underside of the tail nearest the tip is naked and patterned with fine grooves, resembling human fingertip patterns. These increase friction and thus aid grip. Spider monkeys frequently swing through the trees, using their hands like hooks to hang on to the branches. The hands are, accordingly, modified, with long, curved digits and only vestigial thumbs. While this structure makes the hands ideal for swinging in trees, it impedes delicate manipulation of food, but the monkey often uses its highly sensitive tail to gather food and to hold items, such as fruit, while it takes off the skin with its teeth.

Black spider monkeys rarely come to the ground and feed in trees, mainly on fruit and some nuts. They live in groups of 15 to 30 animals in a home range, but a group may split into smaller parties while foraging during the day. Most feeding is done in the early morning and the afternoon.

The female gives birth to 1 young after an average gestation of about 20 weeks. The young is dependent on its mother for 10 months or so.

NAME: **Woolly Spider Monkey,** *Brachyteles arachnoides*
RANGE: **S.E. Brazil**
HABITAT: **coastal forest**
SIZE: **body: about 61 cm (24 in)**
tail: about 67 cm (26¼ in) Ⓔ

The woolly spider monkey usually has a yellowish-grey to brown or reddish coat, and the naked facial skin is often red, especially when the animal is excited. Its body is powerful and its limbs long and slender, and, like the spider monkeys, it has a highly efficient prehensile tail, which it uses as a fifth limb. The underside of the tail near the tip is naked and extremely sensitive. Woolly spider monkeys often move by swinging from branch to branch, and their thumbs, of little use for such locomotion, are vestigial.

Active in the daytime, woolly spider monkeys feed in trees, mainly on fruit. They are gregarious, but little more is known of their habits in the wild.

NAME: **Common Woolly Monkey,** *Lagothrix lagothricha*
RANGE: **upper Amazon basin**
HABITAT: **forest up to 2,000 m (6,600 ft)**
SIZE: **body: 50–68 cm (19¾–26¾ in)**
tail: 60–72 cm (23½–28¼ in) Ⓥ

Heavier in build than the spider monkeys, the common woolly monkey has short, thick hair. Its head is rounded, its body robust, and it has a prominent belly. Fast and agile in the trees, it moves on all fours and by swinging hand over hand, but it is less graceful than the spider monkeys. Its thumbs and toes are well developed for grasping branches, and it has a strong, prehensile tail, with a sensitive naked area near the tip.

Woolly monkeys are highly gregarious and live in troops of up to 50. They forage in the daytime for plant material, mostly fruit, but are less active than many other New World monkeys. Primarily arboreal, they do, nevertheless, often come down to the ground, where they walk upright, using their long tails as counterbalances.

The female gives birth to 1 offspring after a gestation of 18 to 20 weeks. The young monkey holds on to the fur of its mother's belly or back at first and is carried around, but after a few weeks it is able to clamber about the branches unaided. The female suckles her young for 12 months or more. Common woolly monkeys are sexually mature at about 4 years old.

RED HOWLER
(male)

BLACK HOWLER (male)

BLACK SPIDER MONKEY

WOOLLY SPIDER MONKEY

COMMON WOOLLY MONKEY

Old World Monkeys I

NAME: **Barbary Ape, *Macaca sylvanus***
RANGE: **Gibraltar; Africa: Morocco, N. Algeria**
HABITAT: **rocky areas, forest clearings on mountains**
SIZE: **body: 55–75 cm ($21\frac{1}{2}$–$29\frac{1}{2}$ in)**
tail: absent Ⓥ

The barbary ape is a robust, tail-less monkey, with a rounded head and short muzzle. Males are larger than females and have longer hair on the crown. Formerly found elsewhere in southwest Europe, barbary apes now occur outside Africa only on Gibraltar, where the population is reinforced with animals from North Africa. Barbary apes live in troops of 10 to 30 males, females and young in a defined territory. The monkeys sleep in trees or among rocks and feed in the early morning and afternoon, taking a rest at midday. They climb well and forage in trees and on the ground for grass, leaves, berries, fruit, roots, insects and spiders. They will also plunder gardens and crops.

The female gives birth to a single young, rarely twins, after a gestation of about 7 months. Births take place at any time of year but peak from May to September. Males help females to look after and carry young in the first few days. The offspring suckles for about 3 months and stays with its mother for up to 6 months.

NAME: **Stump-tailed Macaque, *Macaca arctoides***
RANGE: **Burma, S. China to Malaysia**
HABITAT: **forest, cultivated land**
SIZE: **body: 50–70 cm ($19\frac{3}{4}$–$27\frac{1}{2}$ in)**
tail: 4–10 cm ($1\frac{1}{2}$–4 in)

Distinguished by its pink-tinged face, shaggy hair and short tail, the stump-tailed macaque is an aggressive, fearless monkey that often invades gardens and cultivated fields. It spends much of its time on the ground but also climbs up into trees to sleep or to find food or a safe refuge, although it is not a particularly agile animal. Leaves, fruit, roots and crops, such as potatoes, are its main foods, and it usually picks up the items with its hands. Two cheek pouches are used for storing food, which is later removed and chewed at leisure. Stump-tailed macaques are active in the daytime and live in groups of 25 to 30, led by a dominant individual. Members of the group continually chatter and squeal to each other, and they also communicate by means of a wide range of facial expressions. Males are larger than females.

Little is known about the reproduction of these macaques in the wild, but females are thought to produce an infant every other year.

CERCOPITHECIDAE: Old World Monkey Family

The monkeys and apes of the Old World are usually grouped together as the catarrhine primates: those with closely spaced nostrils that face forward or downward. The Old World monkeys themselves are the largest group of catarrhines, with about 76 species known, in Africa, Asia and Indonesia. There are a few general differences between Old and New World monkeys. The Old World species are generally larger and often have bare buttock pads, which may be brightly coloured, and their tails, although often long, are seldom, if ever, fully prehensile.

The family includes the macaques, baboons, mandrills, mangabeys, guenons, langurs, colobus and leaf monkeys and many other forms. Almost all are daytime-active animals, with excellent vision, hearing and sense of smell. Most are arboreal, but baboons are ground-feeding specialists, and the macaques are found both in the trees and on the ground. Generally Old World monkeys live in family or larger groups and communicate by a variety of visual and vocal signals. Males are often considerably larger than females.

NAME: **Japanese Macaque, *Macaca fuscata***
RANGE: **Japan**
HABITAT: **high-altitude forest**
SIZE: **body: 50–75 cm ($19\frac{3}{4}$–$29\frac{1}{2}$ in)**
tail: 25–30 cm ($9\frac{3}{4}$–$11\frac{3}{4}$ in)

The only monkey found in Japan, the Japanese macaque is the sole primate other than man able to withstand a cold, snowy winter and near-freezing temperatures. In some parts of its range, it spends long periods immersed up to the neck in thermal pools. It is a medium-sized, well-built monkey, with dense fur and long whiskers and beard. Active both on the ground and in trees, it feeds mainly on nuts, berries, buds, leaves and bark.

Social groups of up to 40 individuals live together, led by an older male. The relationship between females and their mothers is extremely important; as long as their mother lives, females remain in association with her, even when they have their own young, and such female groups are the core of a troop. Males stay with their mothers and kin until adolescence, when they may join a peripheral group of males that drifts between troops. After a period in such a group, or alone, the male joins a troop, usually not that of his birth.

Females give birth to 1 young after a gestation of between 6 and 7 months.

NAME: **Bonnet Macaque, *Macaca radiata***
RANGE: **S. India**
HABITAT: **forest, scrub, cultivated and suburban areas**
SIZE: **body: 35–60 cm ($13\frac{3}{4}$–$23\frac{1}{2}$ in)**
tail: 48–65 cm ($18\frac{3}{4}$–$25\frac{1}{2}$ in)

The common name of this long-tailed macaque is derived from the unruly cap of dark hairs on its crown. Its face is normally pale pink, but the faces of lactating females are dark red. Males are much larger than females. An agile, active monkey, the bonnet macaque spends the bulk of its time in the trees, although it moves readily on the ground and also swims well. It has a voracious appetite and feeds on leaves, fruit, nuts seeds, insects, birds' eggs and, occasionally, lizards.

The female gives birth to 1 young, sometimes twins, after a gestation of about 150 days.

NAME: **White-cheeked Mangabey, *Cercocebus albigena***
RANGE: **Africa: Cameroon to Uganda, Kenya, Tanzania**
HABITAT: **forest**
SIZE: **body: 45–70 cm ($17\frac{3}{4}$–$27\frac{1}{2}$ in)**
tail: 70 cm–1 m ($27\frac{1}{2}$ in–$3\frac{1}{4}$ ft)

A slender, elegant monkey, the white-cheeked mangabey is distinguished by its conspicuous eyebrow tufts and the mane of long hairs on its neck and shoulders. Its tail is immensely long and mobile and covered with rather shaggy hairs; it is semi-prehensile. Males are larger than females and have longer tails. These mangabeys sleep and spend nearly all their time in trees; they feed during the day on fruit, nuts, leaves, bark and insects. Troops of 10 to 30 animals live together and are extremely noisy, constantly chattering and shrieking to one another.

The female gives birth to 1 young after a gestation of 174 to 180 days. The young is suckled for up to 10 months.

NAME: **Agile Mangabey, *Cercocebus galeritus***
RANGE: **Africa: S.E. Nigeria, Zaire to E. Kenya**
HABITAT: **rain and swamp forest**
SIZE: **body: 45–65 cm ($17\frac{3}{4}$–$25\frac{1}{2}$ in)**
tail: 45–75 cm ($17\frac{3}{4}$–$29\frac{1}{2}$ in) Ⓔ

A slender, but strongly built, monkey, the agile mangabey has long legs and tail and a fringe of hairs on its forehead. There are several races, which vary slightly in coloration; one has no fringe on the forehead. The details of this mangabey's daily habits are poorly known, but it is thought to be active on the ground and in trees and to feed on leaves, fruit, crops and insects. It lives in troops of 12 to 20, consisting of several old males, mature females and young.

STUMP-TAILED MACAQUE

JAPANESE MACAQUE

BONNET MACAQUE

BARBARY APE

WHITE-CHEEKED MANGABEY

AGILE MANGABEY

Old World Monkeys 2

NAME: **Olive Baboon,** *Papio anubis*
RANGE: **Africa: Senegal, east to N. Zaire,**
Ethiopia, Kenya, Uganda, N. Tanzania
HABITAT: **savanna**
SIZE: **body: up to 1 m (3¼ ft)**
tail: 45–75 cm (17¾–29½ in)

The olive baboon is a large, heavily built baboon, with a sloping back and a well-developed, doglike muzzle, equipped with powerful teeth. Males have a mane of long hair around neck and shoulders and are larger than females. The tail has a tuft at its end, and the buttock area is naked, with broad callosities.

These baboons live in troops of 20 to 150 individuals, organized in a strict hierarchy. Although mainly ground-living animals, olive baboons sleep at night in the safety of trees or rocks and emerge in the morning to travel to feeding grounds in a well-ordered procession. Older juveniles lead the way, followed by the females and younger juveniles; the older males with the mothers and infants come next, and young males bring up the rear and scout ahead of the party to warn of any danger. Olive baboons are omnivorous, feeding on grass, seeds, roots, leaves, fruit, bark, insects, invertebrates, birds' eggs and young, lizards and young mammals such as antelope and lambs.

The female gives birth to 1 young, rarely 2, after a gestation of about 187 days. At first the young clings to its mother's belly but at 4 or 5 weeks begins to ride on her back. It takes its first solid food at 5 or 6 months and is weaned and independent at 8 months, although it is guarded by the mother until it is about 2 years old.

NAME: **Hamadryas Baboon,** *Papio*
hamadryas
RANGE: **Africa: Ethiopia, Somalia;**
S. Saudi Arabia
HABITAT: **dry rocky country, savanna,**
semi-desert
SIZE: **body: 50–95 cm (19¾–37½ in)**
tail: 40–60 cm (15¾–23½ in)

The male hamadryas baboon is as much as twice the size of the female and has a heavy mane around its neck and shoulders. Females and younger males lack the mane and have brownish hair. Like all baboons, this species has a dog-like muzzle and a sloping back. Family troops of an old male and several females and their young live together, sleeping in trees or among rocks at night and wandering in search of food during the day. They eat almost any plants, insects and small animals.

The peak breeding season is May to July, and the female produces 1, rarely 2, young after a gestation of between 170 and 175 days.

NAME: **Chacma Baboon,** *Papio ursinus*
RANGE: **Africa: Angola, Zambia to South**
Africa
HABITAT: **savanna, rocky areas**
SIZE: **body: up to 1 m (3¼ ft)**
tail: 40–75 cm (15¾–29½ in)

A large, yet slender, baboon, with a prominent muzzle and a sloping back, the chacma baboon carries its tail in a characteristic posture as if "broken" near the base. It lives in troops of about 30 to 100 individuals, sleeping among rocks or in trees at night and searching for food during the day. These baboons feed early and late in the day and rest up at midday. Almost any plant and animal matter, such as leaves, fruit, insects, small invertebrates, lizards, birds and young mammals, is included in their diet, but they are predominantly vegetarian.

When on heat, the bare skin around the female's genital region swells, and at the peak of her fertile period only high-ranking males may mate with her. This is common to all savanna baboons.

The female gives birth to a single young, very rarely twins, after a gestation of 175 to 193 days, and suckles her offspring for about 8 months.

NAME: **Drill,** *Mandrillus leucophaeus*
RANGE: **Africa: S.E. Nigeria, Cameroon**
HABITAT: **forest**
SIZE: **body: 45–90 cm (17¾–35½ in)**
tail: 6–12 cm (2¼–4¾ in) Ⓔ

A powerfully built forest baboon, with a large head and a short, stumpy tail, the drill has a long muzzle, a ridged face and large nostrils. Males are much larger than females, sometimes twice the size, and have heavy manes on neck and shoulders. The skin of the buttock pads and the area around them is brightly coloured, and the hue becomes more pronounced when the animal is excited. Although it climbs well and sleeps in low branches, the drill is ground-dwelling and moves on all fours. It lives in family troops of 20 or more animals, which may join with other troops to form bands of as many as 200. The baboons communicate with a variety of deep grunts and sharp cries, as well as with up and down movements of the head with mouth closed, to express threat, or side to side movements with teeth exposed, to express friendship. Old males dominate the troop and guard its safety; they are formidable animals, well equipped for fighting, with their sharp teeth and strong limbs. They feed on plant matter, insects and small invertebrate and vertebrate animals.

Young are born at all times of year. The female produces 1 young after a gestation of about 7 months.

NAME: **Mandrill,** *Mandrillus sphinx*
RANGE: **Africa: Cameroon, Gabon**
HABITAT: **forest**
SIZE: **body: 55–95 cm (21½–37½ in)**
tail: 7–10 cm (2¾–4 in)

The male mandrill is a large, heavily built forest baboon, with an elongate snout marked with deep, often colourful ridges. Adult females are much smaller and have less pronounced ridges and facial coloration. Although mandrills sleep in trees, they live and feed on the ground in troops of about 20 to 50 animals, led and protected by 1 or more old males. They feed on fruit, nuts, leaves, insects and small invertebrate and vertebrate animals.

Births occur at any time of year but peak from December to February. The female produces a single young after a gestation of about 7½ months.

NAME: **Gelada,** *Theropithecus gelada*
RANGE: **Africa: Ethiopia**
HABITAT: **mountains: rocky ravines,**
alpine meadows
SIZE: **body: 50–75 cm (19¾–29½ in)**
tail: 45–55 cm (17¾–21½ in)

The impressive gelada has a distinctive head, with a somewhat upturned muzzle marked with ridges, and the nostrils are located well back, not at the end of the muzzle as in other baboons. Long side-whiskers project backward and upward, and there is a heavy mane over the neck and shoulders which may reach almost to the ground in old males. On the chest and throat are three areas of bare red skin, which the male expands and brings into full view in his aggressive, threat posture. Females are about half the size of males and have much lighter manes.

The gelada is a ground-dweller and even sleeps on rocky ledges and cliffs. It lives in family groups of several females and their young, led by a large mature male; these family groups may sometimes gather into large troops of several hundred animals. Young and unattached males may form their own units. In the morning, the geladas leave the cliffs where they have slept and move off to alpine meadows, where they feed largely on plant material, such as grasses, seeds and fruit, and on insects and other small animals. They do not have a specific territory, and the male keeps his family together by means of a variety of calls and facial gestures. Geladas have excellent vision, hearing and sense of smell.

Most young are born between February and April. The female produces a single offspring, rarely twins, after a gestation of 147 to 192 days and suckles it for up to 2 years. After giving birth, she does not have a period of heat for 12 to 18 months.

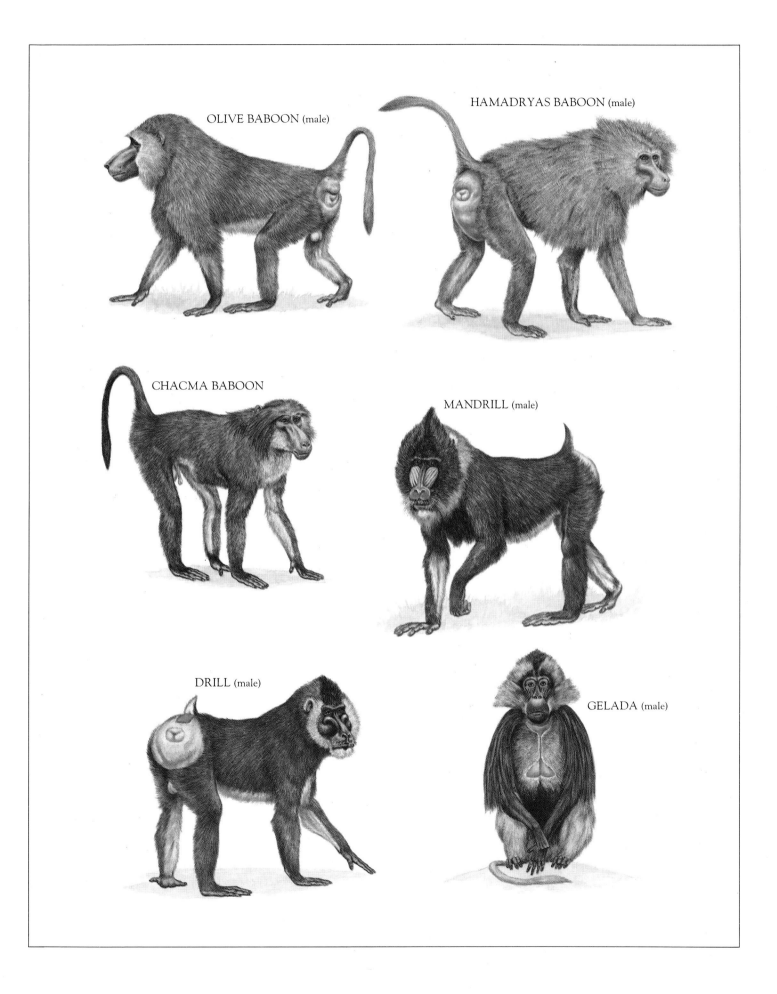

OLIVE BABOON (male)

HAMADRYAS BABOON (male)

CHACMA BABOON

MANDRILL (male)

DRILL (male)

GELADA (male)

Old World Monkeys 3

NAME: Vervet Monkey, *Cercopithecus aethiops*
RANGE: Africa: Senegal to Somalia, south to South Africa;
HABITAT: savanna, woodland edge
SIZE: body: 40–80 cm (15¾–31½ in)
tail: 50–70 cm (19¾–27½ in)

There are many races of this medium- to large-sized monkey, which vary in their facial markings and whiskers. Generally, however, they have black faces, white whiskers and greyish or yellowish-olive hair. Although they sleep and take refuge in trees, these adaptable monkeys forage in open country and will run some distance on the ground. They also climb, jump and swim well. Family troops of an old male and several females and young live together and may join with other troops during the day. Generally rather quiet monkeys, males utter a harsh cry, and others may scream when frightened. Mainly vegetarian animals, they feed on leaves, shoots, fruit, flowers, seeds and bark but also eat some insects, spiders, lizards, birds' eggs and young birds.

Breeding occurs at any time of year. The female gives birth to a single young after a gestation of 175 to 203 days and suckles it for about 6 months. For the first few weeks, the baby clings to its mother's belly but starts to leave her at 3 weeks and to climb at 4 weeks old. When males become sexually mature, their scrotums adopt a blue-green hue.

NAME: De Brazza's Monkey, *Cercopithecus neglectus*
RANGE: Africa: Cameroon, south to Angola, east to Uganda
HABITAT: rain and swamp forest, dry mountain forest near water
SIZE: body: 40–60 cm (15¾–23½ in)
tail: 53–85 cm (20¾–33½ in)

The robust, heavily built De Brazza's monkey has a conspicuous reddish-brown band, bordered with black, on its forehead, and a well-developed white beard. Its back slopes upward to the tail so that the rump is higher than the shoulders. Females look similar to males but are smaller. Active during the day, this monkey is a good climber and swimmer and also moves with speed and agility on the ground, where it spends a good deal of its feeding time. Leaves, shoots, fruit, berries, insects and lizards are its main foods, and it will also raid crops. It lives in small family groups of an old male and several females with young or sometimes in larger troops of 30 or more animals.

The female gives birth to 1 young after a gestation of 177 to 187 days. After a week, the baby first starts to leave go of its mother's body, and by 3 weeks, it is starting to climb and run.

NAME: Red-bellied Monkey, *Cercopithecus erythrogaster*
RANGE: Africa: Nigeria
HABITAT: forest
SIZE: body: about 45 cm (17¾ in)
tail: about 60 cm (23½ in)

This apparently rare monkey has a dark face with a pinkish muzzle, fringed with white side-whiskers. The breast and belly are usually reddish-brown, hence the common name, but can be grey in some individuals. Females look similar to males but have greyish underparts, arms and legs. Few specimens of this monkey have been found, and its habits are not known.

NAME: Diana Monkey, *Cercopithecus diana*
RANGE: Africa: Sierra Leone to Ghana
HABITAT: rain forest
SIZE: body: 40–57 cm (15¾–22½ in)
tail: 50–75 cm (19¾–29½ in)

A slender, elegant monkey, the diana monkey has most striking coloration, with its black and white face, white beard and chest and distinctive patches of bright chestnut on its back and hind limbs. There are also conspicuous white stripes on the otherwise dark hair of each thigh. Females are smaller than males but in other respects look similar. An excellent climber, the diana monkey spends virtually all its life in the middle and upper layers of the forest and is noisy and inquisitive. Troops of up to 30 animals live together, led by an old male, and are most active in the early morning and late afternoon, when they feed on leaves, fruit, buds and other plant matter, and also on some insects and birds' eggs and young.

The female bears a single young after a gestation of about 7 months and suckles it for 6 months.

NAME: Talapoin, *Miopithecus talapoin*
RANGE: Africa: Gabon to W. Angola
HABITAT: rain forest, mangroves, always near water
SIZE: body: 25–40 cm (9¾–15¾ in)
tail: 36–52 cm (14¼–20½ in)

One of the smallest African monkeys, the talapoin has a slender body, a round head, which is large relative to body size, and prominent ears. Its legs are longer than its arms, and its tail exceeds its head and body length. Talapoins live in family troops of 12 to 20 or so individuals, and each troop has its own territory, although several troops may sometimes unite into a larger group. These monkeys sleep in bushes and mangroves and are active in the daytime, particularly in the early morning and late afternoon. Good climbers and runners, they enter water readily and swim and dive well. Leaves, seeds, fruit,

water plants, insects, eggs and small animals are all included in their diet, and they will raid plantation crops.

Most births occur between November and March. After a gestation of about 6½ months, the female produces a single young, which is born well haired and with its eyes open. The young talapoin develops quickly and takes its first solid food at 3 weeks; it is largely independent at 3 months, although it continues to suckle until it is 4 or 5 months old.

NAME: Patas Monkey, *Erythrocebus patas*
RANGE: Africa: Senegal, east to Ethiopia, south to Tanzania
HABITAT: grassland, dry savanna, forest edge, rocky plateaux
SIZE: body: 50–75 cm (19¾–29½ in)
tail: 50–75 cm (19¾–29½ in)

The slender, long-legged patas monkey is among the fastest moving of all primates on the ground: it can attain speeds of up to 50 km/h (31 mph). The male may be as much as twice the size of the female, and the lower parts of his limbs are pure white; the limbs of females are fawn or yellowish-white. Patas monkeys live in troops, consisting of an old male and up to 12 females and their young, and occupy a large territory. Although they sleep in trees, usually at the edge of forest, these monkeys spend virtually all of their day on the ground, searching for fruit, seeds, leaves, roots, insects, lizards and birds' eggs. While the troop feeds, the male leader keeps a look-out for danger and warns his harem of the approach of any enemies.

Most births occur from December to February, and females produce 1 young after a gestation of 170 days. The baby takes its first solid food at about 3 months old.

NAME: Allen's Swamp Monkey, *Allenopithecus nigroviridis*
RANGE: Africa: E. Congo, Zaire
HABITAT: swampy forest
SIZE: body: 40–50 cm (15¾–19¾ in)
tail: 45–55 cm (17¾–21½ in)

Allen's swamp monkey is a sturdily built primate with relatively short limbs and tail. Its head is rounded, and there are rufts of whiskers from the ears to the mouth. Males are slightly larger than females, but otherwise the two look similar. Little is known about the habits of these monkeys. They live in troops and feed mainly on leaves, fruit and nuts, but apparently also eat snails, crabs, fish and insects. They climb and jump well and enter water readily.

The female produces a single young, which clings to her belly and suckles for 2 or 3 months.

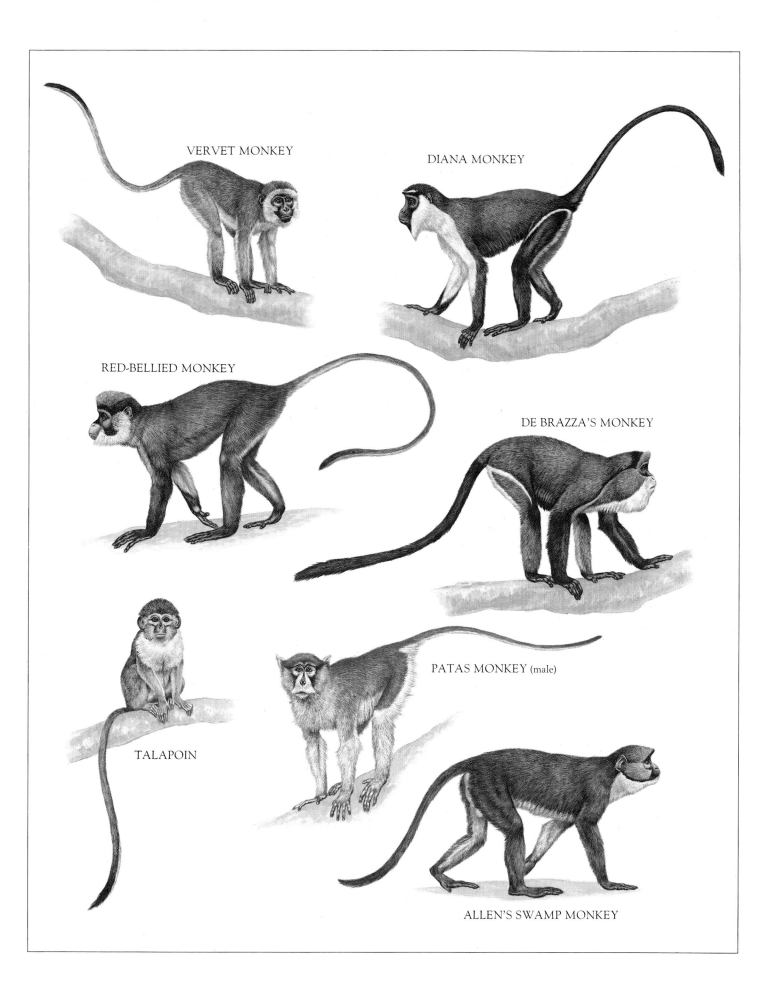

VERVET MONKEY

DIANA MONKEY

RED-BELLIED MONKEY

DE BRAZZA'S MONKEY

PATAS MONKEY (male)

TALAPOIN

ALLEN'S SWAMP MONKEY

Old World Monkeys 4

NAME: **Angolan Black and White Colobus,** *Colobus angolensis*
RANGE: **Africa: Angola to Kenya**
HABITAT: **forest**
SIZE: body: 50–67 cm (19¾–26¼ in)
tail: 63–90 cm (24¾–35½ in)

All colobus monkeys have long limbs and tails and robust bodies. They have only four fingers on each hand, their thumbs being vestigial or absent. The Angolan colobus is one of several black and white species and, with its sturdy body and rounded head, is typical of its genus. It is identified by the characteristic long white hairs on its shoulders, but the many races of this species differ slightly in the extent of the white on shoulders and tail.

These monkeys live in family troops of several females and their young, led and guarded by an old male. As young males mature, they either go off alone or found their own troops. Each troop has its own territory, with feeding areas and sleeping trees, but may sometimes join with other troops to form a group of 50 or so. The animals are active in the daytime, with a period of rest or grooming at midday. Much of their food, such as leaves, fruit, bark and insects, is found in the trees, where they run and leap with astonishing agility, so they rarely need to descend to the ground.

Breeding takes place all year round, females giving birth to 1 young after a gestation of 147 to 178 days. The baby starts to climb at 3 weeks but suckles and stays with its mother for well over a year. Females will suckle young other than their own.

NAME: **Red Colobus,** *Colobus badius*
RANGE: **Africa: Senegal to Ghana, Cameroon, Zaire, Uganda, Tanzania**
HABITAT: **rain, swamp and secondary forest, usually near water**
SIZE: body: 46–70 cm (18–27½ in)
tail: 42–80 cm (16½–31½ in) Ⓔ

There are many races of this slender, long-tailed colobus, with coloration ranging from orange-red to reddish-brown, often with black on the back and shoulders; the underparts are reddish-yellow to grey or white. Females are smaller than males but otherwise look similar. The red colobus lives in a troop of 50 to 100 animals made up of many small family groups, each with a male and several females and young. Active during the day, they feed among the branches on flowers, shoots, fruit and leaves, leaping acrobatically from tree to tree.

The female produces 1 young after a gestation of 4 to 5½ months. She alone nurses the infant until it is weaned at 9 to 12 months.

NAME: **Olive Colobus,** *Procolobus verus*
RANGE: **Africa: Sierra Leone to Ghana**
HABITAT: **rain, swamp and secondary forest**
SIZE: body: 43–50 cm (17–19¾ in)
tail: 57–64 cm (22½–25¼ in) Ⓡ

The smallest of the colobus monkeys, the olive colobus has a little rounded head, a short muzzle and rather subdued coloration. Male and female are about the same size, but the female lacks the crest of upright hairs the male has on his crown. This colobus lives in a family troop of an old male and several females and their young, usually 6 to 10 in all. Sometimes the group may be bigger, up to 20, with several adult males. The monkeys sleep and take refuge in the middle layers of the forest but feed on the lowest branches. They do not climb into the tree-tops and only rarely come to the ground. Leaves and some flowers are their staple diet. They are rather quiet monkeys and make few sounds.

Reproductive details are not known except that the mother carries her baby in her mouth for the first few weeks after birth — a habit shared only with other species of colobus.

NAME: **Proboscis Monkey,** *Nasalis larvatus*
RANGE: **Borneo**
HABITAT: **mangroves, river banks**
SIZE: body: 53–76 cm (20¾–30 in)
tail: 55–76 cm (21½–30 in) Ⓥ

The sturdily built proboscis monkey lives in the tree-tops of mangrove swamp jungles, where the trees are strong and rigid and do not attain enormous heights. It is an agile animal and runs and leaps in the branches, using its long tail as a counterbalance; its long fingers and toes aid grip. The male is considerably larger than the female, perhaps twice her weight, and has an extraordinarily long, bulbous nose. When he makes his loud, honking call, the nose straightens out. The female has a much smaller nose and a quieter cry.

Proboscis monkeys live in small groups of 1 or 2 adult males and several females and their young. They are most active in the morning, when they feed on leaves and shoots of mangrove and pedada trees, as well as on the fruit and flowers of other trees. Much of the rest of the day is spent basking in the tree-tops, where they also sleep at night.

Mating takes place at any time of year, and the female produces a single young after a gestation of about 166 days. These monkeys are difficult to keep in captivity and are becoming increasingly rare in the wild.

NAME: **Snub-nosed Monkey,** *Pygathrix roxellana*
RANGE: **S.W. and S. China, Tibet, E. India**
HABITAT: **mountain forest; winters in lower valleys**
SIZE: body: 50–83 cm (19¾–32½ in)
tail: 51 cm–1 m (20 in–3¼ ft) Ⓡ

This large, long-tailed monkey has a distinctive upturned nose, hence its common name, and golden hairs on its forehead, throat and cheeks. Because of their remote, mountainous range, few snub-nosed monkeys have yet been observed or caught. They are said to live in troops of 100 or more and to feed on fruit, buds, leaves and bamboo shoots.

NAME: **Langur,** *Presbytis entellus*
RANGE: **India, Sri Lanka**
HABITAT: **forest, scrub, arid rocky areas**
SIZE: body: 51 cm–1 m (20 in–3¼ ft)
tail: 72 cm–1 m (28¼ in–3¼ ft)

This large, long-limbed monkey has a black face and prominent eyebrows. It adapts well to many different habitats and will live near human habitation and raid shops and houses for food. In the Himalayas, the langur is believed to make regular migrations, moving up the mountains in summer and down again in winter. Although an agile climber, the langur spends more than half of its time on the ground, where it finds much of its food. It is almost entirely vegetarian, feeding on leaves, shoots, buds, fruit and seeds; very rarely, it eats insects.

Langurs live in groups of 1 or more adult males with females and juveniles, usually about 15 to 35 individuals in all, although in some areas groups of 80 or 90 have been observed. There are also, however, smaller all-male groups containing 4 to 15 animals. In groups with only one male, he is in sole charge of the movements and daily routine, and life is peaceful; in larger groups, the males contest and squabble, even though there is a dominant male.

Breeding takes place at any time of year except in areas with marked seasonal changes, where births are concentrated into the 2 or 3 most climatically favourable months. The female gives birth to 1, occasionally 2, young after a gestation of 6 or 7 months. Other females in the troop show a great interest in the newborn infant, and the mother allows them to touch it shortly after birth. For the first few weeks, the infant clings tightly to its mother, but at 4 weeks it starts to move short distances independently and, at 3 months, is allowed to play with other infants and to take some solid food. It continues to suckle for 10 to 15 months but nevertheless undergoes much stress during the weaning period.

ANGOLAN BLACK AND WHITE COLOBUS

RED COLOBUS

OLIVE COLOBUS (male)

SNUB-NOSED MONKEY

PROBOSCIS MONKEY (male)

LANGUR

Apes I

NAME: **Black Gibbon,** *Hylobates concolor*
RANGE: **Indo-China, Hainan**
HABITAT: **rain forest**
SIZE: **body: 45–63 cm (17¾–24¾ in)**
   tail: absent   Ⓘ

There are several subspecies of black, or crested, gibbon, occurring in various parts of Indo-China, which differ in details of fur coloration. The black male is slightly larger than the female and has a tuft of hair on the crown; the female is buff coloured. Like most gibbons, the male of this species has a throat sac, which amplifies his voice.

Black gibbons feed on a variety of foods, mostly fruit, buds and insects, although occasionally small vertebrates may be eaten.

NAME: **Siamang,** *Hylobates syndactylus*
RANGE: **Malaysia, Sumatra**
HABITAT: **mountain forest**
SIZE: **body: 75–90 cm (29½–35½ in)**
   tail: absent

The largest of the gibbons, the siamang has entirely black fur and no pale brow band. It is also distinguished by the web that unites the second and third toes of each foot. Siamangs are tree-dwelling and extremely agile, despite their large size. They swing from branch to branch and may walk upright on strong boughs. At night, they sleep on high, strong branches. Fruit, especially figs, is a staple food of the siamangs, and they also eat flowers, leaves and shoots, some insects and even birds' eggs.

Adult siamang pairs live together in a territory with their offspring of different ages, but unmated adults live alone. Families communicate by short barks and a distinctive whooping call, which is amplified by the inflatable throat sac. The female siamang gives birth to a single young after a gestation of 230 to 235 days. The newborn baby is almost hairless and clings to its mother for safety and warmth.

NAME: **Kloss's Gibbon,** *Hylobates klossi*
RANGE: **Mentawi Islands west of Sumatra**
HABITAT: **hill and lowland rain forest**
SIZE: **body: 65–70 cm (25½–27½ in)**
   tail: absent   Ⓥ

Kloss's gibbon looks like a small version of *H. syndactylus* and is very similar in its habits. It is the smallest of all the gibbons and is thought to represent the form and structure of ancestral gibbons. Strictly tree-dwelling, it leaps and swings in the trees with tremendous dexterity and feeds on fruit, leaves, shoots and, perhaps, insects.

Family groups, each consisting of an adult pair and up to 3 offspring, occupy a territory and sleep and feed together. Females give birth to a single young.

PONGIDAE: Ape Family

The 10 species of primate in this family are found in Africa, Southeast Asia, Sumatra, Java and Borneo. All members of the family lack an external tail and have protruding jaws, but they divide into two distinct groups: the agile, slender gibbons and the robust great apes.

The gibbons are specialized arboreal forms, very different externally from the great apes and probably having the most remarkable adaptations of all mammals for rapid locomotion in trees. They use their extremely elongate arms and hooked hands to swing through the forest canopy and are spectacularly skilful climbers. All 6 species of gibbon have long, slender hands, with the thumb deeply divided from the index finger, which gives additional flexibility. When standing upright, the gibbon's long arms touch the ground, so they are often carried above its head. Gibbons are primarily vegetarian but also eat insects, young birds and eggs.

While the slender gibbons weigh from 5 to 13 kg (11 to 28 lb), the great apes are much larger, adult male gorillas weighing as much as 270 kg (595 lb), and even chimpanzees between 48 and 80 kg (106 and 176 lb). All great apes are able to move on their hind legs, if briefly, although they are normally on all fours. Gorillas and chimpanzees are largely terrestrial, but the orang-utan spends much of its time in trees, where it swings from branch to branch with surprising agility. While gorillas are vegetarian animals, the chimpanzees and orang-utans are more omnivorous.

Members of this family are generally gregarious and live in family-based groups, which forage together during the day and build sleeping nests at night. Male great apes are considerably larger than females and may have other special characteristics such as the fatty chin and facial flaps of the mature male orang-utan. Male and female gibbons are mostly similar in size.

NAME: **Lar Gibbon,** *Hylobates lar*
RANGE: **S. Burma, Malaysia, Thailand, Kampuchea, Sumatra**
HABITAT: **rain forest, dry forest**
SIZE: **body: 42–58 cm (16½–22¾ in)**
   tail: absent

The lar gibbon may be black or pale buff, but the hands, feet, brow band and sides of the face are always pale. Like all gibbons, it is tree-dwelling and rarely descends to the ground. It moves in the trees by swinging from branch to branch by means of its long arms or by running upright along large branches. Largely vegetarian, the lar gibbon feeds on fruit, leaves, shoots, buds and flowers and occasionally on insects. Lar gibbons live in family groups of 2 to 6 individuals: an adult male and female and their young of different ages.

Females give birth to 1 young at intervals of 2 to 4 years. The gestation period is 7 to 7½ months. Young gibbons remain with their mothers for at least 2 years and are suckled throughout this period.

NAME: **Pileated Gibbon,** *Hylobates pileatus*
RANGE: **S.E. Thailand**
HABITAT: **forest**
SIZE: **body: 43–60 cm (17–23½ in)**
   tail: absent   Ⓔ

The pileated, or capped, gibbons derive their name from the black cap on the heads of both sexes. Like all gibbons, they are born white and pigmentation gradually spreads down from the head. By the time they are sexually mature, male pileated gibbons are completely black, while females are buff with a black cap.

Pileated gibbons are strongly territorial and males shout abuse to one another across territory boundaries; they seldom fight, however. They feed on a mixed diet of leaves, buds, resin and insects.

NAME: **Hoolock Gibbon,** *Hylobates hoolock*
RANGE: **Bangladesh, E. India, S. China, N. Burma**
HABITAT: **hill forest**
SIZE: **body: 46–63 cm (18–24¾ in)**
   tail: absent

The hoolock has the long limbs and the shaggy fur characteristic of gibbons. Males and females are thought to be about the same size, but adult males are blackish-brown in colour and females yellow-brown. Newborn hoolock gibbons are greyish-white and gradually darken, to become black at a few months old. The female's colour fades at puberty — 6 or 7 years of age. Almost entirely arboreal, the hoolock gibbon sleeps in the trees and, in the daytime, swings itself from branch to branch and tree to tree, searching for food. Fruit, leaves and shoots are its staple foods, but it also eats spiders, insects, larvae and birds' eggs.

Family groups of a mated pair and their young live together, but several families may gather in feeding areas, though normally each family feeds in its own territory. The hoolock's loud calls are important, both for communication with its own group and between groups. Mating occurs at the start of the rainy season, and females bear a single young between November and March.

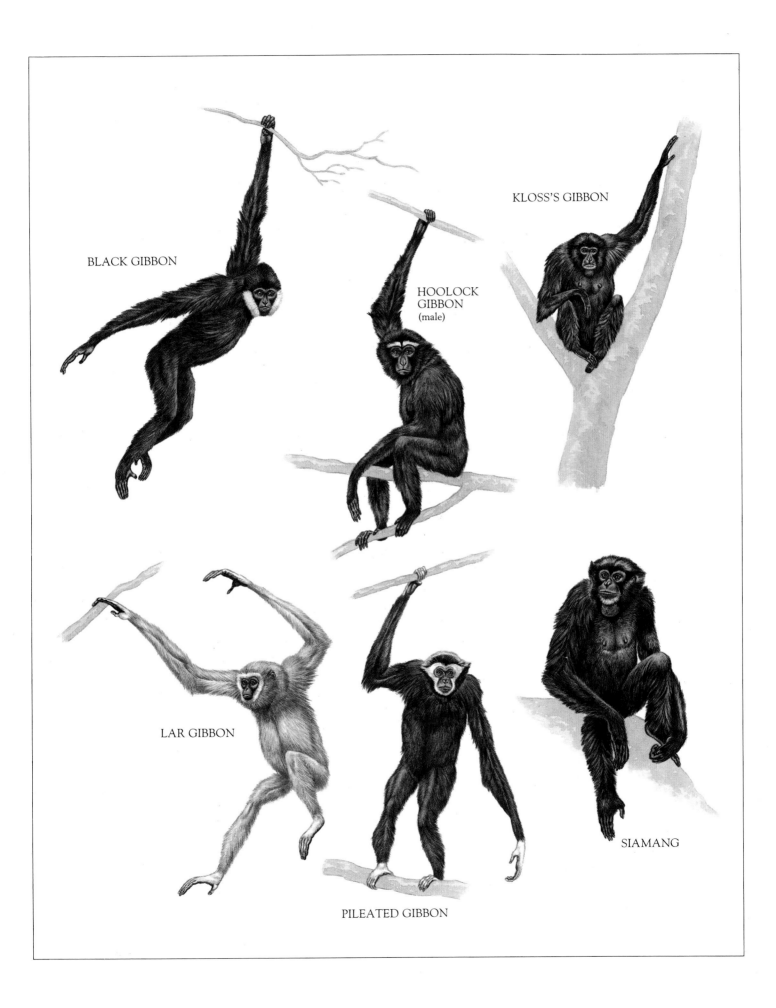

BLACK GIBBON

HOOLOCK
GIBBON
(male)

KLOSS'S GIBBON

LAR GIBBON

PILEATED GIBBON

SIAMANG

Apes 2

NAME: **Orang-utan,** *Pongo pygmaeus*
RANGE: **Sumatra, Borneo**
HABITAT: **rain forest**
SIZE: **height: 1.2–1.5 m (4–5 ft)**
    tail: absent   Ⓔ

The orang-utan, with its reddish-brown, shaggy hair, has a strong, heavily built body and is the second-largest primate. The arms are long and powerful and reach to the ankles when the animal stands erect; there is a small thumb on each broad hand that is opposable to the first digit. The orang-utan's legs are relatively short and are weaker than the arms. Males are much larger and heavier than females and are also identified by the cheek flaps that surround the face of the mature adult. All adults have fatty throat pouches.

Orang-utans live alone, in pairs or in small family groups and are active in the daytime at all levels of the trees. They walk along large branches on all fours or erect and sometimes swing by their hands from branch to branch. On the ground, they walk on all fours or stand erect. Fruit is their staple diet, but they also feed on leaves, seeds, young birds and eggs. The orang-utan sleeps in the trees in a platform nest made of sticks; it may make a new nest every night.

After a gestation period of more than 9 months, the female gives birth to a single young. She cares for her offspring for some time — one captive young was suckled for 6 years — and it clings to her fur as she moves around in the trees.

NAME: **Gorilla,** *Gorilla gorilla*
RANGE: **Africa: S.E. Nigeria to W. Zaire;**
  E. Zaire into adjacent countries
HABITAT: **rain forest up to 3,000 m**
  (10,000 ft)
SIZE: **male height: 1.7–1.8 m (5½–6 ft)**
  female height: 1.4–1.5 m (4½–5 ft)
  tail: absent   Ⓥ

The largest and most robust of the primates, the gorilla is also a gentle, intelligent and sociable animal, which lives a peaceful, quiet existence if undisturbed. Its body, covered with coarse black hair, is massive, with a short, broad trunk and wide chest and shoulders. The head is large, with a short muzzle, and the eyes and ears small; old males have high crowns. The muscular arms are longer than the short, thick legs, and the broad hands are equipped with short fingers and thumbs. Males are bigger and heavier than females, and those over 10 years old have silvery-grey hair on their backs — hence the name silverback, given to old males. There are two races: the lowland and the mountain gorilla, also referred to as the eastern and western races.

On the ground, gorillas normally move in a stooped posture, with the knuckles of the hands resting on the ground, but they do stand erect on occasion. Females and juveniles climb trees, but males rarely do so because of their great bulk. Gorillas live in a close-knit group of a dominant male, 1 or 2 other males, several females and young; some groups may contain only the dominant male, 2 or 3 females and young. The group wanders in a home range of 10 to 40 sq km (4 to 15½ sq mls), which is not defended or marked at the boundaries. There may be some conflict with neighbouring groups, but encounters are generally avoided by communications such as drumming on the ground from a distance. Old males will threaten rivals by standing erect and beating the chest, while roaring and barking, and sometimes by tearing up and throwing plants. When the leader of a troop dies, younger males contest for dominance.

Gorillas are active in the daytime. The troop rises between 6 am and 8 am, feeds for a while on plant material, such as leaves, buds, stalks, berries, bark and ferns, and then has a period of rest and relaxation. Gorillas do not appear to drink but get the water they need from their juicy diet. They feed again in the afternoon and then retire for the night in nests made of twigs and leaves. Young gorillas under 3 years old sleep with their mothers, but all others have their own nests.

Breeding appears to take place at any time of year. The female gives birth to a single young after a gestation of more than 9 months. The young is completely dependent and clings to its mother's fur at first, but it is able to sit up at 3 months and to walk and climb at 5 months. It suckles for 12 to 18 months and remains with its mother for about 3 years.

NAME: **Pygmy Chimpanzee,** *Pan*
  paniscus
RANGE: **Africa: Zaire**
HABITAT: **rain forest**
SIZE: **body: 55–60 cm (21½–23½ in)**
  tail: absent;
  height: up to 1 m (3¼ ft)   Ⓥ

Similar to the chimpanzee, this species (sometimes regarded as only a subspecies), has longer, thinner legs, a more slender body and a narrow face. Its hair and facial skin are black. It moves in the trees and on the ground, feeding mainly on fruit but also on leaves and shoots. A gregarious animal, the pygmy chimpanzee lives in a family group, and several families may gather in a good feeding area. Otherwise its habits are much the same as those of the chimpanzee.

The female bears 1 young, after a gestation of 227 to 232 days, which stays with her for up to 3 years.

NAME: **Chimpanzee,** *Pan troglodytes*
RANGE: **Africa: Guinea to Zaire, Uganda**
  and Tanzania
HABITAT: **rain forest, savanna with**
  woodland
SIZE: **body: 68–94 cm (26¾–37 in)**
  tail: absent;
  height: 1.2–1.7 m (4–5½ ft)   Ⓥ

The intelligent, social chimpanzee has a wide range of sounds and gestures for communication and is probably one of the most expressive of all animals. Thickset and robust, but more lightly built than the gorilla, the chimpanzee has a strong body and long limbs, the powerful arms being longer than the legs. Its hands and feet are narrow and long, with opposable thumbs on the hands. Males are slightly larger than females. There is great variability in the colour of hair and facial skin, but the hair is generally blackish and the face light, darkening in older individuals. The rounded head bears broad, prominent ears, and the lips are mobile and protrusible.

Chimpanzees climb well but spend most of the time on the ground, where they generally walk on all fours, even though they stand erect on occasion, as when their hands are full of food. Their social structure is more variable than that of the gorilla. Rain forest animals live in troops of males; of females with young; of males and females with young, or of adults of both sexes without young. The composition of the troop often changes. Savanna chimpanzees generally live in more stable troops of 1 or more males, several females and their young. They occupy a home range, the size of which depends on the size of the troop and on the food supply. Neighbouring troops meet with much noise and communication, but there is usually little aggression.

Active in the daytime, chimpanzees rise at dawn and feed mainly on plant material, such as fruit, nuts, leaves, shoots and bark, and on eggs and insects. They will use stems or twigs as tools, to winkle termites or ants from their hiding places. Savanna chimpanzees will kill young animals for food by holding them by the hind limbs and striking their heads on the ground. At night, chimpanzees usually sleep in the trees, each making its own nest with interwoven, broken and bent branches. Young under 3 years old sleep with their mothers.

Females have regular periods of heat, with swelling of the genital region, and may be mated by all the males in the troop. Usually 1 young is born, sometimes twins, after a gestation of 227 to 232 days. The young animal lives closely with its mother for 2 to 3 years.

ORANG-UTAN

CHIMPANZEE

PYGMY CHIMPANZEE

GORILLA

Dogs I

NAME: **Coyote**, *Canis latrans*
RANGE: **Alaska, Canada, USA (formerly absent from S.E. but now spreading there); Mexico, Central America**
HABITAT: **prairies, open woodland**
SIZE: **body: 85–95 cm (33½–37½ in)**
    tail: 30–38 cm (11¾–15 in)

A highly adaptable animal, the coyote has managed to thrive, even to increase its population, despite being trapped and poisoned for many years. Although the coyote probably does kill some sheep, young cattle and poultry, its diet consists mainly of rodents and rabbits, and in this way it is of service to the farmer. It also eats snakes, insects, carrion, fruit, berries and grasses and will enter water to catch fish, frogs and crustaceans.

Coyote pairs mate in late winter, and a litter of 5 to 10 young is born after a gestation of 63 to 65 days. The male brings all the food for the female and young at first, but later both parents hunt for food. The young leave when about 6 or 7 months old to find their own home range.

The coyote is the North American equivalent of the jackals, which occur in Asia and Africa. Together with wolves, jackals are the ancestors of domestic dogs.

NAME: **Grey Wolf, *Canis lupus***
RANGE: **E. Europe (isolated populations in Spain and Italy), east to India, USSR; Canada, USA: now only N. Michigan and Wisconsin; Mexico**
HABITAT: **tundra, steppe, open woodland and forest**
SIZE: **body: 1–1.4 m (3¼–4½ ft)**
    tail: 30–48 cm (11¾–18¾ in) Ⓥ

One of the ancestors of the domestic dog, the grey wolf is a powerful, muscular animal, with a thick, bushy tail. Wolves vary in colour from almost white in the Arctic to yellowish-brown or nearly black farther south. Intelligent, social animals, wolves live in family groups or in packs that sometimes include more than one family or other individuals besides the family. The pack members hunt together, cooperating to run down prey such as deer, caribou and wild horses, and they also eat small animals such as mice, fish and crabs. Social hierarchy in the pack is well-organized and is maintained by ritualized gestures and postures; the leading male signals his rank by carrying his tail higher than the others do. Pairs remain together for life.

The female gives birth to 3 to 8 pups after a gestation of about 63 days. Born blind and hairless, the pups first venture outside the den at 3 weeks, and the whole pack then helps to care for and play with them.

ORDER CARNIVORA

There are 8 families in this order: the dogs and foxes, bears, raccoons, pandas, mustelids, viverrids, hyenas and cats. Typically these animals are flesh-eating predators, but not all are totally carnivorous, and a specialized few feed only on plant matter.

CANIDAE: Dog Family

Dogs and their close relatives the jackals, wolves, coyotes and foxes represent one of the most familiar groups of carnivorous mammals. This familiarity is partly due to the fact that dogs were the first animals to be fully domesticated by man.

The family contains about 35 recognized species and is distributed almost world-wide. Domesticated versions aside, dogs are absent only from New Zealand, New Guinea, Madagascar and some other islands. The dingo was introduced into Australia by aboriginal man. All dogs have the well-known, muscular, long-legged body, generally with a bushy tail. The ears are usually large, triangular in outline and erect, and the muzzle is long.

Canids are excellent runners, able to sustain a high speed for considerable distances, and long pursuits are an important part of the hunting technique of many species. Some canids hunt down large prey animals in packs, while others, such as foxes, are, typically, solitary hunters. Males and females generally look alike, although males are often slightly larger than females.

NAME: **Dingo, *Canis dingo***
RANGE: **Australia**
HABITAT: **sandy desert to wet and dry sclerophyll forest**
SIZE: **body: about 1.5 m (5 ft)**
    tail: about 35 cm (13¾ in)

The dingos are descended from domesticated dogs introduced by the aboriginal human inhabitants of Australia many thousands of years ago. In anatomy and behaviour dingos are indistinguishable from domestic dogs, but the two have interbred for so long that there are now few pure dingos. They live in family groups but may gather into bigger packs to hunt large prey. Originally they fed on kangaroos, but when white settlers started to kill off the kangaroos, dingos took to feeding on introduced sheep and rabbits.

A litter of 4 or 5 young is born in a burrow or rock crevice after a gestation of about 9 weeks. The young are suckled for 2 months and stay with their parents for at least a year.

NAME: **Arctic Fox, *Alopex lagopus***
RANGE: **Arctic regions of Europe, Asia and N. America**
HABITAT: **tundra, open woodland**
SIZE: **body: 46–68 cm (18–26¾ in)**
    tail: up to 35 cm (13¾ in)

One of the few truly arctic mammals, the arctic fox has well-furred feet and small, rounded ears. It feeds on ground-dwelling birds, lemmings and other small rodents and also eats the leftovers from polar bear kills and carrion such as stranded marine animals. Burrows, usually in the side of a hill or cliff, provide shelter, but arctic foxes do not hibernate and can withstand temperatures as low as −50°C (−58°F).

A litter of 4 to 11 young is born in May or June after a gestation of 51 to 57 days. They are cared for by both parents.

NAME: **Red Fox, *Vulpes vulpes***
RANGE: **Canada, USA (not Florida or Rockies); Europe (except Iceland); Asia to Japan and Indo-China; introduced in Australia**
HABITAT: **woodland, open country; recently increasing in urban areas**
SIZE: **body: 46–86 cm (18–33¾ in)**
    tail: 30.5–55.5 cm (12–21¾ in)

The versatile, intelligent red fox adapts well to different conditions and has excellent senses and powers of endurance. Although sometimes about at all hours, it is typically active at night, resting during the day in a burrow abandoned by another animal or dug by itself. It lives alone outside the breeding season and is a skilful hunter, preying largely on rodents but also on rabbits, hares, birds, insects and invertebrates. Fruit and berries are also eaten in autumn, and the red fox has taken to scavenging on refuse in urban areas.

A litter of 4 young is born after a gestation of 51 to 63 days. The male brings food for his family until the female is able to leave the cubs or take them out foraging.

NAME: **Fennec Fox, *Vulpes zerda***
RANGE: **N. Africa: Morocco to Egypt, south to N. Niger, Sudan; east to Sinai Peninsula and Kuwait**
HABITAT: **desert, semi-desert**
SIZE: **body: 37–41 cm (14½–16 in)**
    tail: 19–21 cm (7½–8¼ in)

The smallest of the foxes, the fennec fox is identified by its relatively huge ears. It shelters in burrows it digs in the sand and is generally active at night, when it preys on small rodents, birds, insects and lizards.

Fennec foxes are sociable animals which mate for life; each pair or family has its own territory. A litter of 2 to 5 young is born in spring after a gestation of 50 to 51 days.

COYOTE

GREY WOLF

ARCTIC FOX

RED FOX

FENNEC FOX

DINGO

Dogs 2

NAME: Dhole, *Cuon alpinus*
RANGE: central and E. Asia, south to Sumatra and Java
HABITAT: forest, woodland; open country in north of range
SIZE: body: 76 cm–1 m (30 in–3¼ ft)
tail: 28–48 cm (11–18¾ in) Ⓥ

Dholes, or Asiatic wild dogs, are gregarious animals, which live in family groups or in packs of up to 30 that contain several families. Most of their hunting is done in the daytime, and although they are not particularly fast runners, dholes pursue their prey in a steady, relentless chase, finally exhausting the victim. Both smell and sight are important to them when tracking their prey. Because they hunt in packs, dholes are able to kill animals much larger than themselves such as deer, wild cattle, sheep and pigs, water buffalo and the banteng.

A litter of 2 to 6 young is born, after a gestation of about 9 weeks, in a sheltered spot among rocks or a hole in a bank. Several females may breed near one another. Dholes are now becoming rare after many years of persecution by man, and they are also affected by the greatly reduced populations of many of their prey animals. They are now protected in some parts of their range.

NAME: Bush Dog, *Speothos venaticus*
RANGE: Central and South America: Panama to Peru, Brazil and Paraguay
HABITAT: forest, savanna
SIZE: body: 57.5–75 cm (22½–29½ in)
tail: 12.5–15 cm (5–6 in) Ⓥ

Stocky and terrierlike, the bush dog has short legs and tail. It is now rare throughout its range, and little is known of its habits in the wild. A nocturnal dog, it is believed to hunt in packs, preying mainly on rodents, including large species such as pacas and agoutis. Bush dogs swim well, probably better than any other wild canid, and readily pursue prey into water. During the day, they take refuge in a hole or crevice, often the abandoned burrow of an armadillo. They mark the boundaries of their territory with urine and secretions from anal glands.

A litter is thought to contain 4 or 5 young, which are born in a burrow, and both parents tend the young.

NAME: Maned Wolf, *Chrysocyon brachyurus*
RANGE: South America: Brazil, Bolivia, Paraguay, Uruguay, N. Argentina
HABITAT: grassland, swamp edge
SIZE: body: 1.2 m (4 ft)
tail: about 30 cm (11¾ in) Ⓥ

Similar to a red fox in appearance, with its long legs and muzzle, the maned wolf has a yellowish-red coat of fairly long hair, with an erectile mane on the neck and shoulders. The tail may be white or white-tipped. A wary, solitary animal, the maned wolf lives in remote areas and is active mainly at night. It runs fast, with a loping gallop, but has less stamina than many canids and does not usually run down its prey. Large rodents, such as pacas and agoutis, birds, reptiles and frogs are all caught by the maned wolf, and it also feeds on insects, snails and some fruit.

A litter of up to 5 young is born after a gestation of about 2 months. At first, the young have short legs and muzzles, which gradually grow longer during their first year of life.

NAME: Crab-eating Fox, *Dusicyon thous*
RANGE: South America: Colombia to N. Argentina
HABITAT: open woodland, grassland
SIZE: body: 60–70 cm (23½–27½ in)
tail: 30 cm (11¾ in)

The first specimen of this fox ever examined had a crab in its mouth, hence the common name, but, in fact, crabs are only one item in a wide-ranging diet. This fox is also known as the common zorro. Mainly nocturnal and solitary, the crab-eating fox spends the day in a shelter, often a burrow abandoned by another animal. It hunts small rodents, such as mice and rats, lizards, frogs and crabs and also feeds on insects and fruit and digs for turtle eggs. Poultry also figures in the diet of the crab-eating fox.

The female gives birth to a litter of 2 to 6 young.

NAME: Raccoon-dog, *Nyctereutes procyonoides*
RANGE: E. Siberia, N.E. China, Japan, N. Indo-China; introduced in E. and central Europe
HABITAT: forest and rocky banks near rivers and lakes
SIZE: body: 50–55 cm (19¾–21½ in)
tail: 13–18 cm (5–7 in)

Foxlike in build, but with shorter legs and tail, the raccoon-dog has a dark patch on each side of its face, reminiscent of the raccoon's black mask. It lives alone or in family groups of 5 or 6 and is primarily nocturnal. During the day, it shelters in a den among rocks or bushes, in a hollow tree or in a burrow abandoned by another animal. A raccoon-dog may sometimes dig its own burrow. Usually found near water, the raccoon-dog is an excellent swimmer, and frogs and fish are major food items; it also eats rodents, acorns, fruit and berries and scavenges on carrion and refuse around human habitation.

A litter of 6 to 8 young is born after a gestation of about 2 months. The pups are independent at about 6 months old.

NAME: Hunting Dog, *Lycaon pictus*
RANGE: Africa, south of the Sahara to South Africa: Transvaal; (not in rain forest areas of W. and central Africa)
HABITAT: savanna, plains, semi-desert, mountains up to 3,000 m (10,000 ft)
SIZE: body: 80 cm–1.1 m (31½ in–3½ ft)
tail: 30–40 cm (11¾–15¾ in) Ⓥ

Recognized by its dark-brown, black or yellowish coat, well mottled with light patches, the hunting dog has long legs and a short, extremely powerful muzzle. Hunting dogs live in packs of 6 to 30 or more, sometimes up to 90, with a high degree of social cooperation and interaction between individuals in the pack; they communicate by means of gestures and body postures and a few calls. Nomadic animals, hunting dogs roam over a wide area looking for their prey and only remain in one place for more than a few days when the young are too small to travel.

During much of the day, the dogs rest and groom themselves in the shade; most hunting is done in the early morning and evening or on bright, moonlit nights. After a mass greeting ceremony between pack members, the dogs move off to search for prey such as gazelle, impala and zebra. Once prey is located by sight, the dogs follow it slowly for a while before starting the final chase. They may concentrate on one victim or follow several members of a herd before all switching to one particular animal. When close enough, the dogs start to bite the prey wherever they can, often seizing its legs and tail and causing it to fall. They disembowel it and immediately start tearing it to pieces and feeding. The pack shares the kill without aggression, allowing young animals to feed first and disgorging meat to latecomers. Some pack members return to the den and disgorge meat for the adults guarding the young. By expert cooperation, by taking turns in the chase and by combined attack, hunting dogs can successfully bring down prey much bigger than themselves, even wildebeest.

The female gives birth to 2 to 16 young, usually 7, after a gestation of 69 to 72 days. The litter is born in a burrow, such as an abandoned aardvark or warthog hole, and more than one female may share the den. The young are blind at birth, but their eyes open at about 2 weeks, and they soon begin to venture out of the den. They are suckled for about 3 months and fed on regurgitated food by pack members from the age of 2 weeks. The whole pack takes an interest in the young and will feed any motherless pups. At 6 months the young begin to learn to hunt and accompany the pack.

DHOLE

HUNTING DOG

MANED WOLF

BUSH DOG

CRAB-EATING FOX

RACCOON-DOG

Bears

NAME: **Spectacled Bear,** *Tremarctos ornatus*
RANGE: **South America: Venezuela, Colombia, Ecuador, Peru, W. Bolivia**
HABITAT: **forest, savanna, mountainous areas up to 3,000 m (10,000 ft)**
SIZE: **body: 1.5–1.8 m (5–6 ft)**
    tail: 7 cm (2¾ in) Ⓥ

The only South American bear, this species is generally black or dark brown in colour, with white markings around the eyes and sometimes on the neck. Mainly a forest dweller, although it also ranges into open country, this bear feeds largely on plant material such as leaves, fruit and roots. It is also thought to prey on animals such as deer and vicuna. It is a good climber and sleeps in a tree in a large nest it makes from sticks.

The spectacled bear lives alone or in a family group. The female produces a litter of up to 3 young after a gestation of 8 to 8½ months.

NAME: **Big Brown Bear/Grizzly Bear,** *Ursus arctos*
RANGE: **Europe: Scandinavia to Balkans, scattered populations in France, Italy and Spain; USSR; Asia, north of Himalayas; Alaska, W. Canada, mountainous areas of W. USA**
HABITAT: **forest, tundra**
SIZE: **body: 1.5–2.5 m (5–8¼ ft)**
    tail: absent

Many subspecies are covered by the scientific name *U. arctos*, including the Kodiak bear and the grizzly bears. At least 2 races of grizzly are rare or endangered. The races vary in colour from pale yellowish-fawn to dark brown or nearly black, but all are large, immensely strong bears, and they are among the biggest carnivores. They live alone or in family groups and are active night or day, although in areas where bears have been persecuted, they are nocturnal. The diet varies greatly from area to area but may include plant material, such as fruit, nuts, roots and seaweed, as well as insects, fish, small vertebrates and carrion. Alaskan brown bears feed heavily on migrating salmon. Most individuals are too slow to catch wild, hoofed mammals, although they have been known to kill bison, and many are too heavy to climb trees. In late summer and autumn, the bears fatten up on vast quantities of fruit and berries in preparation for the winter sleep — a period of torpor, not true hibernation.

Females breed every 2 or 3 years and produce litters of 1 to 4 young after a gestation of 6 to 8 months. The newborn young are blind and tiny, usually weighing only about 300 to 700 g (10½ to 25 oz). They remain with the mother for a year, sometimes longer.

URSIDAE: Bear Family

The 7 species of bear are an offshoot of doglike ancestors. Dogs and their relatives consume a certain amount of plant material in their largely carnivorous diet, and this omnivorous tendency is increased in the bears, which are adaptable consumers of a wide range of foods, including insects, small vertebrates, grass, leaves, fruit and nuts. Bears' teeth lack the shearing blades of those of cats and dogs.

Bears are large, sturdily built animals, with big heads, short limbs and exceptionally short tails. They have flat, five-toed feet with long, curving claws. Only the polar bear inhabits Arctic regions; the others all occur in temperate and tropical areas of the northern hemisphere, and 1 species lives in northern South America. In cold, winter weather, many bears undergo a period of torpor; this winter sleep is not true hibernation, since temperature and respiration rate do not fall drastically, as they do in true hibernators. Male and female look alike.

NAME: **American Black Bear,** *Ursus americanus*
RANGE: **Alaska, Canada, USA: patchy distribution in New England, through Pennsylvania to Tennessee, Florida, S. Georgia, Mississippi, Louisiana, mountainous areas of the west; N. Mexico**
HABITAT: **wooded areas, swamps, national parks**
SIZE: **body: 1.5–1.8 m (5–6 ft)**
    tail: 12 cm (4¾ in)

American black bears actually vary in colour from glossy black to dark brown, reddish-brown or almost white. There is often a small white patch on the chest. Originally found throughout much of the USA, this bear now lives only in the wilder, uninhabited areas and in national parks, where it is thriving and on the increase. Occasionally about in daytime, black bears are usually active at night, when they roam for long distances in search of food such as fruit, berries, nuts, roots and honey. They also feed on insects, rodents and other small mammals, stranded fish and even carrion and refuse. Their sense of smell is good, but their hearing and eyesight are only fair. In autumn, black bears gorge on the ample supplies of fruit to fatten themselves for their long sleep during the coldest weather.

After mating, pairs separate, and except for females with cubs, black bears are usually solitary. A litter of 1 to 4 young is born in January or February after a gestation of about 7 months.

NAME: **Asiatic Black Bear,** *Selenarctos thibetanus*
RANGE: **Afghanistan to China, Siberia, Japan, Korea, Taiwan, Hainan, S.E. Asia**
HABITAT: **forest and brush up to 3,600 m (11,800 ft)**
SIZE: **body: 1.3–1.6 m (4¼–5¼ ft)**
    tail: 7–10 cm (2¾–4 in)

Although usually black, with some white markings on the snout and chest, some Asiatic black bears may be reddish or dark brown. They feed largely on plant matter and will raid crop fields for food or climb trees to obtain fruit and nuts, as well as eating insect larvae and ants. However, they can also be aggressive predators and sometimes kill cattle, sheep and goats.

The female has a litter of 2 cubs, which are blind and extremely small at birth. They stay with their mother until almost fully grown.

NAME: **Polar Bear,** *Thalarctos maritimus*
RANGE: **Arctic Ocean to southern limits of ice floes**
HABITAT: **coasts, ice floes**
SIZE: **body: 2.2–2.5 m (7¼–8¼ ft)**
    tail: 7.5–12.5 cm (3–5 in) Ⓥ

A huge bear, with an unmistakable creamy-white coat, the polar bear is surprisingly fast and can easily outrun a reindeer over a short distance. It wanders over a larger area than any other bear and, of course, swims well. Seals, fish, seabirds, arctic hares, reindeer and musk-oxen are the polar bear's main prey, and in the summer it also eats berries and leaves of tundra plants.

Normally solitary animals outside the breeding season, polar bears mate in midsummer. A litter of 1 to 4 young is born after a gestation of about 9 months, and the young bears remain with their mother for about a year. Thus females breed only every other year.

NAME: **Sun Bear,** *Helarctos malayanus*
RANGE: **S.E. Asia, Sumatra, Borneo**
HABITAT: **mountain and lowland forest**
SIZE: **body: 1.1–1.4 m (3½–4½ ft)**
    tail: absent

The smallest bear, the sun bear nevertheless has a strong, stocky body and powerful paws, with long, curved claws that help it climb trees. It spends the day in a nest in a tree, sleeping and sunbathing, and searches for food at night. Using its strong claws, the sun bear tears at tree bark to expose insects, larvae and the nests of bees and termites; it also preys on junglefowl and small rodents. Fruit and coconut palm, too, are part of its diet.

There are usually 2 young, born after a gestation of about 96 days and cared for by both parents.

POLAR BEAR

AMERICAN BLACK BEAR

SPECTACLED BEAR

ASIATIC BLACK BEAR

SUN BEAR

BIG BROWN BEAR

Raccoons, Pandas

PROCYONIDAE: Raccoon Family

There are about 18 species in this family, all of which inhabit temperate and tropical areas of the Americas. They are all long-bodied, active animals, thought to be closely linked to the dog-bear line of carnivore evolution. They are good climbers and spend much of their life in trees. Males are usually longer and heavier than females, but otherwise the sexes look alike.

NAME: **Raccoon**, *Procyon lotor*
RANGE: **S. Canada, USA, south to Panama**
HABITAT: **wooded areas, often near water, swamps**
SIZE: **body: 41–60 cm (16–23½ in)**
   tail: 20–40 cm (7¾–15¾ in)

Familiar, adaptable mammals, raccoons have coped well with the twentieth century and today are even spotted in cities, scavenging for food. The raccoon is stocky but agile, with thick, greyish fur and a bushy tail ringed with black bands; its pointed face has a characteristic "bandit" mask across the eyes. On the forepaws are long, sensitive digits, which the raccoon uses with great dexterity when handling food. Mainly active at night, the raccoon is a good climber and can swim if necessary. Its wide-ranging diet includes aquatic animals, such as frogs and fish, small land animals, such as rodents, birds, turtle eggs, nuts, seeds, fruit and corn.

A litter of 3 to 6 young is born in spring after a gestation of about 65 days. The young raccoons' eyes are open at about 3 weeks, and they start to go out with their mother at about 2 months old, remaining with her until autumn.

NAME: **Olingo**, *Bassaricyon gabbi*
RANGE: **Central America, south to Venezuela, Colombia and Ecuador**
HABITAT: **forest**
SIZE: **body: 35–47.5 cm (13¾–18¾ in)**
   tail: 40–48 cm (15¾–18¾ in)

An expert climber, the olingo spends much of its life in trees and rarely descends to the ground. Using its long tail to help it balance, it leaps from tree to tree and runs along the branches. It is primarily nocturnal, and although it lives alone or in pairs, joins in groups with other olingos and with kinkajous to search for food. Fruit is its staple diet, but it also eats insects, small mammals and birds.

Breeding takes place at any time of year, and there is usually only 1 young in a litter, born after a gestation of 73 or 74 days. The female chases her mate away shortly before the birth and rears her offspring alone.

NAME: **Coati**, *Nasua nasua*
RANGE: **South America (not Patagonia)**
HABITAT: **woodland, lowland forest**
SIZE: **body: 43–67 cm (17–26¼ in)**
   tail: 43–68 cm (17–26¾ in)

The coati is a muscular, short-legged animal, with a long, banded tail and a pointed, mobile snout. It lives in groups of up to 40 individuals, which hunt together day and night, resting in the heat of the day. With its mobile snout, the coati probes holes and cracks in the ground, rocks or trees, searching for the insects, spiders and other small ground-dwelling invertebrates that are its staple diet. Fruit and larger animals, such as lizards, are also eaten.

After mating, the group splits up and females go off alone to give birth. A litter of 2 to 7 young is born after a gestation of about 77 days, usually in a cave or a nest in a tree. Once the young are about 2 months old, the females and their offspring regroup with yearlings of both sexes. Males over 2 years old are only allowed to accompany the group for the few weeks of the mating period and even then are subordinate to females.

NAME: **Kinkajou**, *Potos flavus*
RANGE: **E. Mexico, through Central and South America to Brazil**
HABITAT: **forest**
SIZE: **body: 41–57 cm (16–22½ in)**
   tail: 40–56 cm (15¾–22 in)

The tree-dwelling kinkajou is an agile climber, using its prehensile tail as a fifth limb to hold on to branches, which leaves its hands free for picking food. During the day, the kinkajou rests, usually in a hole in a tree, and then emerges at night to forage in the trees. It feeds mainly on fruit and insects and occasionally on small vertebrates, and its long tongue is adapted for extracting the soft flesh from fruit, such as mangoes, avocados and guavas, and for licking up nectar, insects and even honey from wild bees' nests. The nocturnal kinkajous often feed in the same fruit trees that are used by monkeys during the daytime.

The female gives birth to 1 young, rarely 2, in a tree hollow after a gestation of 112 to 118 days. The young kinkajou takes its first solid food at about 7 weeks and is independent when about 4 months old. It is able to hang by its prehensile tail after only 8 weeks or so.

AILUROPODIDAE: Panda Family

The giant panda and the lesser, or red, panda are the only members of this small family, which has evolutionary links with both bears and raccoons.

NAME: **Giant Panda**, *Ailuropoda melanoleuca*
RANGE: **mountains of central China**
HABITAT: **bamboo forest**
SIZE: **body: 1.2–1.5 m (4–5 ft)**
   tail: 12.5 cm (5 in)   ®

One of the most popular, newsworthy mammals, the giant panda is a rare, elusive creature, and surprisingly little is known of its life in the wild. It is large and heavily built, with a massive head, stout legs and a thick, woolly black and white coat, often with a brownish tinge to the black. The panda's forepaw is specialized for grasping bamboo stems, its main food: it has an elongated wrist bone that effectively provides a sixth digit, against which the first and second digits can be flexed. The panda consumes huge amounts of bamboo in order to obtain sufficient nourishment and thus spends 50 to 75 per cent of its day feeding. It is also thought to eat some other plants and, occasionally, small animals. Normally solitary, unless breeding or caring for young, pandas are primarily ground-dwelling but regularly climb trees for shelter or refuge.

The male leaves his territory to find a mate and courts her by uttering whines and barks. He will drive off any rival males before mating. The female gives birth to 1 blind, helpless young, which weighs only about 140 g (5 oz) — tiny in comparison to its mother, which may weigh as much as 115 kg (253 lb). The cub grows rapidly, however, and by 8 weeks is more than 20 times its birth weight.

NAME: **Lesser/Red Panda**, *Ailurus fulgens*
RANGE: **Nepal to W. Burma, S.W. China**
HABITAT: **bamboo forest**
SIZE: **body: 51–63.5 cm (20–25 in)**
   tail: 28–48.5 cm (11–19 in)

The lesser panda, with its beautiful rusty-red coat and long, bushy tail, resembles a raccoon more than it does its giant relative. Primarily nocturnal, it spends the day sleeping, curled up on a branch with its tail over its head or its head tucked on to its chest. It feeds on the ground, on bamboo shoots, grass, roots, fruit and acorns and may occasionally eat mice, birds and birds' eggs. A quiet creature unless provoked, when it rears up on its hind legs and hisses, it lives in pairs or family groups.

In the spring, the female gives birth to 1 to 4 young, usually only 1 or 2, after a gestation of 90 to 150 days. The longer gestations recorded are thought to include a period of delayed implantation, during which the fertilized egg lies dormant in the womb, only starting to develop at the time that will ensure birth at the optimum season for the young's survival. The young stay with their mother for up to a year.

OLINGO

KINKAJOU

LESSER PANDA

RACCOON

GIANT PANDA

COATI

Mustelids I
Weasels, Martens

NAME: **Stoat**, *Mustela erminea*
RANGE: **Europe, Asia, N. USA, Greenland;**
    introduced in New Zealand
HABITAT: **forest, tundra**
SIZE: **body: 24–29 cm (9½–11½ in)**
    tail: 8–12 cm (3–4¾ in)

The stoat is a highly skilled predator. It kills by delivering a powerful and accurate bite to the back of the prey's neck. Rodents and rabbits are the stoat's main diet, but it will also kill and eat other mammals –– including some bigger than itself — as well as birds, eggs, fish and insects. At the beginning of winter, in the northern part of its range, the stoat loses its dark fur and grows a pure white coat, only the black tail-tip remaining. This white winter pelt is the ermine prized by the fur trade.

Stoats produce a litter of 3 to 7 young in April or May. The male assists in caring for and feeding the young which are helpless at birth. Their eyes do not open until they are about 3 weeks old, but at 7 weeks young males are already larger than their mother

There are 15 species of *Mustela*, including the minks now farmed for their dense fur.

NAME: **Black-footed Ferret**, *Mustela*
    nigripes
RANGE: **N. America: Alberta to N. Texas**
HABITAT: **prairie**
SIZE: **body: 38–45 cm (15–17¾ in)**
    tail: 12.5–15 cm (5–6 in) Ⓔ

The black-footed ferret feeds mainly on prairie dogs, but these animals are considered farm pests and large numbers are poisoned. This destruction of their natural prey has caused a drastic decline in the numbers of ferrets — and their indirect poisoning. The black-footed ferret is now protected by law, but it is still in great danger of extinction and its survival depends either on the conservation of prairie dogs or on its ability to adapt to other areas and prey. It is generally a nocturnal animal. In June it produces a yearly litter of 3 to 5 young.

NAME: **Least Weasel**, *Mustela nivalis*
    (rixosa)
RANGE: **Europe, N. Africa, Asia, N.**
    America; introduced in New Zealand
HABITAT: **farmland, woodland**
SIZE: **body: 18–23 cm (7–9 in)**
    tail: 5–7 cm (2–2¾ in)

This weasel is the smallest carnivore. It undergoes the same winter colour change as the ermine in the northern part of its range. The least weasel preys mostly on mice and is small enough to pursue them into their burrows. It is most active by night but will hunt in the daytime. One or two litters a year, of 4 or 5 young each, are born in an underground nest.

MUSTELIDAE: Mustelid Family

The mustelid family of carnivores is an extremely successful and diverse group of small to medium-sized mammals. There are about 67 species in 23 genera from all regions of the world except Australia and Madagascar; 2 species have been introduced into New Zealand to control rodents. Although there is a moderate range of physique, most mustelids conform to the pattern of long, supple body, short legs and longish tails. Males are almost invariably larger than females.

This blueprint of shape has proved adaptable in terms of niche utilization. Mustelids have become expert burrowers, tree climbers and swimmers. One species, the sea otter, is almost entirely marine.

Anal scent glands are well developed in mustelids, and the secretions are used for marking territory boundaries. Some species, notably polecats and skunks, have particularly foul-smelling secretions which they spray as a defensive technique.

NAME: **Western Polecat**, *Mustela putorius*
RANGE: **Europe**
HABITAT: **forest**
SIZE: **body: 38–46 cm (15–18 in)**
    tail: 13–19 cm (5–7½ in)

The western polecat is a solitary, nocturnal creature. It hunts rodents, birds, reptiles and insects, mostly on the ground; it only rarely climbs trees. Like all mustelids, the polecat has anal scent glands, but the secretions are particularly offensive and are used as a defence, as well as for marking territory. It breeds once or twice a year, bearing litters of 5 to 8 young. The domestic ferret is probably descended from this polecat.

NAME: **American Marten**, *Martes*
    americana
RANGE: **Canada, N. USA**
HABITAT: **forest, woodland**
SIZE: **body: 35.5–43 cm (14–17 in)**
    tail: 18–23 cm (7–9 in)

An agile, acrobatic creature with a bushy tail, the marten spends much of its time in trees, where it preys on squirrels. It also hunts on the ground and eats small animals and insects, fruit and nuts. Martens den in hollow trees and produce a yearly litter of 2 to 4 young, usually in April. The young are blind and helpless at birth; their eyes open at 6 weeks and they attain adult weight at about 3 months.

The closely related fisher, M. *penanti*, is one of the few creatures that preys on American porcupines.

NAME: **Sable**, *Martes zibellina*
RANGE: **Siberia; Japan: Hokkaido**
HABITAT: **forest**
SIZE: **body: 38–45 cm (15–18 in)**
    tail: 12–19 cm (5–7½ in)

The sable, one of the 7 species of marten, has long been hunted for its luxuriant fur. Conservation measures have now been introduced in Russia to save the decreasing wild population, and sables are farmed for pelts. The sable is a ground-dweller and feeds on small rodents and other small mammals, as well as on fish, insects, honey, nuts and berries. A yearly litter of 2 to 4 is produced, usually in April.

NAME: **Zorilla**, *Ictonyx striatus*
RANGE: **Africa: Senegal and Nigeria to**
    South Africa
HABITAT: **savanna, open country**
SIZE: **body: 28.5–38.5 cm (11–15 in)**
    tail: 20.5–30 cm (8–12 in)

Also known as the striped polecat, the single species of zorilla can eject a nauseating secretion from its anal glands when alarmed. It is primarily a nocturnal animal and feeds on rodents, reptiles, insects and birds' eggs. By day, it rests in a burrow which it digs itself or takes over from another animal, or in a rock crevice or pile of stones. The litter of 2 or 3 young is born in a burrow.

NAME: **Grison**, *Galictis vittata*
RANGE: **S. Mexico to Peru and Brazil**
HABITAT: **forest, open land**
SIZE: **body: 47–55 cm (18½–21½ in)**
    tail: 16 cm (6 in)

The grison is an agile animal, good at climbing and swimming. It feeds on frogs and worms as well as other ground-living creatures. The abandoned burrow of another animal or a rock or tree-root crevice serves it as a den, and a litter of 2 to 4 young is produced in October. There are 3 species of grison, all living in Central and South America. The local population uses grisons in the same way as ferrets, for flushing out chinchillas.

NAME: **Tayra**, *Eira barbara*
RANGE: **S. Mexico to Argentina;**
    Trinidad
HABITAT: **forest**
SIZE: **body: 60–68 cm (24–27 in)**
    tail: 38–47 cm (15–18½ in)

The single species of tayra runs, climbs and swims well. It preys on small mammals, such as tree squirrels and rodents, and also feeds on fruit and honey. Tayras move in pairs or small family groups and are active at night and in the early morning. They are believed to produce a yearly litter of 2 to 4 young.

ZORILLA

BLACK-FOOTED FERRET

WESTERN POLECAT

LEAST WEASEL

AMERICAN MARTEN

GRISON

SABLE

STOAT

TAYRA

Mustelids 2
Wolverine, Badgers

NAME: **Wolverine**, *Gulo gulo*
RANGE: **Scandinavia, Siberia, Alaska, Canada, W. USA**
HABITAT: **coniferous forest, tundra**
SIZE: **body: 65–87 cm (25½–34¼ in)**
tail: 17–26 cm (6½–10 in)

The single species of wolverine is a heavily built animal, immensely strong for its size and capable of killing animals larger than itself. Although largely carnivorous, wolverines also feed on berries. They are solitary animals, mainly ground-dwelling, but they can climb trees. Each male holds a large territory with 2 or 3 females and mates in the summer. The female wolverine bears 2 or 3 young in the following spring, usually after a period of delayed implantation.

Delayed implantation is an interesting phenomenon allowing animals to mate at the ideal time and bear young at the ideal time, even though the intervening period is longer than their actual gestation. The fertilized egg remains in a suspended state in the womb and development starts only after the required period of dormancy. The young suckle for about 2 months and remain with their mother for up to 2 years, at which time they are driven out of her territory. They become sexually mature at about 4 years of age.

NAME: **Ratel**, *Mellivora capensis*
RANGE: **Africa; Middle East to N. India**
HABITAT: **steppe, savanna**
SIZE: **body: 60–70 cm (23½–27½ in)**
tail: 20–30 cm (7¾ to 11¾ in)

The stocky ratel is also known as the honey badger because of its fondness for honey, and in Africa a honey-eating association has developed between the ratel and a small bird, the honey guide. Calling and flying just ahead, the bird leads the ratel to a wild bees' nest. With its powerful foreclaws, the ratel then breaks open the nest, and both partners share the spoils. The unusually tough hide of the ratel protects it from bee stings and is also a good defence in another way: the skin is so loose on the body that the animal can twist about in its skin and even bite an attacker which has a hold on the back of its neck. In addition to honey and bee larvae, the ratel eats small animals, insects, roots, bulbs and fruit and will occasionally attack large animals such as sheep and antelope.

Although sometimes active in the daytime, ratels are generally nocturnal animals. They live singly or in pairs and produce a litter of 2 young in an underground burrow or a nest among rocks. The gestation period is 6 or 7 months.

NAME: **Eurasian Badger**, *Meles meles*
RANGE: **Europe to Japan and S. China**
HABITAT: **forest, grassland**
SIZE: **body: 56–81 cm (22–32 in)**
tail: 11–20 cm (4–7¾ in)

The gregarious Eurasian badger lives in family groups in huge burrows with networks of underground passages and chambers and several entrances. A burrow system, or sett, may be used by successive generations of badgers, each making additions and alterations. Bedding material of grass, hay and leaves is gathered into the sleeping chambers and occasionally dragged out to air in the early morning. Around the sett are play areas, and the boundaries of the group's territory are marked by latrine holes.

Badgers are generally nocturnal and emerge from the sett around dusk. They are playful creatures, and at this time the young and adults will indulge in boisterous romping. Such play helps to strengthen the social bonds, crucial to group-living animals. Badgers feed on large quantities of earthworms as well as on small animals, bulbs, fruit and nuts. They mate in the summer when social acitivity is at its height, but gestation of the fertilized eggs does not start until such a time as to ensure that the 2 to 4 young are not born until the following spring.

NAME: **American Badger**, *Taxidea taxus*
RANGE: **S.W. Canada to central Mexico**
HABITAT: **open grassland, arid land**
SIZE: **body: 42–56 cm (16½–22 in)**
tail: 10–15 cm (4–6 in)

The single species of *Taxidea* is the only New World badger. It has a rather flattened body shape but is otherwise similar to other badgers. A solitary creature, the American badger is generally active at night, although it will come out of its burrow during the day. It is an excellent digger and burrows rapidly after disappearing rodents — its main food. It sometimes buries a large food item for storage. Birds, eggs and reptiles make up the rest of its diet.

The badgers mate in late summer, but the 6-week gestation period does not begin until February, so the litter is born only the following spring. The 1 to 5 young, usually 2, are born on a grassy bed in the burrow and are covered with silky fur. Their eyes open at 6 weeks and they suckle for several months.

In northern parts of its range and at high altitudes, the American badger sleeps for much of the winter, surviving on its stored fat. However, it does not truly hibernate and becomes active in mild spells.

NAME: **Hog Badger**, *Arctonyx collaris*
RANGE: **N. China, N.E. India, Sumatra**
HABITAT: **wooded regions in uplands and lowlands**
SIZE: **body: 55–70 cm (21½–27½ in)**
tail: 12–17 cm (5–6½ in)

Similar in shape and size to the Eurasian badger, the hog badger is distinguished by its white throat and mostly white tail. The common name refers to the badger's mobile, piglike snout, used for rooting for plant and animal food. A nocturnal animal, it spends its day in a rock crevice or deep burrow; its habits are much the same as those of the Eurasian badger. Like all badgers, the hog badger has anal scent glands with potent secretions, and its black and white markings constitute a warning that it is a formidable opponent.

The breeding habits of the hog badger are not well known, but in one observation the female gave birth to 4 young in April.

NAME: **Stink Badger**, *Mydaus javanensis*
RANGE: **Sumatra, Java, Borneo**
HABITAT: **dense forest**
SIZE: **body: 37.5–51 cm (15–20 in)**
tail: 5–7.5 cm (2–3 in)

The 2 species of stink badger have particularly powerful anal gland secretions, said to be as evil-smelling as those of skunks. When threatened or alarmed, the stink badger raises its tail and ejects a stream of the fluid. However, like musk, although foul in concentration, the secretion can be sweet-smelling in dilution and was formerly used in the making of perfume. This nocturnal creature lives in a burrow and feeds on worms, insects and small animals. The second species, the Palawan badger, *M. marchei*, inhabits the Philippine and Calamian islands.

NAME: **Chinese Ferret Badger**, *Melogale moschata*
RANGE: **N.E. India, S. China, Indo-China, Java, Borneo**
HABITAT: **grassland, open forest**
SIZE: **body: 33–43 cm (13–17 in)**
tail: 15–23 cm (6–9 in)

Distinctive masklike face markings distinguish the Chinese ferret badger from other oriental mustelids. This badger lives in burrows or crevices and is active at dusk and at night. It is a good climber and feeds on fruit, insects, small animals and worms. Ferret badgers are savage when alarmed and their anal secretions are foul-smelling. A litter of up to 3 young is born in May or June. There are 2 other species of ferret badger, both in Southeast Asia.

WOLVERINE

RATEL

EURASIAN BADGER

HOG BADGER

STINK BADGER

AMERICAN BADGER

CHINESE FERRET BADGER

Mustelids 3
Skunks, Otters

NAME: Striped Skunk, *Mephitis mephitis*
RANGE: S. Canada to N. Mexico
HABITAT: semi-open country, woods, grassland
SIZE: body: 28–38 cm (11–15 in)
tail: 18–25 cm (7–10 in)

Notorious for its pungent anal gland secretions, the striped skunk is one of the most familiar mustelids. Like others of its family, it does not use these secretions against rival skunks, only against enemies. The fluid is an effective weapon because the smell temporarily stops the victim's breathing. The striped skunk is a nocturnal animal, spending the day in a burrow or in a den beneath old buildings, wood or rock piles. It feeds on mice, eggs, insects, berries and carrion. A litter of 5 or 6 young is born in early May in a den lined with vegetation. The hooded skunk, M. *macroura*, is a similar and closely related species; the black and white markings of both skunks are highly variable and constitute a warning display.

NAME: Hog-nosed Skunk, *Conepatus mesoleucus*
RANGE: S. USA to Nicaragua
HABITAT: wooded and open land
SIZE: body: 35–48 cm (13¾–19 in)
tail: 17–31 cm (6½–12 in)

This nocturnal, solitary, slow-moving animal has the coarsest fur of all skunks. The common name derives from the animal's long piglike snout, which it uses to root in the soil for insects and grubs. Snakes, small mammals and fruit also feature in its diet. The hog-nosed skunk dens in rocky places or abandoned burrows and produces a litter of 2 to 5 young. There are 7 species of hog-nosed skunk, all found in the southern USA and South America.

NAME: Western Spotted Skunk, *Spilogale gracialis*
RANGE: W. USA to central Mexico
HABITAT: wasteland, brush and wooded areas
SIZE: body: 23–34.5 cm (9–13½ in)
tail: 11–22 cm (4¼–8½ in)

The white stripes and spots of the western spotted skunk are infinitely variable; no two animals have quite the same markings. A nocturnal, mainly terrestrial animal, this skunk usually dens underground, but it is a good climber and sometimes shelters in trees. Rodents, birds, eggs, insects and fruit are the main items in its diet.

In the south of the spotted skunk's range, young are born at any time of year, but farther north, the 4 or 5 young are produced in spring. The gestation period is about 4 months.

NAME: Eurasian Otter, *Lutra lutra*
RANGE: Europe, N. Africa, Asia
HABITAT: rivers, lakes, sheltered coasts
SIZE: body: 55–80 cm (21½–31½ in)
tail: 30–50 cm (12–19½ in) (V)

Although agile on land, otters have become well adapted for an aquatic life. The Eurasian otter has the slim mustelid body, but its tail is thick, fleshy and muscular for propulsion in water. All four feet are webbed and the nostrils and ears can be closed when the otter is in water. Its fur is short and dense and keeps the skin dry by trapping a layer of air around the body. An excellent swimmer and diver, the otter moves in water by strong undulations of its body and tail and strokes of its hind feet. It feeds on fish, frogs, water birds, voles and other aquatic creatures.

Otters are solitary, elusive creatures, now rare in much of their range. They den in a river bank in a burrow called a holt and are most active at night. Even adult otters are playful animals and enjoy sliding down a muddy bank. A litter of 2 or 3 young is born in the spring — or at any time of year in the south of the otter's range. There are 8 species of *Lutra*, all with more or less similar habits and adaptations.

NAME: Giant Otter, *Pteronura brasiliensis*
RANGE: Venezuela to Argentina
HABITAT: rivers, slow streams
SIZE: body: 1–1.5 m (1¼–5 ft)
tail: 70 cm (27½ in) (E)

The giant otter is similar in appearance to *Lutra* species but is larger and has a flattened tail with crests on each edge. It generally travels in a group and is active during the day. Giant otters feed on fish, eggs, aquatic mammals and birds. They den in holes in a river bank or under tree roots and produce yearly litters of 1 or 2 young. Now endangered, the giant otter is protected in some countries, but enforcement of the law in the vast, sparsely inhabited areas of the animal's range is extremely difficult and numbers are still decreasing.

NAME: African Clawless Otter, *Aonyx capensis*
RANGE: Africa: Senegal, Ethiopia, South Africa
HABITAT: slow streams and pools; coastal waters, estuaries
SIZE: body: 95–100 cm (37–39 in)
tail: 55 cm (21½ in)

The African clawless otter swims and dives as well as other otters, although its feet have only small connecting webs. As its name suggests, this otter has no claws other than tiny nails on the third and fourth toes of the hind feet. It has less dense fur than most otters so has not been hunted as extensively. Crabs are the most important item of the clawless otter's diet, and it is equipped with large, strong cheek teeth for crushing the hard shells; it also feeds on molluscs, fish, reptiles, frogs, birds and small mammals. Like most otters, it comes ashore to eat and feeds from its hands rather than from the ground. Indeed, clawless otters seem particularly skilful in their hand movements.

Clawless otters do not dig burrows but live in crevices or under rocks in family groups, pairs or alone. The litter of 2 to 5 young stays with the parents for at least a year.

NAME: Sea Otter, *Enhydris lutra*
RANGE: Bering Sea; USA: California coast
HABITAT: rocky coasts
SIZE: body: 1–1.2 m (1¼–4 ft)
tail: 25–37 cm (10–14½ in)

The most highly adapted of all otters for an aquatic existence, the sea otter spends nearly all its life at sea, always in water less than 20 m (66 ft) deep. Its body is streamlined and legs and tail are short. The hind feet are webbed and flipperlike, and the forefeet are small. Unlike most other marine mammals, it does not have an insulating layer of fat but must rely on a layer of air trapped in its fur for protection against the cold water. Much time and effort is spent on grooming the dense, glossy fur for its insulating qualities are lessened if it becomes unkempt.

Sea otters feed on clams, sea urchins, mussels, abalone and other molluscs which they collect from the sea bed and eat while lying in the water. In order to cope with the hard shells of much of its food, the sea otter has discovered how to use rocks as tools. When diving for food, the sea otter also brings up a rock from the sea bed. Placing the stone on its chest as it lies on its back in the water, the otter bangs the prey against the stone until the shell is broken, revealing the soft animal inside. At dusk, the sea otter swims into the huge kelp beds found in its range and entangles itself in the weed so that it does not drift during the night while it is asleep.

Sea otters breed every two years or so and give birth to 1 pup after a gestation period of 8 or 9 months. The pup is born in an unusually well-developed state, with eyes open and a full set of milk teeth. The mother carries and nurses the pup on her chest as she swims on her back. At one time sea otters were hunted for their beautiful fur and became rare, but they have been protected by law for some years.

STRIPED SKUNK

HOG-NOSED SKUNK

EURASIAN OTTER

WESTERN SPOTTED SKUNK

GIANT OTTER

AFRICAN CLAWLESS OTTER

SEA OTTER

Civets I

NAME: **African Linsang,** *Poiana richardsoni*
RANGE: **Africa: Sierra Leone to Zaire**
HABITAT: **forest**
SIZE: **body: 33 cm (13 in)**
  tail: 38 cm (15 in)

A nocturnal animal, the African linsang is a good climber and spends more time in the trees than on the ground. During the day, it sleeps in a nest, built of green vegetation in a tree, and then emerges at night to hunt for insects and young birds; it also feeds on fruit, nuts and plant material. Elongate and slender, this linsang is brownish-yellow to grey, with dark spots on the body and dark bands ringing the long tail.

Little is known of the breeding habits of the African linsang, but females are thought to produce litters of 2 or 3 young once or twice a year.

NAME: **Banded Linsang,** *Prionodon linsang*
RANGE: **Thailand, Malaysia, Sumatra, Borneo**
HABITAT: **forest**
SIZE: **body: 37.5–43 cm (14¾–17 in)**
  tail: 30.5–35.5 cm (12–14 in)

The slender, graceful banded linsang varies from whitish-grey to brownish-grey in colour, with four or five dark bands across its back and dark spots on its sides and legs. It is nocturnal and spends much of its life in trees, where it climbs and jumps skilfully, but it is just as agile on the ground. Birds, small mammals, insects, lizards and frogs are all preyed on, and this linsang also eats birds' eggs.

The breeding habits of this species are not well known, but it is believed to bear two litters a year of 2 or 3 young each. Young are born in a nest in a hollow tree or in a burrow.

NAME: **Masked Palm Civet,** *Paguma larvata*
RANGE: **Himalayas to China, Hainan, Taiwan, S.E. Asia, Sumatra, Borneo**
HABITAT: **forest, brush**
SIZE: **body: 50–76 cm (19¾–30 in)**
  tail: 51–64 cm (20–25¼ in)

The masked palm civet has a plain grey or brownish-red body, with no stripes or spots but with distinctive, white masklike markings on the face. It is nocturnal and hunts in the trees and on the ground for rodents and other small animals, as well as for insects, fruit and plant roots. The secretions of its anal glands are extremely strong-smelling and can be sprayed considerable distances to discourage any attacker.

A litter of 3 or 4 young is born in a hole in a tree. The young are greyer than adults and do not have conspicuous face masks at first.

VIVERRIDAE: Civet Family

There are about 72 species of civet, mongoose, genet and linsang in this family. All are small to medium-sized carnivores which live in the Old World, with a distribution stretching from southern Europe through Africa, Madagascar and Asia. Most viverrids have long, sinuous bodies, elongate skulls, short legs and long tails. There are generally five claws on each foot, and these claws are partly retractable. Many species of viverrid emit a strong-smelling secretion from anal scent glands, the active ingredient of which, civettone, is used in the manufacture of perfumes.

Viverrids have adapted to a variety of habitats, such as open savanna and dense forest, while some species are semi-aquatic in fresh water. Males and females generally look alike, but males are sometimes larger and heavier.

NAME: **African Palm Civet,** *Nandinia binotata*
RANGE: **Africa: Guinea, east to S. Sudan, south to Mozambique**
HABITAT: **forest, savanna, woodland**
SIZE: **body: 44–60 cm (17¼–23½ in)**
  tail: 48–62 cm (18¾–24½ in)

Active at night, the African palm civet is a skilful climber and spends much of its life in trees. Its diet is varied, ranging from insects, lizards, small mammals and birds to many kinds of fruit (which may sometimes be its sole food), leaves, grass and some carrion. A solitary animal, it spends the day resting in the trees. The male is larger and heavier than the female and both have short legs and long, thick tails. The short muzzle is adorned with long whiskers. Usually greyish-brown to dark reddish-brown in colour, this civet has a pale creamy spot on each shoulder — the origin of its other common name of two-spotted palm civet.

The female gives birth to a litter of 2 or 3 young after a gestation of about 64 days.

NAME: **Congo Water Civet,** *Osbornictis piscivora*
RANGE: **Africa: N.E. Zaire**
HABITAT: **rain forest near streams**
SIZE: **body: 45–50 cm (17¾–19¾ in)**
  tail: 35–42 cm (13¾–16½ in)

The rarely seen Congo water civet is thought to lead a semi-aquatic life and to feed largely on fish and aquatic creatures. It has a slender body, short legs, with the hind legs longer than the forelegs, and a thick, bushy tail. Its coat is reddish-brown, with darker hair on the backs of the ears and the middle of the back and tail.

NAME: **African Civet,** *Viverra civetta*
RANGE: **Africa, south of the Sahara to South Africa: Transvaal**
HABITAT: **forest, savanna, plains, cultivated areas**
SIZE: **body: 80–95 cm (31½–37½ in)**
  tail: 40–53 cm (15¾–20¾ in)

Large and doglike, the African civet has a broad head, strong neck and long legs; the hind legs are longer than the forelegs. Its coat is generally grey, with darker legs, chin and throat, and the back and flanks are patterned with dark stripes and patches; the size and spacing of these dark markings is highly variable. By day, the African civet sleeps in a burrow or in cover of vegetation or rocks. It rarely climbs trees except to escape from an enemy but swims well. Mammals up to the size of young antelope, birds, including poultry and their eggs, reptiles, frogs, toads and insects are all hunted, and this civet also takes some carrion, as well as eating fruit and berries.

The female gives birth to 1 to 4 young, usually 2, after a gestation of 63 to 68 days. The young take their first solid food at 3 weeks and are weaned at 3 months. There may be as many as three litters a year.

NAME: **Small-spotted Genet,** *Genetta genetta*
RANGE: **S.W. Europe: S.W. France, Spain and Portugal; Africa, Middle East**
HABITAT: **semi-desert, scrub, savanna**
SIZE: **body: 50–60 cm (19¾–23½ in)**
  tail: 40–48 cm (15¾–18¾ in)

A slender, short-legged animal, this genet is marked with dark spots, which may form lines down its whitish to brownish-grey body. Its head is small and muzzle pointed, and its long tail is encircled with black bands. An agile, graceful animal, the small-spotted genet moves on land with its tail held straight out behind and climbs well in trees and bushes. It spends the day sleeping in an abandoned burrow of another animal, in a rock crevice or on the branch of a tree and starts to hunt at dusk. Sight, hearing and sense of smell are good, and the genet stalks its prey, crouching almost flat before pouncing. Most prey, such as rodents, reptiles and insects, is taken on the ground, but the genet will climb trees to take roosting or nesting birds; it also kills poultry. It normally lives alone or in pairs.

A litter of 2 or 3 young is born in a hole in the ground or in a tree, or among rocks after a gestation of 68 to 77 days. The young are born blind, and their eyes open after 5 to 12 days. They are suckled for up to 3 months and are fully independent by 9 months.

AFRICAN LINSANG

SMALL-SPOTTED GENET

BANDED LINSANG

AFRICAN CIVET

CONGO WATER CIVET

AFRICAN PALM CIVET

MASKED PALM CIVET

Civets 2

NAME: **Binturong**, *Arctictis binturong*
RANGE: **S.E. Asia, Palawan, Sumatra, Java, Borneo**
HABITAT: **forest**
SIZE: **body: 61–96.5 cm (24–38 in)**
   tail: 56–89 cm (22–35 in)

A large viverrid with long, coarse fur, the binturong has distinctive ear-tufts and a prehensile tail, which it uses as a fifth limb when climbing. It is the only carnivore other than the kinkajou, a member of the raccoon family, to possess such a tail. During the day, it sleeps up in the trees and emerges at night to climb slowly, but skilfully, among the branches, searching for fruit and other plant matter, as well as insects, small vertebrates and carrion.

After a gestation of 90 to 92 days, the female produces a litter of 1 or 2 young. Both parents care for the young, which are born blind and helpless.

NAME: **Fanalouc**, *Eupleres goudoti*
RANGE: **N. Madagascar**
HABITAT: **rain forest, swamps**
SIZE: **body: 46–50 cm (18–19¾ in)**
   tail: 22–24 cm (8½–9½ in)   Ⓥ

The fanalouc, also known as the small-toothed mongoose, with its long, slender body, pointed muzzle and short legs, does resemble a mongoose in build. The hind legs are longer than the forelegs, and the rear is consequently higher than the front of the body. The tail is thick and bushy. Active at dusk and during the night, the fanalouc does not climb or jump well but slowly hops along the ground, searching for earthworms (its main food), insects, water snails, frogs and sometimes even small mammals and birds. It readily wades into water in pursuit of prey. When food is abundant, the fanalouc lays down a store of fat near the base of its tail on which it lives during the dry season, when worms are scarce.

Fanaloucs pair for life, and each pair lives in a territory, the boundaries of which they mark with secretions of the anal and head glands. The female bears 1 young, after a gestation of about 12 weeks, which is born with a full covering of hair and with its eyes open; it is weaned at 9 weeks.

These animals are becoming rare outside nature reserves because of the destruction of forest land, competition with an introduced civet, *Viverricula indica*, and overhunting by local people.

NAME: **Otter-civet**, *Cynogale bennetti*
RANGE: **Indo-China, Malaysia, Borneo, Sumatra**
HABITAT: **swamps, near rivers**
SIZE: **body: 57–67 cm (22½–26¼ in)**
   tail: 13–20 cm (5–7¾ in)

The otter-civet spends much of its life in water and has several adaptations for its aquatic habits. Like many aquatic mammals, it has short, dense underfur, which is waterproof, covered by a layer of longer, coarse guard hairs. Its nostrils open upward and can be closed off by flaps, and the ears can also be closed. The otter-civet's feet are supple and have broad webs; these webs are only partial and do not extend to the tips of the digits, so the animal is able to move as well on land as in water.

With only the tip of its nose above the water, the otter-civet is almost invisible as it swims and so is able to ambush creatures that come to the water's edge to drink, as well as taking prey in water. Fish, small mammals, birds and crustaceans are all included in the otter-civet's diet, and it also eats fruit. It has long, sharp teeth for seizing prey and broad, flat molars, which it uses to crush hard-shelled items such as crustaceans. On land, it climbs well and may take refuge in a tree if attacked, rather than making for water.

A litter of 2 or 3 young is born in a burrow or hollow tree. They are not independent until about 6 months old.

NAME: **Meerkat**, *Suricata suricatta*
RANGE: **Africa: Angola to South Africa**
HABITAT: **open country, savanna, bush**
SIZE: **body: 25–31 cm (9¾–12¼ in)**
   tail: 19–24 cm (7½–9½ in)

The meerkat, also known as the suricate, has the long body and short legs typical of many mongooses. Its coat is mainly greyish-brown to light grey in colour, marked with dark bands across the body, but it has dark ears and nose and a light-coloured head and throat. The fur on the belly is thin and helps the meerkat to regulate its body temperature. It sits up sunning itself or lies on warm ground to increase its temperature and reduces it by lying belly-down in a cool, dark burrow.

Gregarious animals, meerkats live in family units, sometimes several families living together in a group of 30 or so. The colony occupies a home range, which contains shelters, such as burrows or rock crevices, and feeding sites. The animals need to move to a different area several times a year, when food supplies dwindle. Meerkats are good diggers and make burrows with several tunnels and chambers. Active in the daytime, they forage in pairs or small groups, often sitting up on their hind legs to watch for prey or for danger. They run fast for short distances but cannot climb or jump well. Insects, spiders, scorpions, centipedes, small mammals, lizards, snakes, birds and their eggs, snails, roots, fruit and other plant material are all included in their varied, wide-ranging diet. Meerkats have good hearing and sense of smell and excellent eyesight; they keep constant watch for birds of prey, their main enemies, and they instantly dive for a burrow or other cover if alarmed.

Breeding takes place mainly between October and April. The female gives birth to 2 to 5 young, usually 2 or 3, in a grass-lined underground chamber after a gestation of about 77 days. The young are born blind, but their eyes open 12 to 14 days after birth, and they take their first solid food at 3 or 4 weeks.

NAME: **Banded Palm Civet**, *Hemigalus derbyanus*
RANGE: **Malaysia, Sumatra, Borneo**
HABITAT: **forest**
SIZE: **body: 41–51 cm (16–20 in)**
   tail: 25–38 cm (9¾–15 in)

The banded palm civet has a slender, elongate body and a tapering, pointed snout. It is usually whitish to orange-buff in colour, with dark stripes on the head and neck, behind the shoulders and at the base of the tail.

A nocturnal animal, the banded palm civet is a good climber, with strong feet well adapted to arboreal life, and it forages for its prey in trees, on the forest floor and beside streams. Worms and locusts are its main foods, but it also eats ants, spiders, crustaceans, land and aquatic snails and frogs.

Little is known of its breeding habits.

NAME: **Salano**, *Salanoia concolor*
RANGE: **N.E. Madagascar**
HABITAT: **rain forest**
SIZE: **body: 35–38 cm (13¾–15 in)**
   tail: 18–20 cm (7–7¾ in)

Salanos are gregarious animals; they pair for life and live in family groups in a territory, marking the boundaries with secretions of the anal glands. At night, the salano rests in a burrow that it digs or takes over from other animals or in a hollow tree. During the day, it searches for insects, its main food, amphibians, reptiles and, occasionally, small mammals and birds. It also eats the contents of eggs, which it cracks by taking them in its hind feet and hurling them backward against a stone or tree.

The female gives birth to 1 young, but other details of the salanos' breeding habits are not known.

BINTURONG

FANALOUC

OTTER-CIVET

MEERKAT

BANDED PALM CIVET

SALANO

Civets 3

NAME: **Indian Mongoose,** *Herpestes auropunctatus*
RANGE: **Iraq to India, south to Malaysia; introduced in West Indies, Hawaii and Fiji**
HABITAT: **desert, open scrub, thin forest, dense forest**
SIZE: **body: 35 cm (13¾ in)**
　　tail: 25 cm (9¾ in)

This widespread mongoose varies in size and appearance according to its environment: desert populations are the smallest and palest. Generally, however, the soft, silky fur is olive-brown, and the tail is shorter than the head and body length.

At night, the Indian mongoose rests in a burrow, which it digs itself, and during the day it hunts for food, treading the same paths consistently and keeping in cover of vegetation. It feeds on almost anything it can catch, such as rats, mice, snakes, scorpions, centipedes, wasps and other insects, and is much appreciated by man for its ability to keep pest species, such as rats, at bay. Indeed, this adaptable mongoose has been introduced into areas outside its native range specifically for the purpose of destroying the rats and snakes that were infesting crops.

Females may produce two litters a year, each of 2 to 4 young, which are born after a gestation of about 7 weeks. The newborn are blind and hairless, and their mother carries them about in her mouth and defends them fiercely.

NAME: **Marsh Mongoose,** *Atilax paludinosus*
RANGE: **Africa, south of the Sahara**
HABITAT: **marshland, tidal estuaries, swamps**
SIZE: **body: 45–60 cm (17¾–23½ in)**
　　tail: 30–40 cm (11¾–15¾ in)

The large, sturdily built marsh mongoose is an expert swimmer and diver and is probably the most aquatic of all the mongooses. Despite its expertise in water, its feet are not webbed but have short, strong claws. Mainly nocturnal, this mongoose swims or roams along stream banks or in marshes, searching for food such as crabs, aquatic insects, fish, frogs and snakes. It often uses its forepaws to feel in mud or under stones for prey. To crush hard-shelled animals, such as crabs, it will dash the creature against a rock or tree. The marsh mongoose shows complicated and balletic scent-marking behaviour. Standing on its forepaws, with its tail over its back, it marks the underside of branches with secretions of its anal glands.

Marsh mongooses live alone or in pairs or small family groups. The female gives birth to 1 to 3 young in a burrow or in a nest amid a pile of vegetation.

NAME: **Banded Mongoose,** *Mungos mungo*
RANGE: **Africa, south of the Sahara**
HABITAT: **savanna, often near water**
SIZE: **body: 30–45 cm (11¾–17¾ in)**
　　tail: 20–30 cm (7¾–11¾ in)

The banded mongoose has a stout body and a rather short snout and tail. It varies in coloration from olive-brown to reddish-grey, and light and dark bands alternate across its back from shoulders to tail. It is active in the daytime — in hot weather only in the morning and evening — and also sometimes emerges on moonlit nights. A good climber and swimmer if necessary, the banded mongoose is a generally bold, adventurous animal in its search for food. It often digs and forages in the ground and in leaf litter, hunting for prey such as insects, spiders, scorpions, centipedes, small frogs, lizards, snakes and small mammals and birds. It also eats fruit, plant shoots and eggs, which it hurls against stones to smash their shells.

Family troops of up to 30 animals live together, moving their range every few days except in the breeding season. They shelter in hollow trees or rock crevices or in burrows, which they dig or take over from other animals. If attacked, they defend themselves courageously, growling and spitting and arching their backs.

A litter of 2 to 6 young is born in a burrow after a gestation of about 8 weeks. Their eyes open at 10 days, and they first venture out about 3 weeks after birth.

NAME: **Cusimanse,** *Crossarchus obscurus*
RANGE: **Africa: Sierra Leone to Cameroon**
HABITAT: **rain forest**
SIZE: **body: 30–40 cm (11¾–15¾ in)**
　　tail: 15–25 cm (6–9¾ in)

The cusimanse has a particularly long, narrow head and snout, the nose protruding beyond the lower lip. Like most mongooses, it is gregarious and lives in family groups of up to 12 animals, which keep in touch with chattering calls while they wander in search of food. They dig in the soil for worms, woodlice, spiders, snails and insects and also prey on crabs, frogs, reptiles, small mammals and birds and their eggs. To break eggs and hard-shelled prey, the cusimanse throws them against a tree or stone. At night, cusimanses sleep in burrows, which they dig, often in an old termite mound. They rarely climb trees, unless forced to do so in order to escape from an enemy.

The female cusimanse bears several litters a year, each of 2 to 4 young. The gestation period is about 70 days.

NAME: **White-tailed Mongoose,** *Ichneumia albicauda*
RANGE: **Africa, south of the Sahara**
HABITAT: **savanna, dense bush, forest edge, often near water**
SIZE: **body: 47–69 cm (18½–27¼ in)**
　　tail: 36–50 cm (14¼–19¾ in)

This large, long-legged mongoose has a distinctive bushy tail, which is grey at its base, usually becoming white or off-white at the tip; some individuals may have an entirely blackish tail. It lives alone or in pairs and is normally only active at night, but in very secluded areas it may emerge in the late afternoon. A poor climber, the white-tailed mongoose is able to swim, although it rarely does so, and most of its hunting is done on the ground. It preys on insects, frogs, reptiles, rodents and ground-living birds and uses its strong teeth to crush snails and crabs. Eggs, berries and fruit are also included in this varied diet.

The female gives birth to a litter of 2 or 3 young.

NAME: **Bushy-tailed Mongoose,** *Bdeogale crassicauda*
RANGE: **E. Africa: Kenya to Zimbabwe and Mozambique**
HABITAT: **coastal forest, savanna**
SIZE: **body: 40–50 cm (15¾–19¾ in)**
　　tail: 20–30 cm (7¾–11¾ in)

A robust mongoose, with a broad muzzle and sturdy legs, this species does indeed have a broad, heavily furred tail. It is an elusive, nocturnal animal, which rests by day in a burrow, often one taken over from another animal, or in a hole in a hollow tree. At night, it hunts for insects, including termites, and lizards, snakes, rodents and other small creatures to eat.

NAME: **Fossa,** *Cryptoprocta ferox*
RANGE: **Madagascar**
HABITAT: **forest**
SIZE: **body: 60–75 cm (23½–29½ in)**
　　tail: 55–70 cm (21½–27½ in) Ⓥ

The largest Madagascan carnivore, the fossa resembles a cat as much as a viverrid and has a rounded, catlike head, but with a longer muzzle. Its body is slender and elongate, and its hind legs are longer than its forelegs. Active at dusk and at night, it is an excellent climber and is equally agile in trees and on the ground. It lives alone and hunts for mammals up to the size of lemurs, as well as for birds, lizards, snakes and insects; it also kills domestic poultry.

The female bears 2 or 3 young in a burrow, a hole in a tree or a den among rocks after a gestation of about 3 months. She cares for them alone. The young are weaned at 4 months and are fully grown and independent at 2 years.

INDIAN MONGOOSE

MARSH MONGOOSE

CUSIMANSE

BANDED MONGOOSE

BUSHY-TAILED MONGOOSE

WHITE-TAILED MONGOOSE

FOSSA

Hyenas

NAME: Aardwolf, *Proteles cristatus*
RANGE: Africa: Sudan, south to South Africa, Angola
HABITAT: open dry plains, savanna
SIZE: body: 65–80 cm (25½–31½ in)
tail: 20–30 cm (7¾–11¾ in)

A smaller and more lightly built version of the hyena, the aardwolf has a pointed muzzle, slender legs and an erectile mane on the neck and along the back. There are dark stripes on its yellowish- to reddish-brown body and legs. Unless in a family group with young, the aardwolf lives alone in a territory centred on a den, which may be the abandoned burrow of an aardvark or a hole it digs itself. The boundaries of the territory are marked by anal-gland secretions. Active at night, the aardwolf has extremely acute hearing and can detect the movements of termites, its main food. These are lapped from the ground or grass by means of the aardwolf's long tongue, which is covered with sticky saliva, making the task easier. It also eats other insects, birds' eggs, small mammals and reptiles.

A litter of 2 to 4 young is born in a burrow. After weaning, both parents feed the young on regurgitated termites.

NAME: Striped Hyena, *Hyaena hyaena*
RANGE: Africa: Senegal to Tanzania; Middle East to India
HABITAT: dry savanna, bush country, semi-desert, desert
SIZE: body: 1–1.2 m (3¼–4 ft)
tail: 25–35 cm (9¾–13¾ in)

The striped hyena is identified by the dark stripes on its grey or yellowish-grey body and by the erectile mane around its neck and shoulders that extends down the middle of its back. Males are usually larger than females, and both have the heavy head and sloping back typical of the hyenas.

Although they live in pairs in the breeding season, striped hyenas are generally solitary. Each has a home range, which must contain some thick cover, and the boundaries are marked by anal-gland secretions rubbed on to grass stems. Active at night, the striped hyena feeds on carrion, such as the remains of the kills of big cats, and preys on young sheep and goats, small mammals, birds, lizards, snakes and insects; it will also eat fruit. Striped hyenas stay well away from the larger spotted hyenas.

After a gestation of about 3 months, a litter of 2 to 4 young is born in a hole in the ground or among rocks. Both parents help to care for the young, which are blind at birth and are suckled for up to a year.

HYAENIDAE: Hyena Family

The hyenas of Africa and southwest Asia and the aardwolf of southern Africa together constitute a small family of land-dwelling carnivores that is related to viverrids and cats. This relationship is somewhat surprising in terms of superficial appearance, since all members of the hyena family have an extremely doglike body form.

The 3 species of hyena, particularly, resemble dogs but are more heavily built in the fore- than the hindquarters. They have massive heads; indeed the jaws of the spotted hyena, the largest member of the group, are the most powerful of any mammal. All hyenas are able to crush the biggest bones of their prey to extract the marrow. Hyenas specialize in feeding on carrion, often the kills of lions and other large carnivores, and are able to drive smaller predators, such as cheetah, away from their kills. They are, however, also predators in their own right, particularly spotted hyenas, and by hunting in cooperative packs can bring down animals as large as zebras. In fact, in one area of Tanzania, spotted hyenas are more successful killers than lions. Near villages and towns, hyenas are useful scavengers and feed on any refuse left out at night.

The aardwolf is a highly adapted offshoot of the hyena stock. It is more lightly built, has a narrow, pointed head, large ears and tiny teeth. It feeds mostly on termites.

Males and females look alike in all members of the family, but males may be larger.

NAME: Brown Hyena, *Hyaena brunnea*
RANGE: Africa: Angola to Mozambique, south to N. South Africa
HABITAT: dry savanna, plains, semi-desert
SIZE: body: 1.1–1.2 m (3½–4 ft)
tail: 25–30 cm (9¾–11¾ in) ⓥ

Typical of its family, with a bulky head and back sloping toward the rear, the brown hyena has long, rough hair over much of its body, with a mane of even longer hair on the neck and shoulders. This hyena is usually dark brown to brownish-black in colour, with a lighter-brown mane and legs.

Unless in a family group, the brown hyena is solitary, sometimes gathering with others in a hunting pack or at a big carcass. It lives in a large territory, which it marks with secretions from anal scent glands and by defecation sites. During the day, it sleeps in a burrow, often one left by another animal such as an aardvark, or among rocks or

tall grass and emerges at night to find carrion or to hunt prey such as rodents, birds, including poultry, reptiles or wounded large animals. Near the coast, brown hyenas also feed on dead fish, mussels and the stranded corpses of seals and whales.

A litter of 2 to 4 young is born in a burrow after a gestation of 92 to 98 days. The young are suckled for about 3 months but remain with their parents for up to 18 months, during which the male brings them food. Although protected in game reserves, brown hyenas are considered pests because of their habit of attacking livestock, and large numbers have been killed by farmers.

NAME: Spotted Hyena, *Crocuta crocuta*
RANGE: Africa, south of the Sahara
HABITAT: semi-desert to moist savanna
SIZE: body: 1.2–1.8 m (4–6 ft)
tail: 25–30 cm (9¾–11¾ in)

The largest of the hyenas, the spotted hyena has a big, powerful head, slender legs and a sloping back. Its tail is bushy, and a short mane covers its neck and shoulders. The head and feet are always a lighter brown than the rest of the body, and irregular dark spots are scattered overall; these vary greatly in colour and arrangement.

The spotted hyena is an inhabitant of open country and does not enter forest. It lives in packs of 10 to 30 or so animals, sometimes as many as 100, each pack occupying a territory. The boundaries of the territory are marked with urine, droppings and anal-gland secretions and carefully guarded to keep out rival packs. Males are dominant in the pack. Hyenas sleep in burrows, which they dig themselves, or among tall grass or rocks and emerge at dusk. They are normally active only at night but may hunt during the day in some areas. As well as feeding on carrion, spotted hyenas hunt large mammals such as antelope, zebra and domestic livestock. The victim is often brought down by a bite in the leg and then torn to pieces by the pack while still alive. Spotted hyenas are extremely noisy animals, making a variety of howling screams when getting ready for the hunt, as well as eerie sounds like laughter when they kill and when mating.

When courting, spotted hyenas eject strong-smelling anal-gland secretions, and the male prances around the female and rolls her on the ground. The gestation period is 99 to 130 days, and the 1 or 2 young are born in a burrow, with eyes open and some teeth already through. The young are suckled for a year to 18 months, by which time they are able to join the hunting pack.

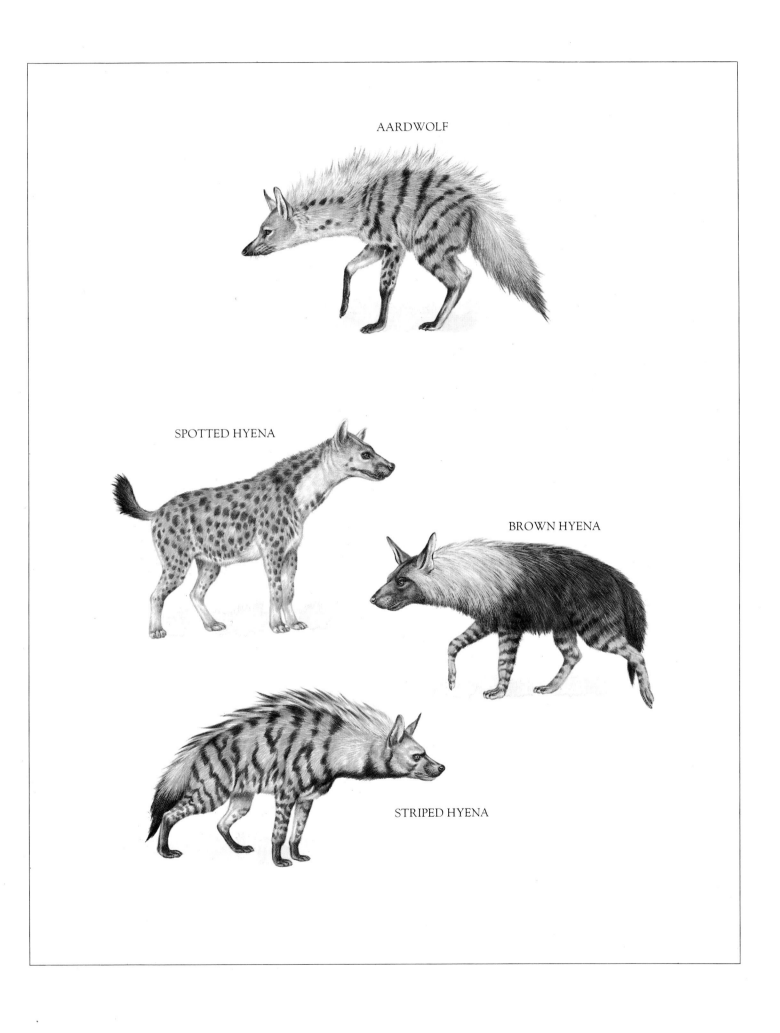

AARDWOLF

SPOTTED HYENA

BROWN HYENA

STRIPED HYENA

Cats I

NAME: Golden Cat, *Felis aurata*
RANGE: Africa: Senegal to Zaire (not Nigeria), Kenya
HABITAT: forest, forest edge
SIZE: body: 72–93 cm (28¼–36½ in)
   tail: 35–45 cm (13¾–17¾ in)

A robust, medium-sized cat, the golden cat has rather short, sturdy legs and small, rounded ears. Coloration varies enormously from brownish-red to slate-grey on the upperparts; some golden cats have distinct spots all over the body, some only on the underside.

An inhabitant of dense forest, this cat spends much of its life in trees and is mainly active at night. During the day it sleeps up in a tree. It is a solitary, elusive creature and little is known of its habits. It preys on mammals up to the size of small antelope and on birds up to the size of guineafowl.

NAME: Leopard Cat, *Felis bengalensis*
RANGE: S.E. Asia, Sumatra, Borneo, Java, Philippines
HABITAT: forest
SIZE: body: about 60 cm (23½ in)
   tail: about 35 cm (13¾ in)

A nocturnal, rarely seen creature, the leopard cat rests during the day in a hole in a tree. It is an agile climber and preys on small birds and on mammals up to the size of squirrels and hares; it may occasionally kill a small deer. The coloration and pattern of the leopard cat are variable, but it is usually yellowish, grey or reddish-brown on the upperparts of the body, with a whitish belly, and is dotted overall with dark spots. These spots are in regular lines and may merge to form bands.

The breeding habits of this cat are not well known, but the female is thought to produce litters of 3 or 4 young in a cave or a den under fallen rocks.

NAME: Pampas Cat, *Felis colocolo*
RANGE: South America: Ecuador, Peru, Brazil to S. Argentina
HABITAT: open grassland, forest
SIZE: body: 60–70 cm (23½–27½ in)
   tail: 29–32 cm (11½–12½ in)

A small, but sturdily built, animal, the pampas cat has a small head and thick, bushy tail. Its long fur is variable in colour, ranging from yellowish-white to brown or silvery-grey. In the north of its range, it lives in forest and, although primarily ground-dwelling, will take refuge in trees. Farther south, it inhabits the vast grasslands, where it takes cover among the tall pampas grass. Active at night, it hunts small mammals, such as cavies, and ground-dwelling birds such as tinamous.

The female gives birth to litters of 1 to 3 young after a gestation thought to be about 10 weeks.

FELIDAE: Cat Family

There are approximately 35 species in the cat family as classified here, but numbers differ according to source, and there is considerable disagreement as to the organization of the family. Of all predators, cats are probably the most efficient killers. Colouring, size and fur-patterning vary within the family, but all species from the smallest to the largest are basically similar in appearance and proportions to the domestic cat — an ideal predatory body form.

Cat bodies are muscular and flexible, and the head is typically shortened and rounded, with large forward-directed eyes. Limbs can be proportionately short or long, but in all species except the cheetah, there are long, sharp, completely retractile claws on the feet for the grasping of prey. The overpowering of prey animals, however, practically always involves a bite from the powerful jaws, which are armed with well-developed, daggerlike canines. Shearing cheek teeth, carnassials, are used for slicing through flesh.

This successful family is distributed almost world-wide, being absent only from Antarctica, Australasia, the West Indies and some other islands, and from Madagascar, which is inhabited by the catlike viverrids, the fossas. Male and female look alike in most species, but males are often slightly larger.

Sadly, the fine fur of the cats has long been coveted by man, and many species have been hunted until they are rare and in danger of extinction.

NAME: Caracal, *Felis caracal*
RANGE: Africa (except rain forest belt); Middle East to N.W. India
HABITAT: savanna, open plains, semi-desert, sand desert
SIZE: body: 65–90 cm (25½–35½ in)
   tail: 20–30 cm (7¾–11¾ in)

The caracal is identified by its long, slender legs, rather flattened head and long, tufted ears. A solitary animal, it occupies a home range, which it patrols in search of prey. It is most active at dusk and at night but may also emerge in the daytime. A wide variety of mammals, ranging from mice to reedbuck, are included in its diet, and it also feeds on birds, reptiles, and domestic sheep, goats and poultry.

The male caracal courts his mate with yowls similar to those of the domestic cat. In a well-concealed den in a rock crevice, tree hole or abandoned burrow, the female bears a litter of 2 or 3 young after a gestation of 69 or 70 days. The young suckle for 6 months and are not independent until 9 to 12 months old.

NAME: Mountain Lion, *Felis concolor*
RANGE: S.W. Canada, W. USA, Mexico, Central and South America
HABITAT: mountainsides, forest, swamps, grassland
SIZE: body: 1–1.6 m (3¼–5¼ ft)
   tail: 60–85 cm (23½–33½ in)   Ⓔ

The widespread mountain lion, also known as the cougar or puma, is now becoming increasingly rare and some subspecies are in danger of extinction. It varies greatly in colour and size over its range, but tawny and greyish-brown are predominant. A solitary creature, the mountain lion occupies a defined territory. A male's home range may overlap with the territories of one or more females but not with the territory of another male. Normally active in the early morning and evening, the mountain lion may emerge at any time. Its main prey are mule deer and other deer, but it also eats rodents, hares and, occasionally, domestic cattle. Having stalked its prey, the mountain lion pounces and kills with a swift bite to the nape of the neck.

Young are born in the summer in the temperate north and south of the range or at any time of year in the tropics. Male and female pair for the season, maybe longer, and during his mate's period of sexual receptivity, or heat, the male fights off any rivals. The litter of 2 to 4 young is born after a gestation of 92 to 96 days in a den, among rocks or in thick vegetation, which the female may use for some years. At 6 or 7 weeks old, the young start to take solid food, brought to them by their mother, and remain with her for 1 or even 2 years.

NAME: Lynx, *Felis lynx*
RANGE: Europe: Scandinavia to Spain and Portugal, east through Asia to Siberia; Alaska, Canada, N. USA
HABITAT: coniferous forest, scrub
SIZE: body: 80 cm–1.3 m (31½ in–4¼ ft)
   tail: 4–8 cm (1½–3 in)

The lynx is recognized by its short tail and its tufted ears and cheeks. Its coat varies in coloration over its wide range, particularly in the degree of spotting, which may be faint or conspicuous. Although strictly protected in most countries, lynx are becoming scarce, and some races are in danger of extinction. A solitary, nocturnal animal, the lynx stalks its prey on the ground or lies in wait for it in low vegetation. Hares, rodents, young deer and ground-living birds, such as grouse, are its main prey.

Breeding normally starts in the spring, and a litter of 2 or 3 young is born in a den among rocks or in a hollow tree after a gestation of about 63 days. The cubs remain with their mother throughout their first winter.

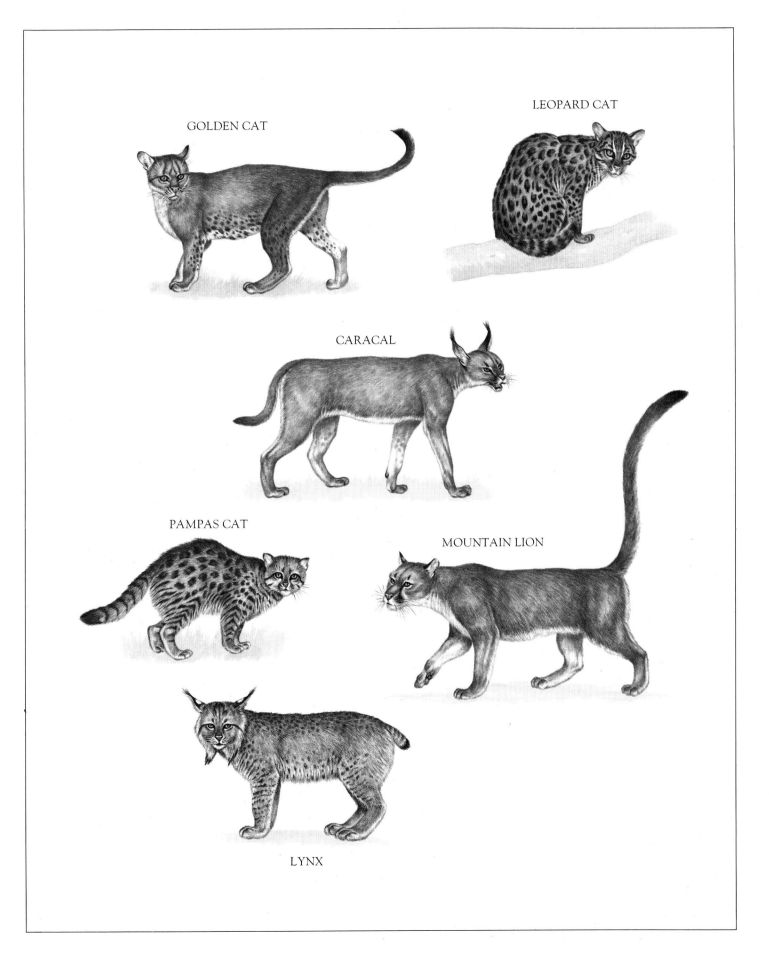

GOLDEN CAT

LEOPARD CAT

CARACAL

PAMPAS CAT

MOUNTAIN LION

LYNX

Cats 2

NAME: Ocelot, *Felis pardalis*
**RANGE: USA: Arizona, Texas; Mexico,
Central and South America to
N. Argentina**
**HABITAT: humid forest, thick bush,
marshy areas**
SIZE: body: 95 cm–1.3 m (37½ in–4¼ ft)
tail: 27–40 cm (10½–15¾ in) Ⓥ

The characteristic dark markings that
pattern the ocelot's coat are so variable
that no two animals are quite alike.
Generally nocturnal, the ocelot sleeps
on a branch or in cover of vegetation
during the day and emerges at night to
hunt for small mammals, such as young
deer and peccaries, agoutis, pacas and
other rodents, as well as birds and
snakes. It is an extremely secretive
animal and rarely shows itself in open
country. Males and females live in pairs
in a territory but do not hunt together.

Ocelots mate at night, and courting
males make loud, screeching calls,
similar to those of domestic cats. A
litter of 2 young, sometimes 4, is born,
after a gestation of about 70 days, in a
safe den in a hollow tree or in thick
vegetation. These beautiful cats have
become rare, both because of the
destruction of their forest habitat and
because they have long been hunted for
their fur. It is now illegal in many
countries to trade ocelot skins, but such
laws are hard to enforce, when the
demand continues, and the black market
price is high.

NAME: Bobcat, *Felis rufus*
**RANGE: S. Canada, USA (mostly western
states), Mexico**
HABITAT: chaparral, brush, swamp, forest
SIZE: body: 65 cm–1 m (25½ in–3¼ ft)
tail: 11–19 cm (4¼–7½ in)

The bobcat is short-tailed, like the lynx,
but is generally smaller than the latter
and has less conspicuous ear-tufts. It
varies considerably in size, the largest
individuals occurring in the north of the
range, and the smallest in Mexico.
Adaptable to a variety of habitats, the
bobcat is ground-dwelling but does
climb trees and will take refuge in a tree
when chased. It is solitary and nocturnal
for the most part but may hunt in the
daytime in winter. Small mammals,
such as rabbits, mice, rats and squirrels,
are its main prey, and it also catches
ground-dwelling birds such as grouse. It
hunts by stealth, slowly stalking its victim until near enough to pounce.

The female bobcat gives birth to a
litter of 1 to 6 young, usually 3, after a
gestation of about 50 days. The young
first leave the den at about 5 weeks and
start to accompany their mother on
hunting trips when they are between
3 and 5 months old.

NAME: Pallas's Cat, *Felis manul*
RANGE: central Asia: Iran to W. China
**HABITAT: steppe, desert, rocky
mountainsides**
SIZE: body: 50–65 cm (19¾–25½ in)
tail: 21–31 cm (8¼–12¼ in)

Pallas's cat has a robust body and short,
stout legs. Its head is broad, and its ears
low and wide apart, protruding only
slightly from the fur. The fur varies in
colour from pale grey to yellowish-buff
or reddish-brown and is longer and
more dense than that of any other wild
cat. An elusive, solitary creature, this
cat lives in a cave or rock crevice or a
burrow taken over from another mammal such as a marmot, usually emerging
only at night to hunt. It preys on small
mammals, such as mice and hares, and
on birds.

Mating occurs in spring, and females
give birth to litters of 5 or 6 young in
summer.

NAME: Serval, *Felis serval*
**RANGE: Africa, south of the Sahara to
South Africa: S. Transvaal**
HABITAT: savanna, open plains, woodland
SIZE: body: 65–90 cm (25½–35½ in)
tail: 25–35 cm (9¾–13¾ in)

A slender, long-legged cat, with a small
head and broad ears, the serval has a
graceful, sprightly air. Coloration varies
from yellowish-brown to dark olive-
brown; lighter-coloured animals tend to
have rows of large black spots on their
fur, while darker individuals are dotted
with many fine spots. The serval is usually active in the daytime and lives in a
small territory, the boundaries of which
are marked with urine. It is generally
solitary, but a female may enter a male's
territory. Mammals, from the size of
rodents up to small antelope, are its
main prey, but it also eats birds, poultry, lizards, insects and some fruit. Ser-
vals have excellent sight and hearing.

A litter of 1 to 4, usually 2 or 3, young
is born in a safe den among rocks or
vegetation or in a burrow taken over
from another mammal. The gestation
period is 67 to 77 days.

NAME: Wild Cat, *Felis silvestris*
**RANGE: Scotland, S. Europe; Africa (not
Sahara), Middle East to India**
**HABITAT: forest, scrub, savanna, open
plains, semi-desert**
SIZE: body: 50–65 cm (19¾–25½ in)
tail: 25–38 cm (9¾–15 in)

One of the ancestors of the domestic
cat, the wild cat is similar in form, but
slightly larger, and has a shorter, thicker
tail, which is encircled with black rings.
Coloration varies according to habitat,
cats in dry sandy areas being lighter than
forest-dwelling cats. Largely solitary
and nocturnal, the wild cat lives in a

well-defined territory. Although it is an
agile climber, it stalks most of its prey
on the ground, catching small rodents
and ground-dwelling birds.

Rival courting males howl and
screech as they vie for the attention of a
female, and it is she who eventually
makes the selection. She bears 2 or 3
young after a gestation of 63 to 69 days.
The young first emerge from the den, in
a cave, hollow tree or fox hole, when
they are 4 or 5 weeks old and leave their
mother after about 5 months.

NAME: Cheetah, *Acinonyx jubatus*
**RANGE: Africa, east to Asia: E. Iran,
Turkmeniya, Afghanistan**
HABITAT: open country: desert, savanna
SIZE: body: 1.1–1.4 m (3½–4½ ft)
tail: 65–80 cm (25½–31½ in) Ⓥ

Superbly adapted for speed, the cheetah
is the fastest of the big cats, able to attain
speeds of 112 km/h (69½ mph). Its body
is long and supple, with high muscular
shoulders, and its legs are long and slender. The tail aids balance during the
cheetah's high-speed sprints.

Cheetahs live in territories in open
country, alone, in pairs or in family
groups. They are active in the daytime,
and sight is the most important sense in
hunting. Having selected its prey when
in hiding, the cheetah stalks its victim
and then attacks with a short, rapid
chase, knocking over the prey and killing it with a bite to the throat. Hares,
jackals, small antelope, the young of
larger antelope, and birds, such as
guineafowl, francolins, bustards and
young ostriches, are the cheetah's main
prey. Several adults may, however,
cooperate to chase and exhaust larger
animals such as zebra.

Rival males compete in bloodless
struggles for the attention of a female.
She bears a litter of 2 to 4 young after a
gestation of 91 to 95 days and brings
them up alone. The young stay with the
mother for up to 2 years.

NAME: Clouded Leopard, *Neofelis
nebulosa*
**RANGE: Nepal to S. China, Taiwan,
Sumatra, Borneo**
HABITAT: forest
SIZE: body: 62–106 cm (24½–41¾ in)
tail: 61–91 cm (24–35¾ in) Ⓥ

The rare, elusive clouded leopard has a
long, powerful body, relatively short
legs and a long tail. It is a good climber
and hunts by pouncing from trees, as
well as by stalking prey on the ground.
Birds, pigs, small deer and cattle are the
clouded leopard's main victims, and it
kills with a single bite from its exceptionally long canine teeth.

A litter of 1 to 5 cubs is born after a
gestation of 86 to 92 days.

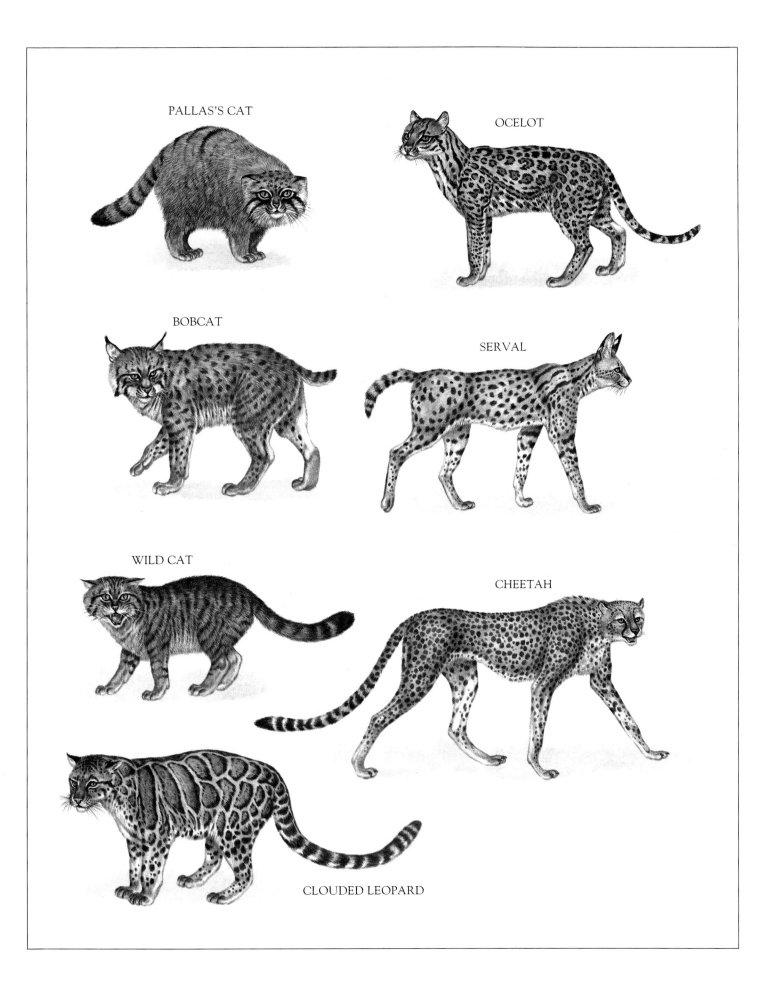

PALLAS'S CAT

OCELOT

BOBCAT

SERVAL

WILD CAT

CHEETAH

CLOUDED LEOPARD

Cats 3

NAME: **Lion,** *Panthera leo*
RANGE: **Africa, south of the Sahara;**
N.W. India; formerly more
widespread in Asia
HABITAT: **open savanna**
SIZE: **body: 1.4–2 m (4½–6½ ft)**
tail: 67 cm–1 m (26¼ in–3¼ ft)

A splendid, powerfully built cat, the lion has a broad head, thick, strong legs and a long tail, tipped with a tuft of hair that conceals a clawlike spine. The male is larger than the female and has a heavy mane on the neck and shoulders. Body coloration varies from tawny-yellow to reddish-brown, and the mane may be light yellow to black.

This impressive creature actually spends 20 or more hours a day resting. Lions normally hunt during the day, but in areas where they themselves are persecuted, they are active only at night. They live in groups, known as prides, consisting of up to 3 adult males and up to 15 females and their young, in a territory that is defended against intruders, particularly other mature male lions. A small group of young males without prides may live together. Lions prey on mammals, such as gazelle, antelope and zebra, and may cooperate to kill larger animals such as buffalo and giraffe. Smaller animals and birds, even crocodiles, may also be eaten. Lionesses do most of the hunting, often in groups, some acting as beaters to drive prey toward other lionesses lying in wait. Lions attack by stalking their prey and approaching it as closely as possible before making a short, rapid chase and pounce. They kill by a bite to the neck or throat.

Breeding occurs at any time of year. A litter of 1 to 6 young, usually 2 or 3, is born after a gestation of 102 to 113 days. They are suckled for about 6 months, but after the first 3 months, an increasing proportion of their food comes from the kills of adults. The cubs are left behind with one or two adults while the rest of the pride goes off to hunt, but if a kill is made, a lioness will return and lead them to it. Once they are over 4 months old, the cubs accompany their mothers everywhere, even following behind on hunting trips. They are not sexually mature until about 18 months old; young males are driven from the pride at about this age, but females remain with their family.

NAME: **Jaguar,** *Panthera onca*
RANGE: **S.W. USA, N. Mexico, Central**
and South America to N. Argentina
HABITAT: **forest, savanna**
SIZE: **body: 1.5–1.8 m (5–6 ft)**
tail: 70–91 cm (27½–35¾ in) Ⓥ

The largest South American cat, the powerful jaguar has a deep chest and massive, strong limbs. Its coloration varies from light yellow to reddish-brown, with characteristic dark spots. Although not quite as graceful and agile as the leopard, the jaguar climbs trees, often to lie in wait for prey, and is also an excellent swimmer. Like lions, jaguars cannot sustain high speeds and depend on getting close to prey for successful kills. Peccaries and capybaras are frequent prey, and jaguars also kill mountain sheep, deer, otters, rodents, ground-living birds, turtles, caimans and fish.

Normally solitary animals, male and female jaguars stay together for a few weeks when breeding. A litter of 1 to 4 young is born, after a gestation of 93 to 105 days, in a secure den in vegetation, among rocks or in a hole in a river bank. The female is aggressive in her protection of the young from any intruder, including the father.

NAME: **Leopard,** *Panthera pardus*
RANGE: **Asia: Siberia to Korea, Sri Lanka**
and Java; Middle East; Africa
HABITAT: **desert to forest, lowland plains**
to mountains
SIZE: **body: 1.3–1.9 m (4¼–6¼ ft)**
tail: 1.1–1.4 m (3½–4½ ft) Ⓥ

Formerly widespread, the leopard is now patchily distributed and many of its subspecies are extinct or endangered. A strong, but elegant, cat, it has a long body and relatively short legs. Most leopards are buff or tawny, with characteristic rosette-shaped black spots, but some are entirely black and are known as panthers. Panthers and leopards are otherwise identical.

Leopards are solitary and normally hunt day or night. In areas where they are persecuted, however, leopards are nocturnal. They swim and climb well and often lie basking in the sun on a branch. Their sight and sense of smell are good, and hearing is exceptionally acute. Prey includes mammals up to the size of large antelope, young apes — particularly baboons and monkeys — birds, snakes, fish and domestic livestock. Large items may be dragged up into a tree for safety while the leopard feeds; it will also feed on carrion.

Females have regular fertile periods, and males may fight over sexually receptive females. The litter of 1 to 6, usually 2 or 3, young is born in a rock crevice or hole in a tree after a gestation of about 90 to 112 days. The young are suckled for 3 months and are independent at 18 months to 2 years. The mother hunts alone, but if she makes a kill, she hides it while she goes to fetch her cubs. Older cubs may catch some small prey for themselves.

NAME: **Tiger,** *Panthera tigris*
RANGE: **Siberia to Java and Bali**
HABITAT: **forest**
SIZE: **body: 1.8–2.8 m (6–9¼ ft)**
tail: about 91 cm (35¾ in) Ⓔ

The largest of the big cats, the tiger has a massive, muscular body and powerful limbs. Males and females look similar, but males have longer, more prominent, cheek whiskers. Coloration varies from reddish-orange to reddish-ochre, and the pattern of the dark, vertical stripes is extremely variable. Tigers of the northern subspecies tend to be larger and paler than tropical subspecies.

Tigers are shy, nocturnal creatures and usually live alone, although they are not unsociable and are on amicable terms with their neighbours. They climb well, move gracefully on land and are capable of galloping at speed when chasing prey. Wild pigs, deer, and cattle, such as gaur and buffalo, are the tiger's main prey, and it also kills other mammals such as the sloth bear.

Male and female associate for only a few days for mating. The female gives birth to a litter of, usually, 2 or 3 young after a gestation of 103 to 105 days. The young may stay with their mother for several years.

Most races of these magnificent animals are now rare, and the Bali and Java tigers may be extinct because of indiscriminate killing earlier this century and the destruction of forest habitats.

NAME: **Snow Leopard,** *Panthera uncia*
RANGE: **Pakistan, Afghanistan, north to**
USSR; Himalayas, east to China
HABITAT: **mountain slopes, forest**
SIZE: **body: 1.2–1.5 m (4–5 ft)**
tail: about 90 cm (35½ in) Ⓔ

In summer, the snow leopard, also known as the ounce, lives in alpine meadows above the tree-line, amid snow and glaciers; but in winter it follows the migrations of prey animals down to forest and scrub at about 2,000 m (6,600 ft). A powerful, agile animal, capable of huge leaps over ravines, the snow leopard stalks prey such as ibex, markhor, wild sheep and goats, boar and ground-dwelling birds, such as pheasants, partridges and snow-cocks; in winter it will also take domestic livestock. Females may be accompanied by young, but otherwise snow leopards live alone, constantly roaming around their huge territories. They are active mainly in the early morning and late afternoon.

The female gives birth to 2 or 3 cubs, rarely 4 or 5, after a gestation of 98 to 103 days. The young start to accompany their mother on hunting trips when about 2 months old.

LION (male)

JAGUAR

LEOPARD

TIGER

SNOW LEOPARD

Sea Lions, Walrus

ORDER PINNIPEDIA

The 3 families in this order — sea lions and fur seals, earless seals and the walrus — are all carnivores, adapted to life in water. Their limbs have become flippers, but pinnipeds are still able to move, albeit awkwardly, on land, where they all spend part of their lives.

OTARIIDAE: Sea Lion Family

There are approximately 14 species of fur seal and sea lion. The main features that distinguish these animals from the seals, family Phocidae, are the presence of external ears and their ability to tuck the hind flippers forward to facilitate locomotion on land. Sea lions occur in the southern Atlantic and Indian oceans and in the North and South Pacific, and come to land, or haul out, on coasts and islands.

Generally gregarious, these animals haul out in large numbers at traditional breeding sites, called rookeries, where males compete for the best territories.

Females give birth to the young conceived the previous year and mate again some days later. Mating and giving birth are synchronized in this convenient manner as a result of the phenomenon of delayed implantation. The embryo lies dormant for a period before development starts, thus ensuring the correct timing of the birth without producing young of undue size. Males are considerably larger than females, and have big, bulbous heads.

NAME: **South American Fur Seal,** *Arctocephalus australis*
RANGE: **South Pacific and Atlantic Oceans, from Brazil around to Peru**
HABITAT: **breeds on coasts and islands**
SIZE: **1.4–1.8 m (4½–6 ft)**

The South American fur seal usually has deep reddish-brown underfur; the male has a coarse mane. It feeds on marine invertebrates, fish, squid and penguins and prefers to haul out on rocky coasts.

These fur seals are extremely territorial in the breeding season. Males take up their territories in November, competing for the prime spots and rigorously enforcing boundaries, since those with the best and largest sites mate with the most females. Males are joined 2 weeks later by the females, who give birth to their young within a few days. The female remains with her pup for up to 12 days, during which period she mates with the male whose territory she is in, then goes to sea to feed, returning at intervals to suckle the pup. The males rarely leave their territories until all the mating is over.

NAME: **Northern Fur Seal,** *Callorhinus ursinus*
RANGE: **Bering Sea, Okhotsk Sea**
HABITAT: **breeds on islands in range such as Aleutian and Pribilof Islands**
SIZE: **1.5–1.8 m (5–6 ft)**

Northern fur seals have larger rear flippers than other otariids, and both male and female have a pale patch on the neck. The male may be four times the female's weight. Usually alone or in pairs at sea, these fur seals feed on fish and squid and rarely come to land outside the breeding season.

Males establish territories on the breeding beaches before females arrive to give birth to their young. The female stays with her single offspring for 7 days before going off on brief feeding trips, returning to suckle it at intervals.

NAME: **California Sea Lion,** *Zalophus californianus*
RANGE: **Pacific coasts: British Columbia to Mexico; Galápagos Islands**
HABITAT: **breeds on coasts and islands in south of range**
SIZE: **1.7–2.2 m (5½–7¼ ft)**

This attractive sea lion takes well to training and is the most commonly seen species in circuses and marine shows. Females and juveniles are tan-coloured when dry, while the larger males are brown; males are also distinguished by the horny crest on their heads. Social animals, these sea lions occur in groups and often come on to land outside the breeding season. They feed on fish, octopus and squid.

Males gather at a breeding site but only establish territories when the females arrive and start to give birth; territories are ill defined and somewhat unstable. The female produces 1 young, and mates again a few days later.

NAME: **Australian Sea Lion,** *Neophoca cinerea*
RANGE: **off S. and S.W. Australia**
HABITAT: **coasts and islands**
SIZE: **up to 2.4 m (7¾ ft)**

A non-migratory species, this sea lion does not travel far from the beach of its birth and often comes out on to land throughout the year. It moves quite easily on land and may travel several kilometres. A gregarious species, it is usually found in small groups. Fish, squid and penguins are its main foods.

The males establish well-defined territories, which they defend vigorously, and usually manage to prevent females on their sites from leaving. The female produces 1 young and remains with it for 14 days, during which time she mates. She then goes to sea, returning at 2-day intervals to suckle the pup, which she feeds for up to 2 years.

NAME: **Steller Sea Lion,** *Eumetopias jubata*
RANGE: **N. Pacific Ocean**
HABITAT: **breeds on Pribilof and Aleutian Islands, Kurile Islands, islands in Okhotsk Sea, coasts of N. America to San Miguel Island off S. California**
SIZE: **2.4–2.8 m (7¾–9¼ ft)**

The largest otariid, the steller sea lion overlaps in its range with the California sea lion but is distinguished by its size and lighter colour. It feeds on fish, squid and octopus, and stomach contents have revealed that it dives to 180 m (600 ft) or more to find food.

Males establish well-defined territories at breeding grounds and maintain the boundaries with ritual threat displays. They remain there throughout the breeding period, mating with the females in their area, and do not feed for about 2 months. Shortly after arriving, the female gives birth to 1 young. She remains with the pup constantly for the first 5 to 13 days before going briefly to sea to feed, leaving her pup in the company of other young.

ODOBENIDAE: Walrus Family

The 1 species of walrus resembles the sea lions in that its hind flippers can be brought forward to help it move on land. It cannot move as fast or adeptly as sea lions, however, and will often just drag itself forward.

NAME: **Walrus,** *Odobenus rosmarus*
RANGE: **Arctic Ocean; occasionally N. Atlantic Ocean**
HABITAT: **pack ice, rocky islands**
SIZE: **male: 2.7–3.5 m (8¾–11½ ft); female: 2.2–2.9 m (7¼–9½ ft)** ①

The largest, heaviest pinniped, the male walrus is a huge animal with heavy tusks, formed from the upper canine teeth which extend downward. Females also have tusks, but they are shorter and thinner. Walruses feed on bottom-living invertebrates, particularly molluscs, of which they consume only the soft muscular foot or siphon. How they extract these is still a mystery, but it is thought to be by a form of suction. They also feed on crustaceans, starfish, fish and even mammals.

Walruses are gregarious throughout the year. During the mating season, they congregate in traditional areas, where males compete for space near potential mates and display. There is a period of delayed implantation of 4 or 5 months, and a gestation of 11 months, so females can breed only every other year at most. Usually 1 young is born, rarely 2, and it may be suckled for up to 2 years.

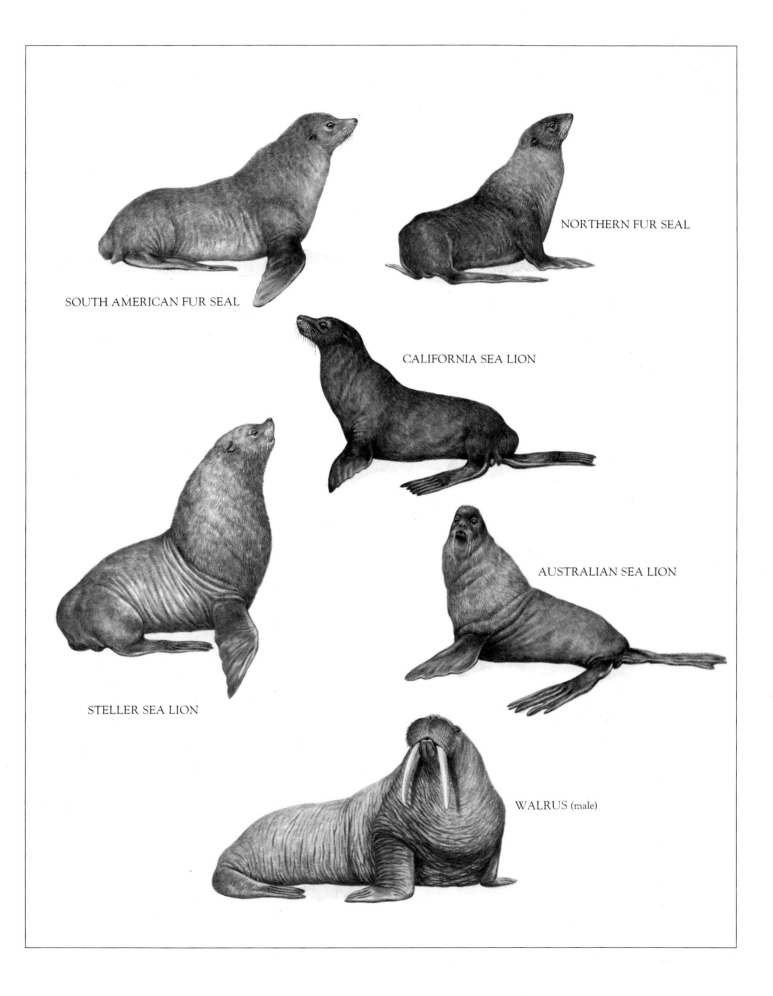

SOUTH AMERICAN FUR SEAL

NORTHERN FUR SEAL

CALIFORNIA SEA LION

STELLER SEA LION

AUSTRALIAN SEA LION

WALRUS (male)

Seals I

NAME: Grey Seal, *Halichoerus grypus*
RANGE: N. Atlantic Ocean
HABITAT: breeds on rocky coasts of Scandinavia and Britain; Iceland and Faroe Islands; Labrador, Gulf of St. Lawrence, Newfoundland
SIZE: 1.6–2.3 m (5¼–7½ ft)

The largest of the seals, excepting the elephant seals, a male grey seal may weigh up to 300 kg (660 lb) and be more than twice as heavy as a female. The male is also identified by his massive shoulders, covered with thick skin, which forms heavy folds and wrinkles, and by the elongated snout, rounded forehead and wide, heavy muzzle. The female has a flatter profile and a daintier, more slender muzzle. Grey seals travel far from breeding sites but stay mostly in coastal waters, feeding on fish and also on some crustaceans, squid and octopus.

The timing of breeding differs in the three areas of its range, but females always arrive at breeding grounds first and give birth before the males appear. The males take up positions on the beach, the older, experienced individuals getting the best places, but there is little fighting, and males may move sites from day to day. Having suckled her pup for about 3 weeks, the female mates and then leaves the area.

NAME: Harp Seal, *Pagophilus groenlandicus*
RANGE: N. Atlantic and Arctic Oceans: N. USSR to Scandinavia and Greenland; Labrador, Newfoundland
HABITAT: subarctic and arctic waters
SIZE: 1.6–1.9 m (5¼–6¼ ft)

The harp seal is identified by its black head and the dark band along its flanks and over its back; the rest of the body is usually pale grey, but this is highly variable. An expert, fast swimmer, the harp seal spends much of the year at sea and makes regular north-south migrations; it can also move fast over ice if necessary. Fish and crustaceans are the harp seal's main foods, and it is renowned for its ability to dive long and deeply for food. It is generally a gregarious species, and only old males live alone.

Females form whelping groups on the ice and give birth to their young in late February and early March. The pups are suckled for 2 to 4 weeks, growing rapidly on the nourishing milk, which is rich in fat. They are then left by the mothers, who go off to feed for a few weeks before migrating north to summer feeding grounds. Courting males fight rivals with their teeth and flippers and probably mate with females 2 or 3 weeks after they have given birth.

PHOCIDAE: Seal Family

Often known as earless seals because they lack external ears, the 19 species in this family have made the most complete transition from a terrestrial to an aquatic way of life of all the pinnipeds. Their hind flippers cannot be turned forward like those of other pinnipeds, and locomotion on land is, therefore, more restricted, since they must drag themselves forward with the fore-flippers. In water, however, they are highly skilful swimmers and divers, moving with undulatory movements of the hind portion of the body and the hind flippers. Seals have particularly sophisticated mechanisms to enable them to dive deeply for food and to stay underwater for long periods. Most impressive are the adaptations related to the blood circulation. During a dive, the heart rate may drop from 120 to about 4 beats a minute, but without any corresponding drop in blood pressure. This is achieved by the restriction of the blood supply to the muscles of the heart itself and to the brain, thus reserving the seal's blood oxygen for only the most vital organs.

The body of a seal is typically torpedo shaped, thick layers of fatty blubber under the skin accounting for much of its weight. The flippers as well as the body are furred, and the seals undergo an annual moult.

In some species, males are much bigger and heavier than females, but in others, females are the larger sex. Some are monogamous, but others, such as the elephant seals, are gregarious and polygamous. In most species there is a delay between fertilization and the actual start of gestation — a phenomenon known as delayed implantation. This ensures that both birth and mating can be accomplished within the short period while the seals are all together on dry land.

NAME: Leopard Seal, *Hydrurga leptonyx*
RANGE: Southern Ocean
HABITAT: pack ice, coasts, islands
SIZE: 3–3.5 m (9¾–11½ ft)

An unusually slender seal, the leopard seal is built for speed and has a large mouth, well suited to grasping the penguins, and sometimes other seals, that are its main prey. It catches penguins under water or as they move off the ice, and, using its teeth, tears away the skin with great efficiency before eating them. Squid, fish and crustaceans are also caught.

Mating is thought to take place from January until March, but there is little information available.

NAME: Common Seal, *Phoca vitulina*
RANGE: N. Atlantic and N. Pacific Oceans
HABITAT: temperate and subarctic coastal waters
SIZE: 1.4–1.8 m (4½–6 ft)

The common seal has a proportionately large head and short body and flippers. Although often grey with dark blotches, these seals vary considerably in coloration and spots may be light grey to dark brown or black. Males are larger than females. Generally non-migratory, the common seal often hauls out on protected tidal rocks and even travels up rivers and into lakes. It feeds mostly during the day on fish, squid and crustaceans and has been known to make dives lasting 30 minutes, although they normally last only 4 or 5 minutes.

Courtship and mating take place in water. The single pup is closely guarded by its mother and suckles for 2 to 6 weeks. Born in an advanced state of development, the pup can swim from birth and dive for 2 minutes when 2 or 3 days old. After weaning, the pup is left, and the mother mates again.

NAME: Crabeater Seal, *Lobodon carcinophagus*
RANGE: Antarctic
HABITAT: edge of pack ice
SIZE: 2–2.4 m (6½–7¾ ft)

Crabeater seals may well be the most abundant of all pinnipeds, and in their remote habitat they have few enemies other than killer whales. Capable of rapid movement over the ice, the crabeater seal thrusts with alternate forelimbs and the pelvis, and it is thought to achieve speeds of as much as 25 km/h (15½ mph). Krill, small shrimplike crustaceans, are its main food, which is strained from the water by means of its trident-shaped teeth.

Births and mating probably occur from October to the end of December. The pup is well developed at birth and suckles for about 5 weeks.

NAME: Bearded Seal, *Erignathus barbatus*
RANGE: Arctic Ocean
HABITAT: shallow waters; breeds on ice floes
SIZE: 2.1–2.4 m (6¾–7¾ ft)

Numerous long bristles on the snout are the identifying feature of the bearded seal and the source of its common name. It is a robust, heavily built species, in which females are slightly longer than males. Bottom-dwelling invertebrates, such as crustaceans and molluscs, and fish are its main foods.

The female gives birth to 1 pup, which can swim immediately. The pup suckles for 12 to 18 days, during which time the female usually mates again.

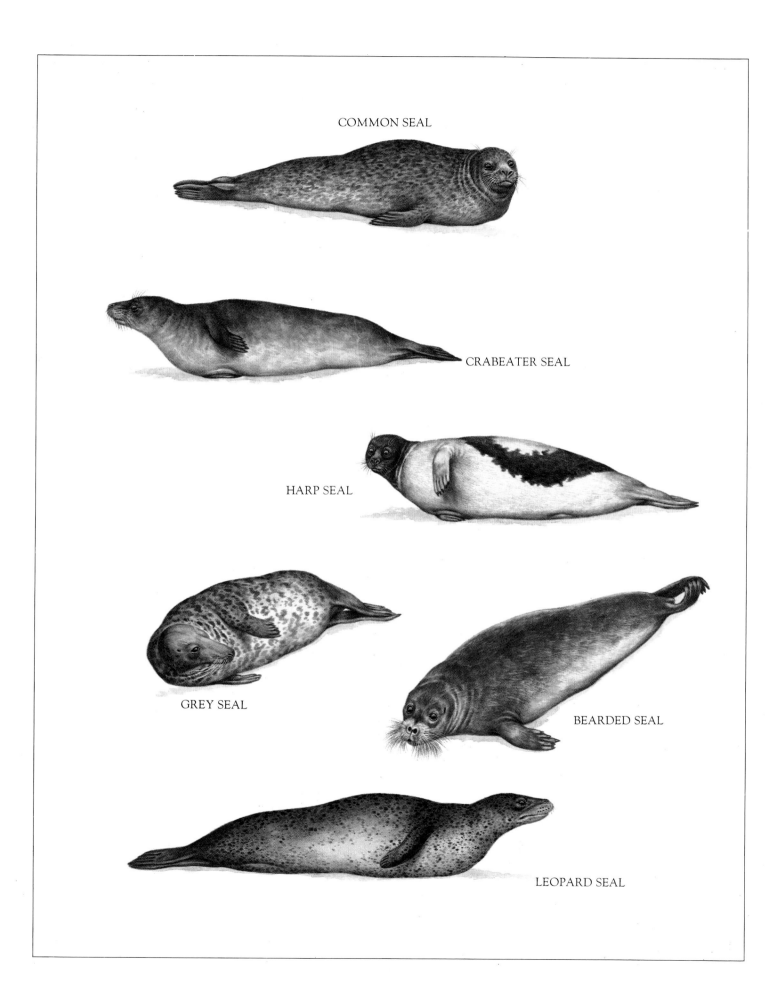

COMMON SEAL

CRABEATER SEAL

HARP SEAL

GREY SEAL

BEARDED SEAL

LEOPARD SEAL

Seals 2,
Dugong, Manatees

NAME: Northern Elephant Seal,
Mirounga angustirostris
RANGE: Pacific coast of N. America:
Vancouver Island to central Baja
California
HABITAT: breeds on offshore islands
SIZE: male: up to 6 m (19¾ ft)
female: up to 3 m (9¾ ft)

The largest pinniped in the northern
hemisphere, the male elephant seal may
weigh a massive 2,700 kg (6,000 lb),
much of it accounted for by the thick
layer of blubber. Females rarely weigh
more than 900 kg (2,000 lb). Because of
its size, this species was a major target
for commercial sealers, and by the end
of the nineteenth century the population was dangerously low. Only one of
the breeding islands appeared to be
used, and that by only a hundred or so
seals. With strict protection, numbers
have increased since then, and between
1957 and 1976, the population tripled to
over 47,000 — a remarkable recovery.

Northern elephant seals feed on fish
and squid and make long, deep dives.
Adult males haul out for breeding in late
November and fight for dominance in
the social hierarchy — high-ranking
males mate with most females. With the
aid of the greatly enlarged nasal chamber, which creates the elephantine
snout, males utter loud, vocal threats
against rivals. Females arrive a couple of
weeks after males and each gives birth to
a single pup, which is suckled for about
a month. The bond between mother and
young is close, and the female defends
the pup from other adults and rarely
leaves the breeding colony, existing on
her blubber until the pup is weaned. She
then mates again and leaves the breeding
ground. Weaned pups gather in a group
on the beach, where they remain for
another month, living on fat reserves
built up while suckling.

NAME: Mediterranean Monk Seal,
Monachus monachus
RANGE: W. Atlantic Ocean: Canary
Islands to Mediterranean Sea; Turkish
coast of Black Sea
HABITAT: breeds on rocky islets and cliffs
SIZE: 2.3–2.7 m (7½–8¾ ft) Ⓔ

The Mediterranean monk seal is becoming rare now that its previously remote
hauling-out spots on islets and cliffs are
accessible to humans with motor boats
and scuba diving equipment; these seals
often become entangled in fishing nets.
They are extremely upset by any disturbance, and mothers and pregnant
females particularly are nervous of any
approach, and pregnancies may be
spontaneously aborted.

Births occur from May to November,
with a peak in September and October;
pups are suckled for about 6 weeks.

NAME: Weddell Seal, *Leptonychotes*
weddelli
RANGE: Antarctic
HABITAT: edge of pack ice
SIZE: up to 2.9 m (9½ ft)

One of the larger seals, the Weddell has
a small head in proportion to its body
and an appealing, short-muzzled face;
the female is longer than the male. It
makes deeper, longer dives than any
other seal, the maximum recorded
depth being 600 m (2,000 ft), and the
longest duration 73 minutes. Dives to
300 and 400 m (1,000 and 1,300 ft) are
common, and antarctic cod, abundant
at these depths, are one of the Weddell
seal's main food species. This seal is able
to reach such depths because, when
diving, its heart rate slows to 25 per cent
of the pre-dive rate.

Weddell seals are normally solitary
outside the breeding season, but young
animals may form groups. In the breeding season, males seem to set up underwater territories, which females can
enter freely. The female gives birth on
land to 1 pup, with which she stays constantly for about 12 days; she then
spends about half her time in water until
the pup is weaned at about 6 weeks.
When only 7 weeks old, the pup can
dive to 90 m (295 ft). Females mate once
their pups are weaned.

NAME: Hooded Seal, *Cystophora cristata*
RANGE: N. Atlantic Ocean: arctic and
subarctic waters
HABITAT: edge of pack ice
SIZE: 2–2.6 m (6½–8½ ft)

Hooded seals spend much of their lives
in open seas, diving deeply for fish and
squid. They make regular migrations to
areas of pack ice in the Denmark Strait
and east of Greenland, where all adults
gather to haul out and moult. After
moulting, the seals disperse to re-
assemble at breeding grounds in different areas the following spring.

Pups are born in March on ice floes
and are suckled for 7 to 12 days. During
this period, the female is courted by a
male, who stays in the water near her
and her pup, chasing away any rivals. If
necessary, he hauls out and displays or
fights, making threat calls that are amplified by the huge, inflatable nasal sac.
The female mates about 2 weeks after
giving birth.

ORDER SIRENIA

The sirenians are the only completely
aquatic herbivorous mammals. There
are 2 families, the dugong' and the
manatees, containing a total of 4 species.
All have streamlined bodies and flipperlike forelimbs. Their hind limbs have
been lost in the formation of a tail with
a horizontal fluke.

DUGONGIDAE: Dugong Family

There is now only 1 species in this
family, the other member, Steller's sea
cow, having been exterminated in the
eighteenth century by excessive hunting, only 25 years after its discovery.

NAME: Dugong, *Dugong dugon*
RANGE: coast of E. Africa, Indian Ocean,
Red Sea to N. Australia
HABITAT: coastal waters
SIZE: up to 3 m (9¾ ft) Ⓥ

The dugong is a large, but streamlined,
animal, identified by its tail, which has
a crescent-shaped fluke. Its head is
heavy, with a fleshy, partially divided
snout. The male has two tusks, formed
from the incisor teeth, but these are
often barely visible under the fleshy
lips. A shy, solitary animal, it leads a
quiet, sedentary life, lying on the sea bed
for much of the time and only rising to
the surface to breathe every couple of
minutes. Its nostrils are placed on the
upper surface of the muzzle, so the
dugong can breathe, while remaining
almost submerged. Seaweed and sea
grass are its main food.

Little is known of the dugong's
reproductive habits, but the gestation is
thought to be about a year. The single
young is born in the water and is helped
to the surface by its mother.

TRICHECHIDAE: Manatee Family

There are 3 species of manatee; 2 live in
fresh water in West Africa and the
Amazon, and the third in coastal waters
of the tropical Atlantic Ocean.

NAME: American Manatee, *Trichecus*
manatus
RANGE: Atlantic Ocean: Florida to
Guyana
HABITAT: coastal waters
SIZE: up to 3 m (9¾ ft) Ⓥ

The manatee has a heavier body than the
dugong and is also distinguished by its
oval tail fluke. There are three nails in
each of its flippers, which it uses to
gather food. The manatee feeds at night,
foraging by touch and smell. Its diet is
quite varied, since it eats any sort of
vegetation and also often takes in small
invertebrate animals with the plants. For
much of the day, the manatee lies
on the sea bed, rising every couple
of minutes to breathe at the surface.
Manatees are social animals and live in
family groups, sometimes gathering in
larger herds.

The gestation period is about a year.
One young is born in water and is helped
by its mother to the surface to breathe.

WEDDELL SEAL

MEDITERRANEAN MONK SEAL

NORTHERN ELEPHANT SEAL (male)

HOODED SEAL

AMERICAN MANATEE

DUGONG

River Dolphins, Porpoises

ORDER CETACEA

There are 76 species of whale, dolphin and porpoise. They are the only aquatic mammals to spend their entire lives in water. All are streamlined animals with strong, horizontally-set tail flukes. Their front limbs are modified into flippers and there are no visible hind limbs. As a general rule, whales produce only one young at a time although twins are known.

There are two groups within the order. First, the toothed whales, with 66 species of small whale, dolphin and porpoise, all of which prey on fish and squid. To help locate their prey, they use a form of ultrasonic sonar: they emit high-frequency clicking sounds which bounce off objects, the echoes informing the whale with astonishing accuracy of the size, distance and speed of travel of the object. All of these whales have teeth, some a pair, some as many as 200.

In the second group are the 10 largest whales, known as the baleen whales. These marine giants feed on tiny planktonic animals, which they extract from the sea by filtering water through plates of fringed horny material hanging from their upper jaws. These baleen plates, as they are called, act as sieves to trap the plankton. There are 3 families of baleen whale: rorquals, grey and right whales.

PLATANISTIDAE: River Dolphin Family

The 5 species in this family all inhabit rivers in South America and Asia. They look alike and are grouped as a family, but the resemblance may be more to do with evolutionary pressures of similar habitat than close relationship. All the river dolphins are small for cetaceans, with long slender beaks and prominent rounded foreheads. The rivers these dolphins inhabit are muddy and full of sediment, and visibility is poor; as a result they rely heavily on echolocation to find food and avoid obstacles, and their eyes have become much reduced.

NAME: **Boutu**, *Inia geoffrensis*
RANGE: **Amazon basin**
HABITAT: **rivers, streams**
SIZE: **1.8–2.7 m (6–9 ft)**

The boutu has a strong beak, studded with short bristles, and a mobile, flexible head and neck. Most boutus have a total of 100 or more teeth. Their eyes, although small, seem to be more functional than those of other river dolphins. Boutus feed mainly on small fish and some crustaceans, using echolocation clicks to find their prey. Boutus live in pairs and seem to produce young between July and September.

NAME: **Ganges Dolphin**, *Platanista gangetica*
RANGE: **India: Ganges and Brahmaputra river systems**
HABITAT: **rivers, streams**
SIZE: **1.5–2.4 m (5–8 ft)**

The Ganges dolphin has a beak, which can be as long as 46 cm (18 in), and has up to 120 teeth. Its forehead curves up steeply from the beak. An agile animal, it generally swims on its side and returns to the normal upright position to breathe. It can dive for a maximum of 3 minutes at a time but usually remains under for about 45 seconds.

The Ganges dolphin is blind — its eyes have no lenses — but it finds its food by skilful use of echolocation signals. It feeds mainly on fish and some shrimps and hunts in the evening and at night. The dolphins are usually seen in pairs and may gather in groups of 6 or so to feed. They mate in autumn, and calves are born the following summer after a gestation of about 9 months.

NAME: **Whitefin Dolphin**, *Lipotes vexillifer*
RANGE: **China: Yangzte River; formerly Lake Tungting**
HABITAT: **muddy-bottomed rivers**
SIZE: **2–2.4 m (6½–8 ft)**   ①

Since 1975 this species has been protected by law in China, but although the total numbers are not known, population still seems to be low. The whitefin dolphin has a slender beak, which turns up slightly at the tip, and a total of 130 to 140 teeth. With little or no vision, it relies on sonar for hunting prey, mainly fish, but may also probe in the mud with its beak for shrimps.

Groups of 2 to 6 dolphins move together, sometimes gathering into larger groups for feeding. In the rainy summer season, they migrate up small swollen streams to breed, but no further details are known of their reproductive behaviour.

PHOCOENIDAE: Porpoise Family

Although the name porpoise is sometimes erroneously applied to members of other families, strictly speaking, only the 6 members of this family are porpoises. They are small beakless whales, rarely exceeding 2.1 m (7 ft) in length, and usually with prominent dorsal fins. They have 60 to 80 spatular teeth and feed mainly on fish and squid.

Porpoises live in coastal waters throughout the northern hemisphere, often ascending the estuaries of large rivers. One species, the spectacled porpoise, *Phocoena dioptrica*, occurs off the coasts of South America.

NAME: **Common Porpoise**, *Phocoena phocoena*
RANGE: **N. Atlantic, N. Pacific Oceans; Black and Mediterranean Seas**
HABITAT: **shallow water, estuaries**
SIZE: **1.4–1.8 m (4½–6 ft)**

Gregarious, highly vocal animals, porpoises live in small groups of up to 15 individuals. There is much communication within the group, and porpoises will always come to the aid of a group member in distress. Porpoises feed on fish, such as herring and mackerel, and can dive for up to 6 minutes to pursue prey, which are pin-pointed by the use of echolocation clicks.

Breeding pairs mate in July and August and perform prolonged courtship rituals, caressing one another as they swim side by side. Gestation is 10 or 11 months, and calves are suckled for about 8 months. While her calf feeds, the mother lies on her side at the surface so that it can breathe easily.

NAME: **Dall's Porpoise**, *Phocoenides dalli*
RANGE: **temperate N. Pacific Ocean**
HABITAT: **inshore and oceanic deeper waters**
SIZE: **1.8–2.3 m (6–7½ ft)**

Dall's porpoise is larger and heavier than most porpoises; its head is small, and the lower jaw projects slightly beyond the upper. It lives in groups of up to 15, which may gather in schools of 100 or more to migrate north in summer and south in winter. It feeds on squid and fish such as hake, and most probably uses echolocation when hunting. Pairs mate at any time of the year, and the young are suckled for as long as 2 years.

NAME: **Finless Porpoise**, *Neophocaena phocaenoides*
RANGE: **E. and S.E. Asia: Pakistan to Borneo and Korea; Yangtze River, E. China Sea**
HABITAT: **coasts, estuaries, rivers**
SIZE: **1.4–1.8 m (4½–6 ft)**

The finless porpoise is different from other porpoises in that it has a prominent rounded forehead, which gives the appearance of a slight beak, and a ridge of small rounded projections just behind where the dorsal fin should be. Finless porpoises dive for less than a minute in search of prey and are quick and agile in the water. They feed largely on crustaceans, squid and fish and are skilful echolocators. Although finless porpoises generally move in pairs, groups of up to 10 are sometimes seen. Little is known of their breeding behaviour, but young calves travel clinging to the projections on their mothers' backs.

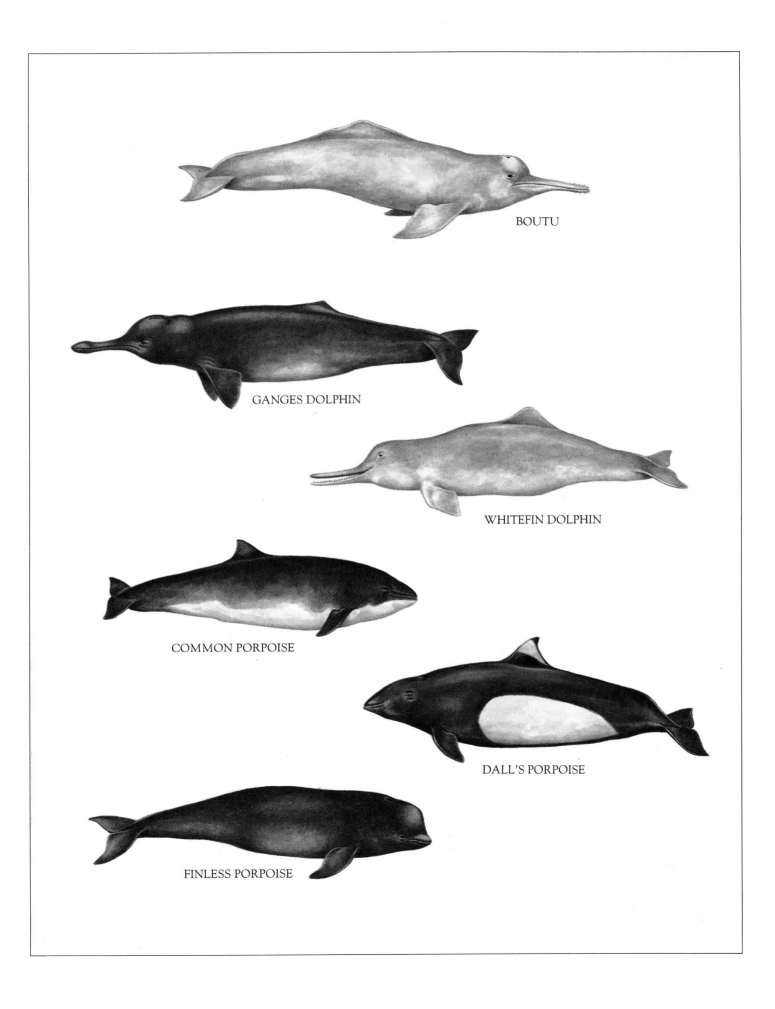

BOUTU

GANGES DOLPHIN

WHITEFIN DOLPHIN

COMMON PORPOISE

DALL'S PORPOISE

FINLESS PORPOISE

Dolphins

NAME: **Indo-Pacific Humpbacked Dolphin**, *Sousa chinensis*
RANGE: **Indian Ocean, S.W. Pacific Ocean, Yangtze River**
HABITAT: **coasts, estuaries, swamps**
SIZE: **2–3 m (6½–10 ft)**

The young of this species all have the normal streamlined body shape, but adults have humps of fatty tissue on the back. The beak is long, and there is a total of at least 120 teeth. These dolphins feed in shallow water on fish, molluscs and crustaceans and use echolocation when searching for prey. They are gregarious creatures, living in groups of up to 20 individuals.

NAME: **Striped Dolphin**, *Stenella coeruleoalba*
RANGE: **Atlantic and Pacific Oceans, temperate and tropical areas**
HABITAT: **deep offshore waters**
SIZE: **2.4–3 m (8–10 ft)**

Colour is variable in this species, but there is always a dark stripe running along the side and usually a dark band curving from the dorsal fin toward the eye. Striped dolphins have between 90 and 100 teeth and feed on small fish, squid and shrimps. They move in large schools of several hundred, even several thousand individuals, which are organized into age-segregated groups. Females breed about every 3 years. The gestation period is 12 months, and calves are nursed for from 9 to 18 months.

NAME: **Common Dolphin**, *Delphinus delphis*
RANGE: **world-wide, temperate and tropical oceans**
HABITAT: **coastal and oceanic waters**
SIZE: **2.1–2.6 m (7–8½ ft)**

The classic dolphin depicted by artists for centuries, the common dolphin is a beautifully marked animal with a long beak and pointed flippers. The markings are the most complex of any whale and are extremely variable. There are a number of geographically recognizable forms of this widespread species.

Dolphins live in hierarchical groups of 20 to 100 or more; groups sometimes join together, forming huge schools. There are many reports of these highly intelligent social animals coming to the aid of injured companions. Active animals, they roll and leap in the water and often swim at the bows of ships. Although they normally breathe several times a minute, they can dive for as long as 5 minutes to depths of 280 m (920 ft) to feed on fish and squid and certainly make good use of echolocation when hunting. Young are born in the summer, after a gestation of 10 or 11 months.

DELPHINIDAE: Dolphin Family

There are 32 species in this, the largest, most diverse cetacean family, which is found in all oceans and some tropical rivers. Most have beaked snouts and slender streamlined bodies and they are among the smallest whales. Typically, the dolphin has a bulging forehead, housing the melon, a lens-shaped pad of fat thought to help focus the sonar beams. A few species, notably the killer whale, are much larger and do not have beaks. Male dolphins are usually larger than females, and in some species the sexes differ in the shape of their flippers and dorsal fins. Dolphins swim fast and feed by making shallow dives and surfacing several times a minute. They are extremely gregarious and establish hierarchies within their social groups.

NAME: **Bottlenose Dolphin**, *Tursiops truncatus*
RANGE: **world-wide, temperate and tropical oceans**
HABITAT: **coastal waters**
SIZE: **3–4.2 m (10–14 ft)**

Now the familiar performing dolphin in zoos and on screen, this dolphin is a highly intelligent animal which is, tragically, still hunted and killed by man in some areas. It is a sturdy creature with a broad, high fin and a short, wide beak. Its lower jaw projects beyond the upper and this, combined with the curving line of the mouth, gives it its characteristic smiling expression.

Bottlenose dolphins live in groups of up to 15 individuals, sometimes gathering into larger schools, and there is much cooperation and communication between the group members. They feed mainly on bottom-dwelling fish in inshore waters but also take crustaceans and large surface-swimming fish. They are highly skilful echolocators, producing a range of click sounds in different frequencies to analyse any object at a distance with great precision. Up to 1,000 clicks a second are emitted.

Breeding pairs perform gentle courtship movements, caressing one another before copulation. Gestation lasts 12 months, and two adult females assist the mother at the birth and take the calf, which is normally born tail first, to the surface for its first breath. The mother feeds her calf for about a year, so there must be at least a 2-year interval between calves. Like all whales, the bottlenose dolphin produces extremely rich milk with a fat content of over 40 per cent to satisfy the high energy demands of her fast-growing youngster.

NAME: **Killer Whale**, *Orcinus orca*
RANGE: **world-wide, particularly cooler seas**
HABITAT: **coastal waters**
SIZE: **7–9.7 m (23–32 ft)**

The largest of the dolphin family, the killer whale is a robust yet streamlined animal, with a rounded head and no beak. The characteristic dorsal fin of an adult male is almost 2 m (6½ ft) high; and while the fins of females and juveniles are much smaller and curved, they are still larger than those of most other cetaceans. Adults have a total of 40 to 50 teeth.

Killer whales are avid predators and feed on fish, squid, sea lions, birds and even other whales. Their echolocation sounds are unlike those of other dolphins and are probably used to find food in turbid water.

Extended family groups of killer whales live together and cooperate in hunting. They have no regular migratory habits but do travel in search of food.

NAME: **Long-finned Pilot Whale**, *Globicephala melaena*
RANGE: **N. Atlantic Ocean; temperate southern oceans**
HABITAT: **coastal waters**
SIZE: **4.8–8.5 m (16–28 ft)**

The long-finned pilot whale has an unusual square-shaped head and long, rather narrow flippers. Pilot whales have a vast repertoire of sounds, some of which are used for echolocation purposes. Squid is their main food, but they also feed on fish such as cod and turbot. Social groups are made up of 6 or more whales and they have particularly strong bonds. Groups may join into larger schools. Gestation lasts about 16 months and the mother feeds her young for well over a year.

The range of this species is unusual in that it occurs in two widely separated areas.

NAME: **Risso's Dolphin**, *Grampus griseus*
RANGE: **world-wide, temperate and tropical oceans**
HABITAT: **deep water**
SIZE: **3–4 m (10–13 ft)**

Most adult Risso's dolphins are badly marked with scars, apparently caused by members of their own species, since the marks correspond to their own tooth pattern. The body of this dolphin is broad in front of the fin and tapers off behind it. It has no beak, but there is a characteristic crease down the centre of the forehead to the lip. There are no teeth in the upper jaw and only three or four in each side in the lower jaw. Squid seems to be its main food.

STRIPED DOLPHIN

INDO-PACIFIC HUMPBACKED DOLPHIN

COMMON DOLPHIN

BOTTLENOSE DOLPHIN

KILLER WHALE (male)

LONG-FINNED PILOT WHALE

RISSO'S DOLPHIN

Sperm Whales, White Whales

PHYSETERIDAE:
Sperm Whale Family

There are 3 species of sperm whale, 1 of which is the largest of all toothed whales, while the other 2 are among the smallest whales. Their characteristic feature is the spermaceti organ, located in the space above the toothless upper jaw; this contains a liquid, waxy substance which may be involved in controlling buoyancy when the whale makes deep dives. All sperm whales have underslung lower jaws but have little else in common.

NAME: **Pygmy Sperm Whale,** *Kogia breviceps*
RANGE: **all oceans**
HABITAT: **tropical, warm temperate seas**
SIZE: **3–3.4 m (10–11 ft)**

Its underslung lower jaw gives the pygmy sperm whale an almost sharklike appearance, belied by its blunt, square head. Unlike its large relative with its disproportionately huge head, the pygmy sperm whale's head accounts for only about 15 per cent of its total length. There are 12 or more pairs of teeth in the lower jaw. The short, broad flippers are located far forward, near the head, and there is a small dorsal fin, behind which the body tapers off markedly.

Pygmy sperm whales are thought to be shy, slow-moving animals. They feed on squid, fish and crabs from both deep and shallow water so may be less of an exclusively deep-water species than the sperm whale. Pygmy sperm whales have often been sighted alone, but evidence suggests that they form social units of 3 to 5 individuals.

Little is known of the reproductive habits of pygmy sperm whales. Gestation is believed to last about 9 months; calves are born in the spring and fed by the mother for about a year.

NAME: **Dwarf Sperm Whale,** *Kogia simus*
RANGE: **all oceans**
HABITAT: **tropical and subtropical seas**
SIZE: **2.4–2.7 m (8–9 ft)**

Superficially similar to the pygmy sperm whale, the dwarf sperm whale tends to have a more rounded head than its relative although there is considerable individual variation in head shape. Its lower jaw is set back, and it has up to 11 pairs of teeth.

Little is known of the biology and habits of this whale, but fish and squid are believed to be its main items of diet. Species found in the stomachs of dwarf sperm whales are all known to live at depths of more than 250 m (800 ft), so there seems little doubt that these whales make prolonged dives for food.

NAME: **Sperm Whale,** *Physeter catodon*
RANGE: **all oceans**
HABITAT: **temperate and tropical waters**
SIZE: **11–20 m (36–66 ft)**

The largest of the toothed whales, the sperm whale has a huge head, as great as one-third of its total body length, and a disproportionately small lower jaw, set well back from the snout. On its back is a fleshy hump and behind this are several smaller humps. Its flippers are short, but the tail is large and powerful and useful for acceleration. Surrounding the nasal passages in the huge snout is a mass of the waxy substance known as spermaceti. When the whale dives, it allows these passages to fill with water, and by controlling the amount and temperature of the water taken in at different depths, it can alter the density of the wax and thus the buoyancy of its whole body. This facility enables the whale to make its deep dives and to remain down, at neutral buoyancy, while searching for prey. Sperm whales are known to dive to 1,000 m (3,300 ft) and may even dive to more than twice this depth. They feed mainly on large, deepwater squid, as well as on some fish, lobsters and other marine creatures. Their sonar system is essential for finding prey in the black depths of the ocean.

All sperm whales migrate toward the poles in spring and back to the Equator in autumn, but females and young do not stray farther than temperate waters. Adult males, however, travel right to the ice caps in high latitudes. They return to the tropics in winter and contest with each other to gather harem groups consisting of 20 to 30 breeding females and young. Males under about 25 years old do not generally hold harems, but gather in bachelor groups.

The gestation period for sperm whales is about 14 to 16 months. When a female gives birth, she is surrounded by attendant adult females, waiting to assist her and to help the newborn to take its first breath at the surface. As with most whales, usually only 1 young is produced at a time, but twins have been known. Mothers suckle their young for up to 2 years.

MONODONTIDAE:
White Whale Family

There are 2 species in this family, both of which live in Arctic waters. They are distinctive whales with many features in common. Both have more flexible necks than is usual for whales, and their tails, too, are highly manoeuvrable. They have no dorsal fins. In both species males are larger than females.

NAME: **White Whale,** *Delphinapterus leucas*
RANGE: **Arctic Ocean and subarctic waters**
HABITAT: **shallow seas, estuaries, rivers**
SIZE: **4–6.1 m (13–20 ft)**

Often known as the beluga, the white whale has a rounded, plump body and just a hint of a beak. There is a short raised ridge on its back where the dorsal fin would normally be. At birth, these whales are a dark brownish-red; they then turn a deep blue-grey and gradually become paler until, at about 6 years old, they are a creamy-white colour. They have about 32 teeth.

White whales feed on the bottom in shallow water, mainly on crustaceans and some fish. They actually swim under pack ice and can break their way up through the ice floes to breathe.

Sexual maturity is attained when the whales are 5 to 8 years old. The whales mate in spring and calves are born in the summer, after a gestation of about 14 months. Since the young are suckled for at least a year, white whales are able to breed only every 3 years or so. All white whales are highly vocal and make a variety of sounds for communication, as well as clicks used for echolocation. Their intricate songs caused them to be known as sea canaries by the nineteenth-century whalers. White whales congregate in herds of hundreds of individuals to migrate south in winter and then return to rich northern feeding grounds in summer.

NAME: **Narwhal,** *Monodon monoceros*
RANGE: **high Arctic Ocean (patchy distribution)**
HABITAT: **open sea**
SIZE: **4–6.1 m (13–20 ft)**

The male narwhal has an extraordinary spiral tusk which is actually its upper left incisor. The tooth grows out of a hole in the upper lip to form the tusk, which may be as much as 2.7 m (9 ft) long. The female may sometimes have a short tusk. It is not known what the purpose of the tusk is, but it is probable that it is a sexual characteristic, used by males to dominate other males and impress females.

Narwhals feed on squid, crabs, shrimps and fish. Groups of 6 to 10 whales form social units and may gather into larger herds when migrating. They do make click sounds, as well as other vocal communications, but it is not certain whether or not these are used for echolocation. Narwhals mate in early spring, and gestation lasts 15 months. Mothers feed their calves for up to 2 years. At birth, calves are a dark blue-grey but as they mature this changes to the mottled brown of adults.

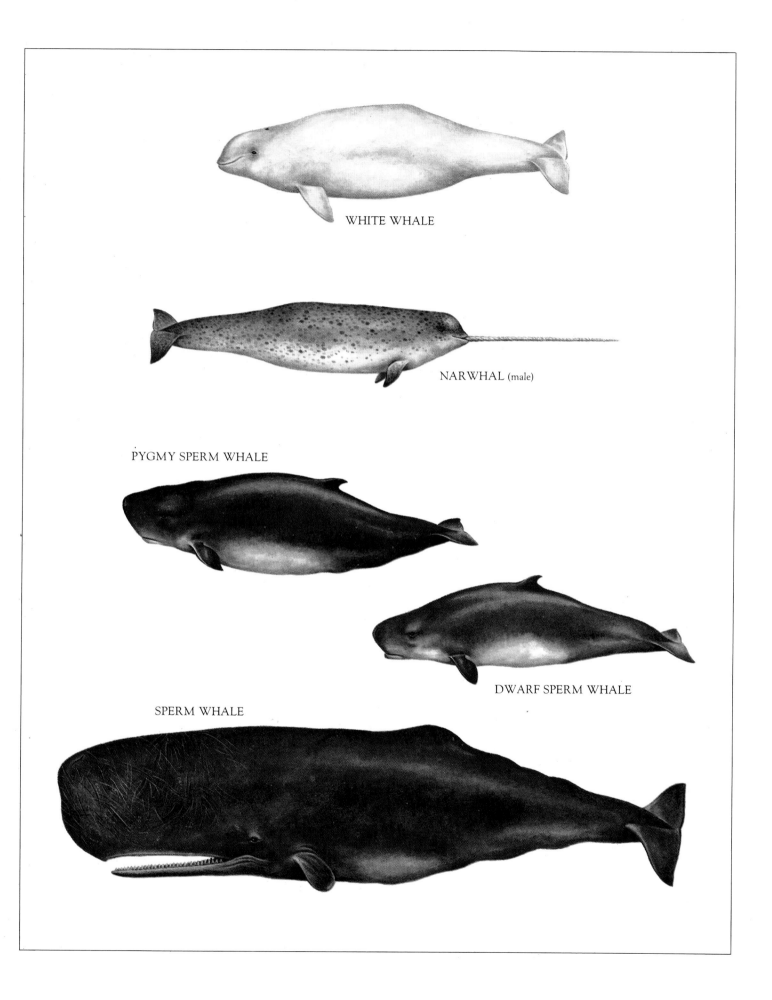

WHITE WHALE

NARWHAL (male)

PYGMY SPERM WHALE

DWARF SPERM WHALE

SPERM WHALE

Beaked Whales

ZIPHIIDAE: Beaked Whale Family

There are 18 species of beaked whale, found in all oceans. Most are medium-sized whales with slender bodies and long narrow snouts; some species have bulging, rounded foreheads. A particular characteristic of the family is the pair of grooves on the throat. Although Shepherd's beaked whale has more than 50 teeth, all other beaked whales have only one or two pairs, and the arrangement and shape of these are a useful means of identification.

Beaked whales feed largely on squid. They are deep divers and are believed to dive deeper and to remain submerged for longer periods than any other marine mammals. They generally move in small groups, but adult males are often solitary.

Although the second-largest family of whales (the dolphin family is the largest with 32 species), beaked whales are a little-known group. Some species are known to exist only from a few skulls and bones. The 12 species in the genus *Mesoplodon* are particularly unresearched but interesting. Only males have functional teeth, with a single pair protruding from the lower jaw. The shape and length of these differ from species to species, culminating in M. *layardi*, in which the backward-pointing teeth grow upward out of the mouth like tusks.

NAME: Northern Bottlenose Whale,
Hyperoodon ampullatus
RANGE: Arctic, N. Atlantic Oceans
HABITAT: deep offshore waters
SIZE: 7.3–10 m (24–33 ft) (V)

A sturdy, round-bodied whale, the northern bottlenose has a prominent, bulbous forehead that is particularly pronounced in older males. Males are generally larger than females. The adult male has only two teeth, which are in the lower jaw, but these are often so deeply embedded in the gums that they cannot be seen. Adult females also have only two teeth, and these are always embedded in the gum. Some individuals have further vestigial, unusable teeth in the gums.

Squid, some fish such as herring and sometimes starfish make up the diet of the northern bottlenose. A member of a deep-diving family, the bottlenose is believed to dive deeper than any other whale and certainly remains under water for longer. These are gregarious whales, and they collect in social units of 4 to 10 individuals, a group usually consisting of a male and several females with young. Pairs mate in spring and summer, and gestation lasts about 12 months. The whales are sexually mature at between 9 and 12 years of age.

Since commercial whaling of this species began in 1887, populations have been seriously depleted.

NAME: Cuvier's Beaked Whale, *Ziphius cavirostris*
RANGE: all oceans, temperate and tropical areas
HABITAT: deep waters
SIZE: 6.4–7 m (21–23 ft)

Cuvier's beaked whale has the typical tapering body of its family and a distinct beak. Adult males are easily distinguished by the two teeth which protrude from the lower jaw; in females these teeth remain embedded in the gums. The colouring of this species is highly variable: Indo-Pacific whales are generally various shades of brown, many with darker backs or almost white heads, while in the Atlantic, Cuvier's whales tend to be grey or grey-blue. All races are marked with scars and with discoloured oval patches caused by the feeding of parasitic lampreys.

Squid and deep-water fish are the main food of Cuvier's whales, and they make deep dives, lasting up to 30 minutes, to find prey. There is no definite breeding season, and calves are born at any time of the year. Groups of up to 15 individuals live and travel together.

NAME: Sowerby's Beaked Whale,
Mesoplodon bidens
RANGE: N. Atlantic Ocean
HABITAT: deep, cool coastal waters
SIZE: 5–6 m (16½–20 ft)

There are 12 closely related species of beaked whale in the genus *Mesoplodon*. Most tend to live in deep water, staying clear of ships, so they are rarely seen and their habits are little known. All have fairly well-rounded bodies, with small flippers in proportion to body size. Males are larger than females. Mature whales are generally marked with many scars; some of these are caused by parasites and others are, perhaps, the result of fights between individuals of the same species.

Sowerby's beaked whale was the first beaked whale to be recognized officially and described as a species, in 1804. The male has a pointed tooth at each side of the lower jaw. Females have smaller teeth in this position or no visible teeth at all. Squid and small fish are the main food of Sowerby's whale. Its breeding habits are not known, but it is thought to migrate south in winter and to give birth in its wintering area.

NAME: Shepherd's Beaked Whale,
Tasmacetus shepherdi
RANGE: New Zealand seas; off coasts of Argentina and Chile
HABITAT: coasts, open ocean
SIZE: 6–6.6 m (20–22 ft)

Shepherd's beaked whale was not discovered until 1933, and very few individuals have since been found or sighted. Until recently, the species was believed to occur only around New Zealand, but in the 1970s identical specimens were found off Argentina and Chile.

Unique in its family for its tooth pattern, Shepherd's beaked whale has a pair of large teeth at the tip of its lower jaw, with 12 or more pairs behind them, and about 10 pairs in the upper jaw. It resembles the rest of the family in habits and appearance. Squid and fish are believed to be its main food.

NAME: Baird's Beaked Whale, *Berardius bairdi*
RANGE: temperate N. Pacific Ocean
HABITAT: deep water over 1,000 m (3,300 ft)
SIZE: 10–12 m (33–39 ft)

The largest of the beaked whales, Baird's beaked whale has a distinctive beak, with the lower jaw extending beyond the upper. A pair of large teeth protrudes at the tip of the lower jaw, and behind these is a pair of smaller teeth. Female Baird's whales are generally larger than males and lighter in colour but have smaller teeth. Adult males are usually marked with scars, caused by their own species, suggesting that there is much rivalry and competition for leadership of groups of breeding females.

The normal social unit is a group of 6 to 30, led by a dominant male. The whales mate in midsummer and gestation lasts for 10 months, sometimes longer. The migration pattern of this species is the exact opposite of the normal migration habits of whales. They spend the summer in warm waters to the south of their range off California and Japan, then move northwards in winter to the cooler waters of the Bering Sea and similar areas. These movements are probably connected with the local abundance of food supplies. Deep divers, Baird's whales feed on squid, fish, octopus, lobster, crabs and other invertebrates.

Arnoux's beaked whale, *B. arnouxii*, which occurs in the temperate South Pacific and South Atlantic, is the closely related southern counterpart of Baird's whale. Although rarely seen, it is believed to be similar in both appearance and habits.

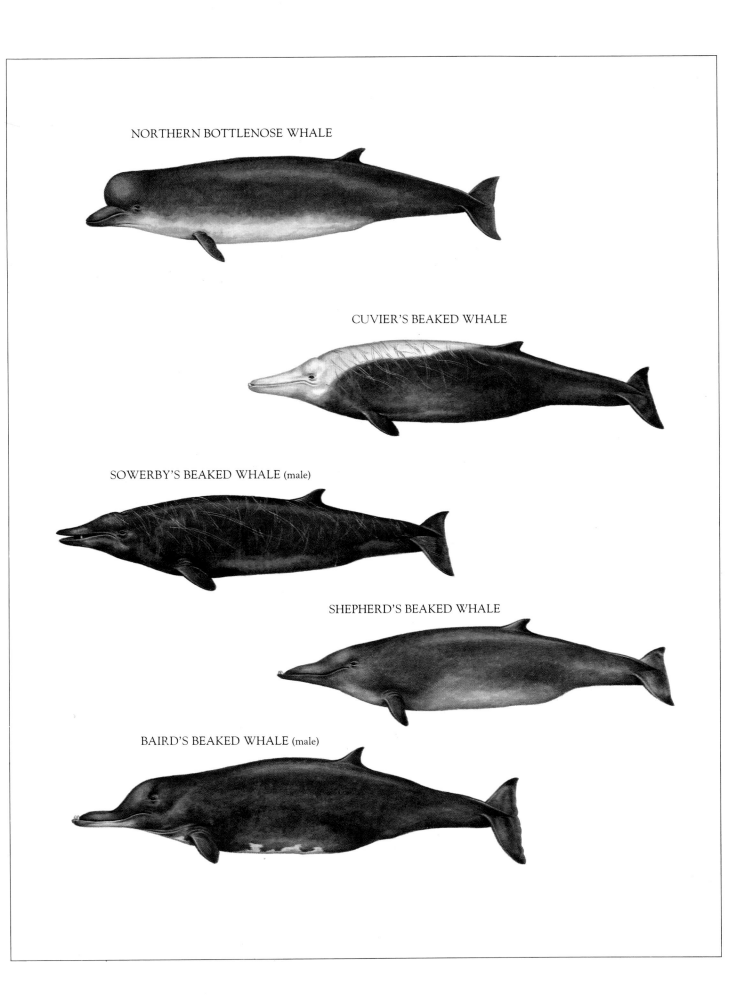

NORTHERN BOTTLENOSE WHALE

CUVIER'S BEAKED WHALE

SOWERBY'S BEAKED WHALE (male)

SHEPHERD'S BEAKED WHALE

BAIRD'S BEAKED WHALE (male)

Grey Whale, Rorquals, Right Whales

ESCHRICHTIDAE:
Grey Whale Family

There is a single species in this family, which is in some ways intermediate between the rorqual and right whales, the other 2 baleen whale families. The grey whale differs from both of these in that it has two or sometimes four throat grooves instead of the 100 or more in the other baleen whales.

NAME: Grey Whale, *Eschrichtius robustus*
RANGE: N.E. and N.W. Pacific Ocean
HABITAT: coastal waters
SIZE: 12.2–15.3 m (40–50 ft) Ⓥ

The grey whale has no dorsal fin, but there is a line of bumps along the middle of its lower back. Its jaw is only slightly arched and the snout is pointed. Males are larger than females. Like all baleen whales, it feeds on small planktonic animals by filtering water through rows of fringed horny plates, suspended from the upper jaw. Any creatures in the water are caught on the baleen plates and the water is expelled at the sides of the mouth. Using its tongue, the whale takes the food from the baleen to the back of its mouth to be swallowed. The grey whale feeds at the bottom of the sea, unlike other baleen whales, stirring up the sediment with its pointed snout then sieving the turbulent water.

Grey whales perform migrations of some 20,000 km (12,500 mls) between feeding grounds in the north and breeding grounds in the south. They spend the summer months in the food-rich waters of the Arctic, when they do most of their feeding for the year. At the breeding grounds, they gather to perform courtship rituals, and breeding animals pair off with an extra male in attendance. They lie in shallow water and, as the pair mate, the second male lies behind the female, apparently supporting her. The gestation period is 12 months, so the calf is born at the breeding grounds a year after mating has taken place and travels north with its mother when it is about 2 months old.

BALAENOPTERIDAE:
Rorqual Family

There are 6 species of rorqual whale. All except the humpback whale are similar in appearance, but they differ in size and colour. Most have about 300 baleen plates on each side of the jaws, which are not highly arched, and there are a large number of grooves on the throat. Females are larger than males.

NAME: Minke Whale, *Balaenoptera acutorostrata*
RANGE: all oceans, temperate and polar areas
HABITAT: shallow water, estuaries, rivers, inland seas
SIZE: 8–10 m (26–33 ft)

The smallest of the rorqual family, the minke whale has a distinctive, narrow pointed snout and 60 to 70 throat grooves. In polar areas, minke whales feed largely on planktonic crustaceans, but temperate populations eat fish and squid more often than any other baleen whale. Out of the breeding season, these whales tend to occur alone or in pairs, but they may congregate in rich feeding areas. Gestation lasts 10 or 11 months and calves are suckled for 6 months.

NAME: Sei Whale, *Balaenoptera borealis*
RANGE: all oceans (not polar regions)
HABITAT: open ocean
SIZE: 15–20 m (49–65½ ft)

A streamlined, flat-headed rorqual, the sei whale is a fast swimmer, achieving speeds of 50 km/h (26 kn). Almost any kind of plankton, as well as fish and squid, are eaten by sei whales and they usually feed near the surface. Family groups of 5 or 6 whales move together, and pair bonds are strong and may last from year to year. The gestation period is 12 months, and the calf is fed by its mother for 6 months.

NAME: Blue Whale, *Balaenoptera musculus*
RANGE: all oceans
HABITAT: open ocean
SIZE: 25–32 m (82–105 ft) Ⓔ

The largest mammal that has ever existed, the blue whale may weigh more than 146 te (144 UK t, 161 US t). Its body is streamlined and, despite its enormous bulk, it is graceful in the water. It has 64 to 94 grooves on its throat. These gigantic whales feed entirely on small planktonic crustaceans and, unlike many baleen whales, are highly selective, taking only a few species. They feed during the summer months, which they spend in nutrient-rich polar waters, and over this period take up to 4.1 te (4 UK t, 4.4 US t) of small shrimps apiece each day.

In autumn, when ice starts to cover their feeding grounds, the blue whales migrate toward the Equator but eat virtually nothing while in the warmer water. They mate during this period and, after a gestation of 11 or 12 months, the calves are born in warm waters the following year.

Even though blue whales have been protected since 1967, populations of this extraordinary animal are still low, and it is in danger of extinction.

NAME: Humpback Whale, *Megaptera novaeangliae*
RANGE: all oceans
HABITAT: oceanic, coastal waters
SIZE: 14.6–19 m (48–62 ft) Ⓔ

The humpback has a distinctly curved lower jaw and an average of 22 throat grooves. Its most characteristic features are the many knobs on the body and the flippers, which are about 5 m (16 ft) long and scalloped at the front edges. Humpback whales are more gregarious than blue whales and are usually seen in family groups of 3 or 4, although they may communicate with many other groups.

In the southern hemisphere, humpbacks feed on planktonic crustaceans, but in the northern hemisphere they eat small fish. Populations in both hemispheres feed in polar regions in summer and then migrate to tropical breeding areas for the winter. The gestation period is 11 or 12 months, and a mother feeds her calf for almost a year.

Humpbacks perform the most extraordinary, complex songs of any animal. The songs may be repeated for hours on end and are specific to populations and areas. They may change from year to year.

BALAENIDAE:
Right Whale Family

It is in the 3 species of right whale that the baleen apparatus is most extreme. Right whales have enormous heads, measuring more than a third of their total body length, and highly arched upper jaws to carry the long baleen plates. They have no throat grooves.

Regarded by whalers as the "right" whales to exploit, they have been killed in such numbers by commercial whalers over the last century that they are rare today.

NAME: Bowhead Whale/Greenland Right Whale, *Balaena mysticetus*
RANGE: Arctic Ocean
HABITAT: coastal waters
SIZE: 15–20 m (49–65½ ft) Ⓔ

The bowhead has a massive head and a body that tapers sharply toward the tail. Its jaws are strongly curved to accommodate the 4.5 m- (15 ft-) long baleen plates, the longest of any filter-feeding whale.

Bowheads feed on the smallest planktonic crustaceans, which they catch on the fine fringes of their baleen. They mate in early spring, the gestation period is 10 to 12 months, and the calf is fed for almost a year. Occasionally twins are produced.

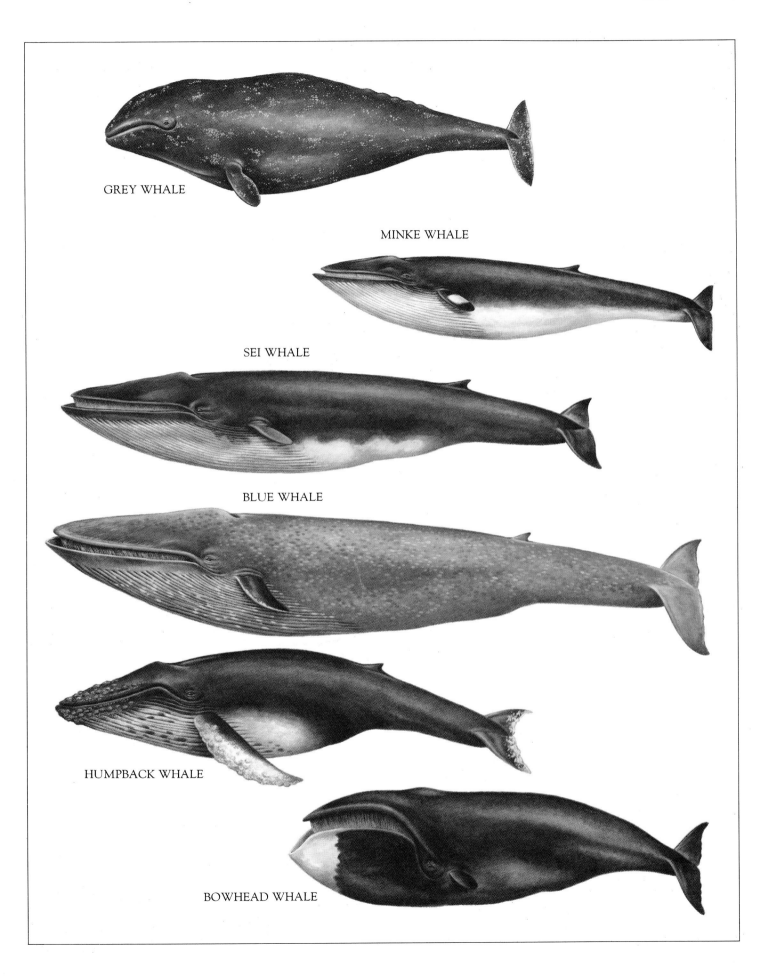

GREY WHALE

MINKE WHALE

SEI WHALE

BLUE WHALE

HUMPBACK WHALE

BOWHEAD WHALE

Elephants, Hyraces

ORDER PROBOSCIDEA

The elephants are the only surviving representatives of this once diverse and widespread group, which formerly contained many species of huge herbivorous mammal.

ELEPHANTIDAE: Elephant Family

The two species of elephant are by far the largest terrestrial mammals; they may stand up to 4 m (13 ft) at the shoulder and weigh as much as 5,900 kg (13,000 lb). One species lives in Africa, the other in India and Southeast Asia. Elephants are quite unmistakable in appearance; they have thick, pillarlike legs, and their feet are flattened, expanded pads. On the head are huge ears, which are fanned to and fro to help dissipate excess body heat, and the most remarkable of adaptations, the trunk. This is an elongated nose and upper lip, which is extremely flexible and has a manipulative tip. An extraordinarily sensitive organ, it is used for gathering food, drinking, smelling and fighting.

Both species of elephant have suffered badly from destruction of forest and vegetation in their range, and large numbers have been killed for their ivory tusks. Although hunting is now strictly controlled, poaching continues, since the demand for ivory persists.

NAME: **African Elephant, *Loxodonta africana***
RANGE: **Africa, south of the Sahara**
HABITAT: **forest, savanna**
SIZE: **body: 6–7.5 m (19¾–24½ ft)**
 tail: 1–1.3 m (3¾–4¼ ft) Ⓥ

The huge, majestic elephant is perhaps the most imposing of all the African mammals. It has larger ears and tusks than the Asian species and two fingerlike extensions at the end of its trunk. Females are smaller than males and have shorter tusks. Elephants rest in the midday heat and have one or two periods of rest at night but are otherwise active at any time, roaming with their swinging, unhurried gait in search of food. Depending on its size, an elephant may consume up to 200 kg (440 lb) of plant material a day, all of which is grasped with the trunk and placed in the mouth. The diet includes leaves, shoots, twigs, roots and fruit from many plants, as well as cultivated crops on occasion.

Elephants are social animals, particularly females, and are known to demonstrate concern for others in distress. A troop centres around several females and their young of various ages. As they mature, young males form separate troops. Old males may be shunned by the herd when they are displaced by younger males.

Breeding occurs at any time of year, and a female on heat may mate with more than one male. The gestation period is about 22 months, and usually only 1 young is born. The female clears a secluded spot for the birth and is assisted by other females. The calf is suckled for at least 2 years and remains with its mother even longer. She may have several calves of different ages under her protection and gives birth only every 2 to 4 years.

NAME: **Asian Elephant, *Elephas maximus***
RANGE: **India, Sri Lanka, S.E. Asia, Sumatra**
HABITAT: **forest, grassy plains**
SIZE: **body: 5.5–6.5 m (18–21¼ ft)**
 tail: 1.2–1.5 m (4–5 ft) Ⓥ

Although an equally impressive animal, the Asian elephant has smaller ears than its African counterpart, a more humped back and only one fingerlike extension at the end of its mobile trunk. The female is smaller than the male and has only rudimentary tusks. The main social unit is a herd led by an old female and including several females, their young and an old male, usually all related. Other males may live alone but near to a herd and will sometimes feed or mate with members of the herd. The herd rests in the heat of the day but spends much of the rest of the time feeding on grass, leaves, shoots, fruit and other plant material, all of which they search out and grasp with highly sensitive trunks. Their hearing and sense of smell is excellent, and eyesight poor.

During the heat period, called musth, which is often accompanied by a profuse secretion of scented liquid from a gland on the side of the head, normally docile animals become excited and unpredictable. The gestation period is about 21 months, and the female usually gives birth to a single young.

ORDER HYRACOIDEA
PROCAVIIDAE: Hyrax Family

Small herbivorous mammals found in Africa and the Middle East, the hyraces, conies or dassies have the general appearance of rabbits with short, rounded ears. There are about 5 species, and the family is the only one in its order. Some hyraces are agile climbers in trees, while others inhabit rocky koppies, or small hills. On the feet are flattened nails, resembling hoofs, and a central moist cup that functions as an adhesive pad when the hyrax is climbing.

NAME: **Tree Hyrax, *Dendrohyrax arboreus***
RANGE: **Africa: Kenya to South Africa: Cape Province**
HABITAT: **forest**
SIZE: **body: 40–60 cm (15¾–23½ in)**
 tail: absent

The tree hyrax is an excellent climber and lives in a tree hole or rock crevice, where it rests during the day. It emerges in the afternoon or evening to feed in the trees and on the ground on leaves, grass, ferns, fruit and other plant material. Insects, lizards and birds' eggs are also eaten on occasion. Tree hyraces normally live in pairs and are extremely noisy animals, uttering a wide range of screams, squeals and grunts.

A litter of 1 or 2 young is born after a gestation of about 8 months.

NAME: **Small-toothed Rock Hyrax, *Heterohyrax brucei***
RANGE: **Africa: Egypt to South Africa: Transvaal; Botswana and Angola**
HABITAT: **open country, plains to mountains, forest, savanna**
SIZE: **body: 40–57 cm (15¾–22½ in)**
 tail: absent

Despite its name, this hyrax lives among rocks or trees, depending on its habitat, sheltering in crevices or holes. Sociable animals, rock hyraces form colonies of up to 30, each colony consisting of several old males, many breeding females and their young. They feed during the day, mainly on leaves of trees but also on small plants and grass.

The female gives birth to 1 or 2 young, rarely 3, after a gestation of 7½ to 8 months.

NAME: **Large-toothed Rock Hyrax, *Procavia capensis***
RANGE: **Arabian Peninsula; Africa: N.E. Senegal to Somalia and N. Tanzania, S. Malawi, S. Angola to South Africa: Cape Province**
HABITAT: **rocky hillsides, rock piles**
SIZE: **body: 43–47 cm (17–18½ in)**
 tail: absent

This hyrax lives among rocky outcrops and is an agile climber. It feeds mostly on the ground on leaves, grass, small plants and berries, but readily climbs to feed on fruits such as figs; in winter bark is eaten. Much of the rest of the day it spends lying in the sun or shade to maintain its body temperature, and at night the hyraces huddle together to minimize loss of body heat. These hyraces are sociable and live in colonies of 50 or more individuals.

Males are aggressive at mating time and reassert their dominance over rivals and younger males. The female gives birth to 1 to 6 young, usually 2 or 3, after a gestation of 7 to 8 months.

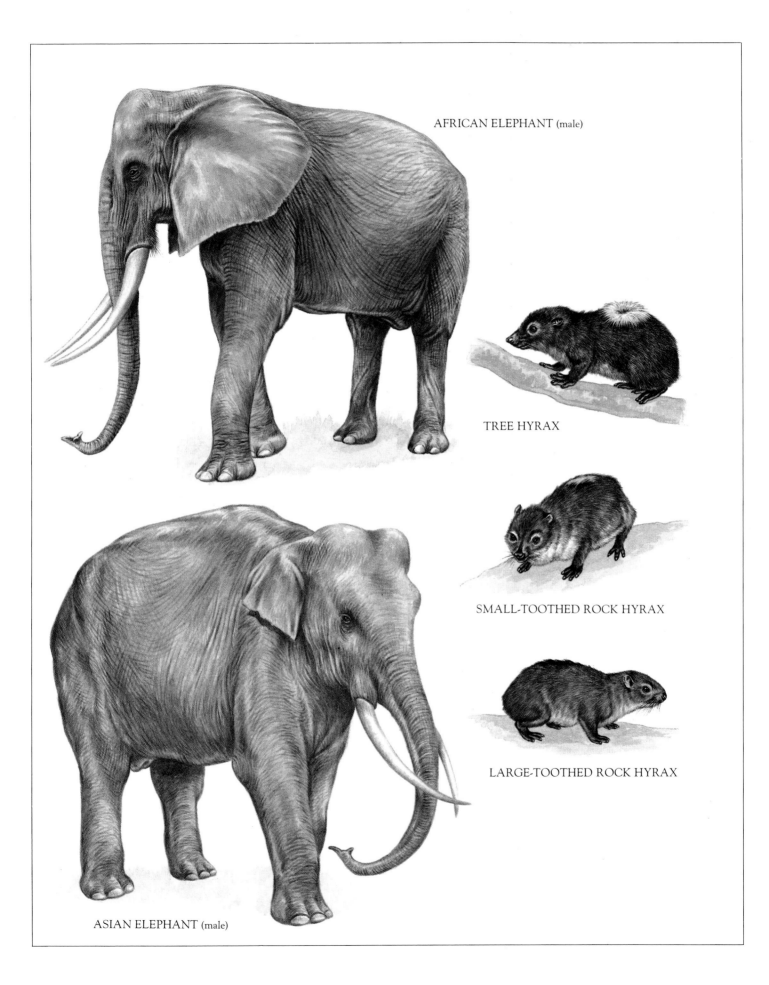

AFRICAN ELEPHANT (male)

TREE HYRAX

SMALL-TOOTHED ROCK HYRAX

LARGE-TOOTHED ROCK HYRAX

ASIAN ELEPHANT (male)

Horses

NAME: **Common Zebra,** *Equus burchelli*
RANGE: **E. and S. Africa**
HABITAT: **grassy plains, lightly wooded savanna, hills**
SIZE: **body: 1.9–2.4 m (6¼–7¾ ft)**
   tail: 43–57 cm (17–22½ in)

Great variation in pattern occurs in these zebras, both between individuals and the various subspecies. Toward the south of the range, the stripes on the hind parts of the body generally become lighter. The body is rounded, and the legs slender, and there is a small erect mane on the back of the neck. The base colour of the body varies from white to yellowish, and stripes may be light to dark brown or black.

Active in the daytime, these zebras leave their resting place at dawn and move to grazing grounds to feed on grass and sometimes on leaves and bark. They must drink regularly. Zebras live in families of up to 6 females and their young, led by an old male. When the male is 16 to 18 years old, he is peacefully replaced by a younger male of 6 to 8 years and then lives alone. Several families share a home range and may join in large herds, but they can always recognize each other by pattern, voice and scent.

The female gives birth to a single young, rarely twins, after a gestation of about a year. Until the foal learns to recognize its mother — in 3 or 4 days — she drives other animals away. It suckles for about 6 months and is independent at about a year old.

NAME: **Grevy's Zebra,** *Equus grevyi*
RANGE: **E. Africa: Kenya, Ethiopia**
HABITAT: **savanna, semi-desert**
SIZE: **body: 2.6 m (8½ ft)**
   tail: 70–75 cm (27½–29½ in) Ⓥ

The largest of its family, Grevy's zebra has a long head, broad, rounded ears and a relatively short, strong neck; an erect mane runs from its crown down the back of its neck. Its body is white with black stripes, which are narrower and more numerous than those of the common zebra. Grevy's zebra grazes during the day, resting in the shade if possible in the noon heat, and likes to drink daily. Mature males live alone, each in his own territory, while males without territories live in troops. Females and their young live in separate troops of a dozen or more. In the dry season, male and female troops migrate, but territorial males stay on unless there is a severe drought.

The female gives birth to a single young after a gestation thought to be about a year. The young is able to recognize its mother after the first few days. It suckles for 6 months and remains with its mother for up to 2 years.

ORDER PERISSODACTYLA

There are only 3 surviving families of perissodactyl, or odd-toed, hoofed mammals: the horses, tapirs and rhinoceroses. Nine other families, now extinct, are known from fossils.

EQUIDAE: Horse Family

The horses, asses and zebras together make up a small family of about 8 species of hoofed mammal, highly adapted for fast, graceful running. In this group the foot has evolved to the point of being a single hoof on an elongate third digit. The family has a natural distribution in Asia and Africa, but the wild horse itself has been domesticated by man and has since spread to other parts of the world.

In the wild, all equids live in herds, make regular migrations and feed in the main on grass. Their teeth are adapted for grass-cropping and grinding, with chisel-shaped incisors and very large premolars and molars, with convoluted surfaces. The skull is elongate to accommodate the large cheek teeth.

NAME: **African Ass,** *Equus africanus*
RANGE: **N.E. Africa**
HABITAT: **open grassy plains, rugged rocky country, semi-desert, mountains**
SIZE: **shoulder height: 1.2 m (4 ft)** Ⓔ

The ancestor of the domestic ass, the African ass is now rare in the wild, and it is thought that many of those which remain are cross-breeds with domestic stock. They have suffered from excessive hunting and from competition from increasing numbers of domestic livestock. Of the 4 subspecies, 1 is now extinct and 2 nearly so.

The African ass has a large head with long, narrow ears, and short, smooth hair, which varies from yellowish-brown to bluish-grey in colour. A good climber, it is adept at moving over rocks and rugged country as it wanders, feeding on grass and herbage. It may sometimes browse on foliage and needs to drink regularly. Most active at dusk and night-time and in the early morning, it spends much of the heat of the day resting, in shade if possible. Female African asses live in loose-knit troops with their young or in mixed troops of young animals of both sexes. Older males live alone or in male troops. The asses' senses are good, and they defend themselves by kicking and biting.

The female gives birth to 1 young after a gestation of 330 to 365 days. She keeps all other animals away from her newborn offspring for several days until it has learned to recognize her.

NAME: **Przewalski's Wild Horse,** *Equus ferus*
RANGE: **Mongolia, W. China**
HABITAT: **plains, semi-desert**
SIZE: **body: 1.8–2 m (6–6½ ft)**
   tail: 90 cm (35½ in) Ⓔ

The ancestor of the domestic horse, the wild horse is distinguished by its erect mane and lack of forelock. It interbreeds with domestic stock, and few pure-bred animals remain. Populations have declined drastically because of hunting, cold winters and competition with domestic livestock for water and pasture, and true wild horses may soon be extinct.

When their numbers were greater, wild horses moved in large herds, but now they live in small herds, each led by a dominant male. Young are born in April or May.

NAME: **Onager,** *Equus hemionus*
RANGE: **Iran, Afghanistan, USSR**
HABITAT: **steppe, gorges, river margins**
SIZE: **body: about 2 m (6½ ft)**
   tail: 42.5 cm (16¾ in) Ⓥ

Also called the Asiatic wild ass, the onager has declined in numbers because of human settlement in its range, grazing by domestic stock and hunting. In summer, it lives on high grassland and migrates in winter to lower levels to find water and pasture. It feeds on many types of grass, and the availability of fresh water is critical to its survival. A sociable animal, it lives in troops of up to 12 females and young, led by a dominant male, but in autumn, troops may gather in larger herds to migrate.

The female gives birth to a single young after a gestation of about a year.

NAME: **Kiang,** *Equus kiang*
RANGE: **Tibet**
HABITAT: **high plateaux**
SIZE: **body: 2.2 m (7¼ ft)**
   tail: 49 cm (19¼ in)

The largest of the wild asses, the kiang is the only one that still occurs in large, although dwindling, numbers. It remains on the high plateaux throughout the year and does not migrate, but in autumn it does put on an enormous amount of fat, as much as 40 kg (88 lb), which helps to sustain it through the winter and also forms an insulating layer against the cold. With their hard lips and horny gums, kiangs are well equipped for feeding on the tough grasses in their habitat, and they will eat snow if they cannot find water.

Kiangs live in herds in which a specific dominance order is maintained, but they are generally aggressive, and in the breeding season, males fight one another over females. The female gives birth to 1 young after a gestation of about a year.

COMMON ZEBRA

GREVY'S ZEBRA

PRZEWALSKI'S WILD HORSE

AFRICAN ASS

KIANG

ONAGER

Tapirs, Rhinoceroses

TAPIRIDAE: Tapir Family
The 4 living species of tapir are thought to resemble the ancestors of the perissodactyls. They are stocky, short-legged animals, with four toes on the forefeet and three on the hind feet. The body is covered with short, bristly hairs, giving the tapir a smooth-coated appearance. The snout and upper lip are elongated to form a short, mobile trunk, with nostrils at its tip.

Tapirs are mainly nocturnal, forest-dwelling animals and feed on vegetation. Of the 4 species, 3 occur in Central and South America and 1 in Southeast Asia.

NAME: **Malayan Tapir, *Tapirus indicus***
RANGE: **S.E. Asia: Burma to Malaysia, Sumatra**
HABITAT: **humid, swampy forest**
SIZE: **body: 2.5 m (8¼ ft)**
tail: 5–10 cm (2–4 in) Ⓔ

The Malayan tapir differs from other tapirs in its striking greyish-black and white coloration. It also has a longer, stronger trunk than the tapirs from South America. A shy, solitary animal, it is active only at night, when it feeds on aquatic vegetation and the leaves, buds and fruit of low-growing land plants. It swims well and, if alarmed, makes for water.

The female gives birth to a single young after a gestation of about 395 days. The body of the young tapir is patterned with camouflaging stripes and spots, which disappear at about 6 to 8 months. Malayan tapirs have been badly affected by the destruction of large areas of forest and the changes wrought by human settlement and are now extremely rare.

NAME: **Brazilian Tapir, *Tapirus terrestris***
RANGE: **South America: Colombia, Venezuela, south to Brazil and Paraguay**
HABITAT: **rain forest, near water or swamps**
SIZE: **body: 2 m (6½ ft)**
tail: 5–8 cm (2–3 in)

Nearly always found near water, the Brazilian tapir is a good swimmer and diver but also moves fast on land, even over rugged, mountainous country. It is dark brown in colour and has a low, erect mane, running from the crown down the back of the neck. Using its mobile snout, this tapir feeds on leaves, buds, shoots and small branches that it tears from trees, fruit, grasses and water plants.

The female gives birth to a single spotted and striped young after a gestation of 390 to 400 days.

RHINOCEROTIDAE: Rhinoceros Family
Of the living perissodactyl mammals, only the rhinoceroses show the massive, heavy-bodied structure that was so common among the earlier, now extinct, families of the order. There are 5 species of rhinoceros, found in Africa and Southeast Asia, and all have huge heads, with one or two horns and a prehensile upper lip, which helps them browse on tough plant material. The legs are short and thick, with three hoofed toes on each foot, and the skin is extremely tough, with only a few hairs. Male and female look much alike, but females tend to have smaller horns.

NAME: **Indian Rhinoceros, *Rhinoceros unicornis***
RANGE: **Nepal, N.E. India**
HABITAT: **grassland in swampy areas**
SIZE: **body: 4.2 m (13¾ ft)**
tail: 75 cm (29½ in) Ⓔ

The largest of the Asian species, the Indian rhinoceros has a thick, dark-grey hide, studded with many small protuberances. The skin falls into deep folds at the joints, giving the rhinoceros an armour-plated appearance. Both sexes have a single horn on the head, but the female's horn is smaller. Generally a solitary animal, the Indian rhinoceros feeds in the morning and evening on grass, weeds and twigs and rests during the remainder of the day.

The female gives birth to a single young after a gestation of about 16 months. The young rhinoceros is active soon after birth and is suckled for up to 2 years.

NAME: **Sumatran Rhinoceros, *Dicerorhinus sumatrensis***
RANGE: **Burma, Thailand, Malaysia, Sumatra, Borneo, possibly Laos**
HABITAT: **dense forest, near streams**
SIZE: **2.5–2.8 m (8¼–9¼ ft)**
tail: about 60 cm (23½ in) Ⓔ

The smallest member of its family, the Sumatran rhinoceros has two horns; those of the female are smaller than the male's. Bristlelike hairs are scattered over the thick skin and fringe the edges of the ears. Sumatran rhinoceroses are usually solitary, although a male and female pair may live together. They feed mostly in the early morning and evening on leaves, twigs, fruit and bamboo shoots and may trample down small trees to browse on their foliage. Like other rhinoceroses, this species has good hearing and sense of smell, but its sight is poor.

The female bears a single young after a gestation of about 7 to 8 months.

NAME: **Square-lipped/White Rhinoceros, *Ceratotherium simum***
RANGE: **Africa: N.W. Uganda and adjacent regions; Zimbabwe to N. South Africa**
HABITAT: **savanna**
SIZE: **body: 3.6–5 m (11¾–16½ ft)**
tail: 90 cm–1 m (35½ in–3¼ ft)

The largest living land animal after the elephant, the square-lipped rhinoceros has a pronounced hump on its neck and a long head, which it carries low. Its muzzle is broad, with a squared upper lip. This rhinoceros is generally greyish in colour but takes on the colour of the mud in which it has been wallowing, so may, in fact, be any shade of greyish- or reddish-brown.

Although so vast, the square-lipped rhinoceros is a placid animal and tends to flee from trouble, rather than attack. Each old male occupies his own territory, which may be shared by younger males, but the female is sociable and is usually accompanied by another female with young or by her own young and several others. They feed only on grass, grazing and resting from time to time throughout the day and night.

The female gives birth to a single young after a gestation of about 16 months. The calf suckles for at least a year and stays with its mother for 2 or 3 years, leaving her only when her next calf is born.

NAME: **Black Rhinoceros, *Diceros bicornis***
RANGE: **Africa: S. Chad and Sudan to South Africa**
HABITAT: **bush country, grassland, woodland**
SIZE: **3–3.6 m (9¾–11¾ ft)**
tail: 60–70 cm (23½–27½ in) Ⓥ

The black rhinoceros is, in fact, grey in colour but varies according to the mud in which it wallows. It has no hump on its neck but has a large head, held horizontally, which bears two horns and sometimes a third small horn. Its upper lip is pointed and mobile, which helps the animal to browse on the leaves, buds and shoots of small trees and bushes. Less sociable than the square-lipped rhinoceros, black rhinoceroses live alone, except for mothers and young. Adults live in overlapping home ranges, with boundaries marked by dung heaps.

Male and female remain together for only a few days when mating. The female gives birth to a single young after a gestation of about 15 months. The young rhinoceros suckles for about a year and stays with its mother for 2 or 3 years, until her next calf is born.

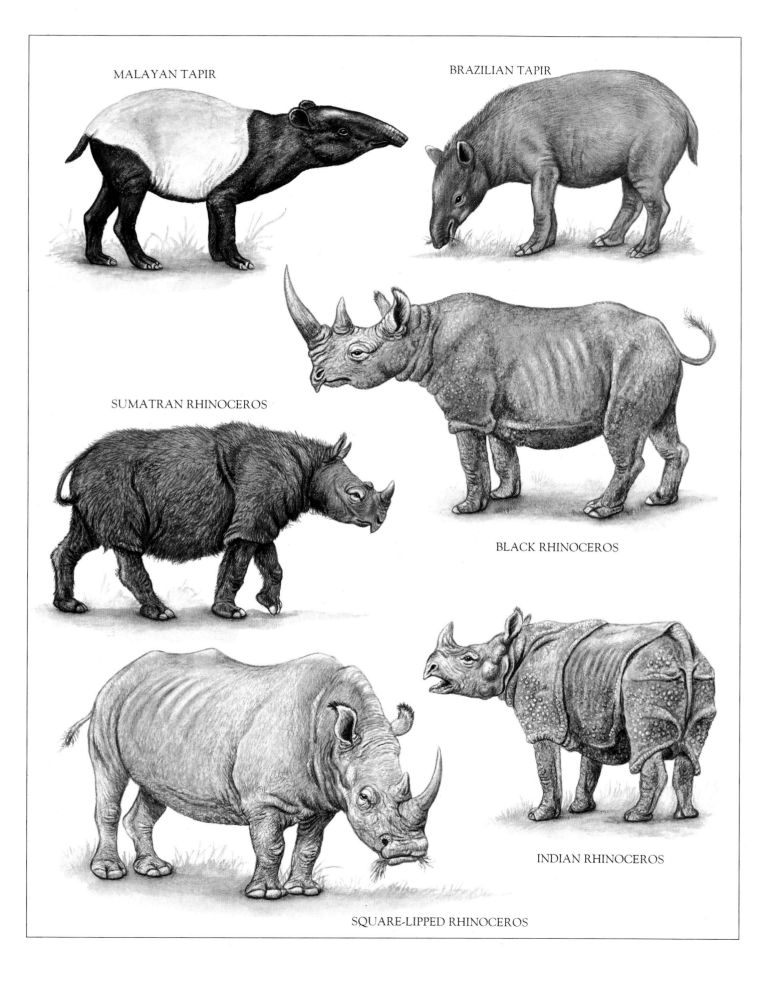

MALAYAN TAPIR

BRAZILIAN TAPIR

SUMATRAN RHINOCEROS

BLACK RHINOCEROS

INDIAN RHINOCEROS

SQUARE-LIPPED RHINOCEROS

Pigs

NAME: **Bush Pig, *Potamochoerus porcus***
RANGE: **Africa, south of the Sahara;
Madagascar**
HABITAT: **forest, bush, swamps, thickets
in savanna**
SIZE: **body: 1–1.5 m (3¼–5 ft)
tail: 30–45 cm (11¾–17¾ in)**

Also known as the red river hog, this pig has an elongate snout and long, tufted ears. Its bristly coat varies from reddish-to greyish-brown, with a white dorsal mane and whiskers. Bush pigs are gregarious animals and live in groups of up to 12 or so, led by an old male. They wander over a wide range to seek food and eat almost anything, including plant matter, such as grass, roots, fruit and grain crops, and small mammals, young birds and carrion. Normally active in the daytime, bush pigs are nocturnal in areas where they are hunted.

Breeding occurs throughout the year, particularly at times of abundant food supplies. The female gives birth to 3 to 6 young in a grassy nest, after a gestation of 120 to 130 days.

NAME: **Warthog, *Phacochoerus
aethiopicus***
RANGE: **Africa: Ghana to Somalia, south
to South Africa: Natal**
HABITAT: **savanna, treeless open plains**
SIZE: **body: 1.1–1.4 m (3½–4½ ft)
tail: 35–50 cm (13¾–19¾ in)**

The warthog has long legs, a large head and a broad muzzle that bears tusks derived from the canine teeth. On each side of the big head are two wartlike protuberances — the origin of the animal's common name. Its bristly coat is sparse, but there is a mane of long bristles running to the middle of the back, and there are whiskers on the lower jaw. The female is smaller than the male and has shorter tusks.

Generally gregarious, warthogs live in family groups in a territory that may be shared by more than one family. They prefer to have water for drinking and wallowing within their range and also some form of shelter, such as aardvark burrows or holes among rocks, where they rest in the heat of the day and at night. As well as grazing on short grass, warthogs feed on fruit and, in dry spells, will probe the ground with their tusks to obtain bulbs, tubers and roots. They occasionally prey on small mammals and will take carrion.

Timing of the breeding season tends to be associated with the local rainy seasons. The female gives birth to 2 to 4 young after a gestation of 170 to 175 days. The young suckle for up to 4 months but, after a week, start to leave the burrow in which they are born to feed on grass.

ORDER ARTIODACTYLA

This, the largest and most diverse order of hoofed herbivorous mammals, contains the pigs, peccaries, hippopotamuses, camels, deer, antelope, cows, sheep and goats. Artiodactyls have an even number of toes, the weight of the body being carried on digits 3 and 4, which are, typically, encased in pointed hoofs. The first digit is always absent, and digits 2 and 5 are always more or less reduced.

Artiodactyls can run rapidly and have specialized teeth for cutting and grinding vegetation. Plant material is digested in a complex, four-chambered stomach with the aid of enzymes and symbiotic micro-organisms.

SUIDAE: Pig Family

Pigs are more omnivorous than other artiodactyls and lack some of their specializations. There are about 8 species, found in Europe, Asia and Africa, usually in forested or brush areas. Pigs are stocky animals, with long heads terminating in mobile snouts, which are tough, sensitive and flattened at the tip. The snout is used for ploughing up forest litter or soil to find food items. The upper canine teeth usually grow outward and upward to form tusks. On each foot are four toes, but only the third and fourth reach the ground and have functional hoofs.

NAME: **Wild Boar, *Sus scrofa***
RANGE: **S. and central Europe, N.W.
Africa; through Asia to Siberia, south
to Sri Lanka, Taiwan and S.E. Asia**
HABITAT: **forest, woodland**
SIZE: **body: 1.1–1.3 m (3½–4¼ ft)
tail: 15–20 cm (6–7¾ in)**

The ancestor of the domestic pig, the wild boar has a heavy body covered with dense, bristly hair, thin legs and a long snout. The male has prominent tusks derived from the canine teeth. Wild boars live alone or in small groups of up to 20, with males separate from, but remaining close to, the females. Active at night and in the morning, they forage over a wide area for food, digging for bulbs and tubers and also eating nuts and a variety of other plant material, as well as insect larvae and, occasionally, carrion. An agile, fast-moving animal, the wild boar is aggressive if alarmed, males using their strong tusks to defend themselves.

The breeding season varies according to regional climate, but in Europe, wild boars mate in winter and give birth to a litter of up to 10 striped young in spring or early summer, after a gestation of about 115 days.

NAME: **Bearded Pig, *Sus barbatus***
RANGE: **Malaysia, Sumatra, Borneo**
HABITAT: **rain forest, scrub, mangroves**
SIZE: **body: 1.6–1.8 m (5¼–6 ft)
tail: 20–30 cm (7¾–11¾ in)**

A large pig, with an elongate head and a narrow body, the bearded pig has abundant whiskers on its chin and a bristly, wartlike protuberance beneath each eye. These warts are more conspicuous in males than in females. Fallen fruit, roots, shoots and insect larvae are the bearded pig's staple foods, and it also invades fields of root crops. It often follows gibbons and macaques and picks up the fruit the monkeys drop.

After a gestation of about 4 months, the female makes a nest of plant material and gives birth to 2 or 3 young, which stay with her for about a year.

NAME: **Giant Forest Hog, *Hylochoerus
meinertzhageni***
RANGE: **Africa: Liberia, Cameroon, east
to S. Ethiopia, Tanzania, Kenya**
HABITAT: **forest, thickets**
SIZE: **body: 1.5–1.8 m (5–6 ft)
tail: 25–35 cm (9¾–13¾ in)**

The largest of the African pigs, the giant forest hog has a huge elongate head, a heavy body and rather long legs for its family. Its muzzle is broad, and there are glandular swellings in the skin under its eyes and across its cheeks. Males are bigger and heavier than females. These pigs live in family groups of up to 12, and pairs remain together for life. They wander over a large, but undefined, range, feeding on grass, plants, leaves, buds, roots, berries and fruit. Mostly active at night and in the morning, they rest during the heat of the day in dense vegetation.

The female gives birth to 1 to 4 young, sometimes up to 8, after a gestation of 4 to 4½ months.

NAME: **Babirusa, *Babyrousa babyrussa***
RANGE: **Sulawesi, Sula Islands**
HABITAT: **moist forest, lake shores and
river banks**
SIZE: **body: 87–107 cm (34¼–42 in)
tail: 27–32 cm (10½–12½ in)**   Ⓥ

The babirusa has unusual upper tusks, which grow upward through the muzzle and curve back toward the eyes. Only males have prominent lower tusks, and these are thought to be a sexual characteristic. Elusive animals, babirusas prefer dense cover near water and are fast runners and good swimmers, even in the sea. They move in small groups, the male doing most of the rooting and unearthing of food, while females and young trail behind, feeding on items such as roots, berries, tubers and leaves.

The female gives birth to 2 young after a gestation of 125 to 150 days.

BUSH PIG (male)

BABIRUSA (male)

WILD BOAR (male)

GIANT FOREST HOG (male)

WARTHOG (male)

BEARDED PIG (female)

Peccaries, Hippopotamuses

TAYASSUIDAE: Peccary Family

The 3 species of peccary occur only in the New World, from the southwestern USA to central Argentina. The New World equivalent of pigs in their habits, peccaries resemble pigs but are smaller and differ from them in a number of ways. First, they have only three toes on each hind foot (pigs have four); second, peccaries have a prominent musk gland on the back about 20 cm (7¾ in) in front of the tail; and third, their tusks are directed downward, not upward like those of pigs.

NAME: Chaco Peccary, *Catagonus wagneri*
RANGE: Bolivia, Argentina, Paraguay
HABITAT: semi-arid thorn scrub, grassland
SIZE: body: about 1 m (3¼ ft)
tail: 87 cm (34¼ in) Ⓥ

Once thought to be extinct, the chaco peccary is now believed to be reasonably abundant in areas where it is undisturbed, although the species as a whole is vulnerable. The animals have suffered from excessive hunting and from the loss of much of their thorn scrub habitat, which has been cleared for cattle ranching.

A long-tailed, long-legged animal, this species is active during the day and has better vision than other peccaries. It moves in small groups of up to 6 animals, among which there are strong social bonds, and feeds largely on cacti and the seeds of leguminous plants.

NAME: White-lipped Peccary, *Tayassu pecari*
RANGE: Mexico, Central and South America to Paraguay
HABITAT: forest
SIZE: body: 95 cm–1 m (37½ in–3¼ ft)
tail: 2.5–5.5 cm (1–2¼ in)

The white-lipped peccary has a heavy body, slender legs and a long, mobile snout. A gregarious animal, it gathers in groups of 50 to 100 individuals of both sexes and all ages. Active in the cooler hours of the day, these peccaries are fast, agile runners, even on rugged ground. Using their sensitive snouts, they dig around on the forest floor, searching for plant material, such as bulbs and roots, and for small animals. Although their sight is poor and hearing only fair, these peccaries have such an acute sense of smell that they can locate bulbs underground by scent alone.

The female gives birth to a litter of 2 young after a gestation period of about 158 days.

NAME: Collared Peccary, *Tayassu tajacu*
RANGE: S.W. USA, Mexico, Central and South America to Patagonia
HABITAT: semi-desert, arid woodland, forest
SIZE: body: 75–90 cm (29½–35½ in)
tail: 1.5–3 cm (½–1¼ in)

Collared peccaries are robust, active animals, able to run fast and swim well. They live in groups of 5 to 15 individuals, and the musky secretions of the gland on each animal's back seem to play a part in maintaining the social bonds of the herd, as well as being used for marking territory. With their sensitive snouts, collared peccaries search the ground for roots, herbs, grass and fruit; they also eat insect larvae, worms and small vertebrates. In summer, they feed only in the morning and evening, but in winter they are active all day long, treading well-worn, regular paths through their home range. Hearing is the most acute of this peccary's senses.

Several males in a herd may mate with a female on heat, and there is rarely fighting or rivalry. After a gestation of 142 to 149 days, the female leaves the herd and gives birth to 2 or 3 young in a sheltered spot. The young are active soon after birth, and the mother returns to the herd with them within a couple of days.

HIPPOPOTAMIDAE: Hippopotamus Family

There are 2 species of hippopotamus, found only in Africa, although fossil evidence shows that the family was once more widely distributed in the southern parts of the Old World. Both species are amphibious, spending much of their lives in water, and they have various adaptations for this mode of life, including nostrils that can be closed and specialized skin glands that secrete an oily, pink substance, which protects their virtually hairless bodies.

NAME: Hippopotamus, *Hippopotamus amphibius*
RANGE: Africa, south of the Sahara to Namibia and South Africa: Transvaal
HABITAT: rivers or lakes in grassland
SIZE: body: (male) 3.2–4.2 m (10½–13¾ ft)
(female) 2.8–3.7 m (9¼–12 ft)
tail: 35–50 cm (13¾–19¾ in)

One of the giants of Africa, the hippopotamus has a bulky body and a massive head and mouth equipped with an impressive set of teeth; the canine teeth form tusks. Its legs are short and thick, and there are four webbed toes on each foot. When the hippopotamus is in water, it lies with much of its vast body submerged; often only the bulging eyes, ears and nostrils are visible. It swims and dives well and can walk along the river or lake bottom. Daytime hours are spent resting in water or on the shore, then, in the evening, the hippopotamus emerges to graze on land, taking short grass and other plants and fallen fruit. Hippopotamuses play a vital role in the ecology of inland waters, both by keeping down bankside vegetation and by excreting tons of fertilizing manure into the water, which encourages the growth of plankton and invertebrates and thus sustains the whole ecosystem.

Hippopotamuses are gregarious animals and live in groups of up to 15 or so, sometimes more, led by an old male. Males are aggressive and will fight for prime positions on the river bank or for dominance of the group. To threaten or challenge a rival, the male opens his mouth in a huge, yawning gape and bellows. All adults are fierce in defence of their young.

Mating takes place in water at any time of year but is generally timed so that births coincide with the rains and, thus, the luxuriant growth of grass. A single young is born on land or in shallow water after a gestation of 233 to 240 days. The young is suckled for about a year, and females usually give birth every 18 months to 2 years.

NAME: Pygmy Hippopotamus, *Choeropsis liberiensis*
RANGE: Guinea to Nigeria
HABITAT: rain forest, swamps and thickets near water
SIZE: body: 1.7–1.9 m (5½–6¼ ft)
tail: 15–21 cm (6–8¼ in) Ⓥ

The pygmy hippopotamus is much less aquatic than its giant relative and has a proportionately smaller head and longer legs; only the front toes are webbed. It lives near water but stays on land for much of the time, feeding at night on leaves, swamp vegetation and fallen fruit and also on roots and tubers, which it digs up. Usually alone, except for breeding pairs or females with young, it occupies a territory, which it defends against rivals. When alarmed, the pigmy hippopotamus seeks refuge in dense cover or in water.

The female gives birth to a single young after a gestation of 180 to 210 days. The young stays with its mother for up to 3 years. Rare over almost all its range, the pygmy hippopotamus may be extinct in some places. The population has suffered from excessive hunting, combined with the destruction of large areas of its forest habitat.

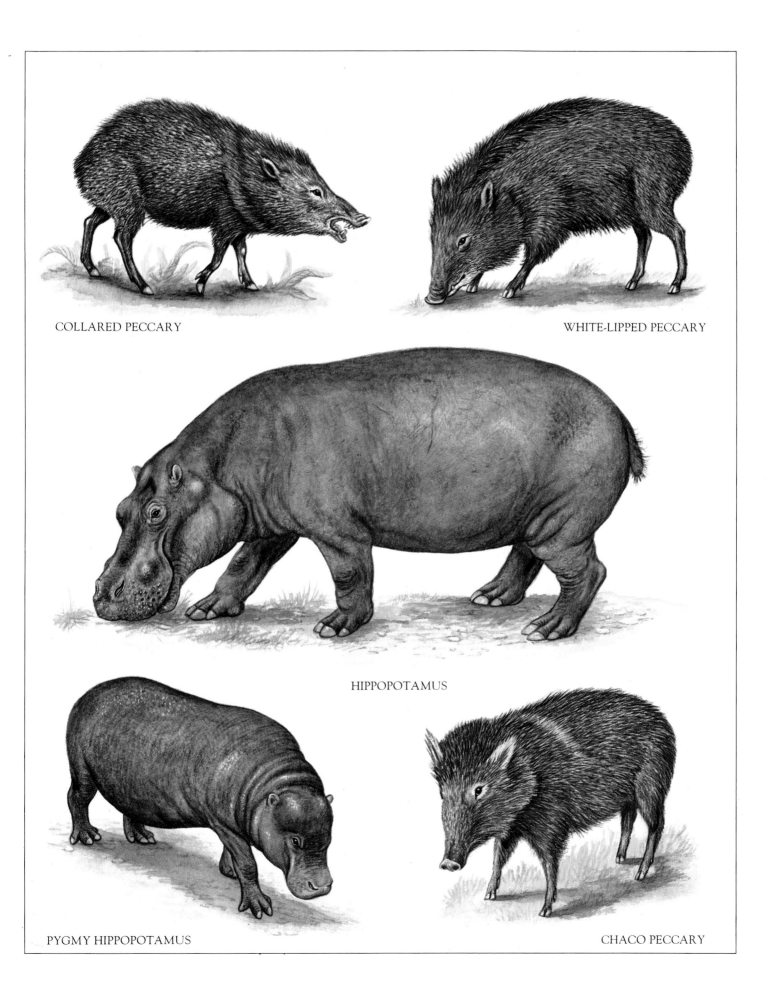

COLLARED PECCARY

WHITE-LIPPED PECCARY

HIPPOPOTAMUS

PYGMY HIPPOPOTAMUS

CHACO PECCARY

Camels

NAME: **Guanaco,** *Lama guanicoe*
RANGE: **South America: Peru to Patagonia**
HABITAT: **semi-desert to about 5,000 m (16,500 ft)**
SIZE: **body: 1.2–1.7 m (4–5½ ft)**
    tail: 25 cm (9¾ in)

The guanaco is a slender, long-limbed animal, capable of fast movement over rugged terrain and able to leap nimbly up mountain trails. Adaptable to heat or cold, it lives in open country and feeds on grass. Males are polygamous·and lead harems of 4 to 10 females with their young, which they defend, fighting off any rivals or intruders that try to steal one of their females. Young males and males without harems also form herds.

The female gives birth every other year, producing a single young after a gestation of 10 or 11 months. The young guanaco is active soon after birth and is able to run with speed and grace.

Llamas and alpacas are domesticated forms of the guanaco and are bred as draft animals and fleece producers respectively. They interbreed readily with one another and with wild guanacos.

NAME: **Vicuna,** *Vicugna vicugna*
RANGE: **South America: Peru to N. Chile**
HABITAT: **semi-arid grassland at altitudes over 4,000 m (13,000 ft)**
SIZE: **body: 1.4–1.6 m (4½–5¼ ft)**
    tail: 15 cm (6 in) Ⓥ

The vicuna's tawny-brown coat is thick and woolly and longest on the sides; it enables the animal to tolerate the cold, snow and ice of its mountain habitat. Gregarious animals, vicunas live in groups of up to 15 females led by a male, or in all-male herds. The harem band lives in a territory which is fiercely guarded by the adult male; at the first sign of any danger he alerts the females so that they can escape. Male troops consist mainly of young animals and do not have a specific territory but wander nomadically. Since most of the best grazing is appropriated by the territorial family males, these nomads are continually trespassing and being driven away. Rival males have a characteristic habit of spitting at each other as they fight. Vicunas are fast, graceful animals, capable of maintaining speeds of 47 km/h (29 mph) over long distances, even at high altitudes. They feed on grass and small plants. Eyesight is their most acute sense, while hearing is fair and smell poor.

The female gives birth to 1 young after a gestation of 10 to 11 months. The young can stand and walk soon after birth and suckles for about 10 months. Vicunas have long been killed for their fine wool and meat, but despite this, a few years ago the vulnerable population was said to be on the increase.

CAMELIDAE: Camel Family

The 4 surviving species in this formerly more diverse family are the most primitive of the ruminants, or cud-chewing animals. Of the 4, dromedaries and most of the bactrian camels are wholly domesticated. There are still wild bactrians in the Gobi Desert, and guanacos and vicunas maintain wild populations in South America.

Camels and their relatives have highly specialized feet. They have evolved to the point of having only two toes on each foot, but the foot bones are expanded sideways to produce the support for two broad, flat pads on each foot, with a nail on the upper surface of each toe. This foot structure is particularly well developed in the camel species and it enables them to walk on soft, sandy soil, where conventional hoofs would sink in deeply.

The head of a camelid is relatively small, with an elongate snout terminating in a cleft upper lip. Vegetation is cropped by using long, forward-pointing lower incisors that work against tough upper gums. Camelids have complex three-chambered stomachs and ruminate, or chew the cud.

The humps of the 2 species of camel are fat stores, which provide food reserves — vital in the unpredictable conditions of the camel's desert habitat.

NAME: **Bactrian Camel,** *Camelus ferus*
RANGE: **central Asia: China, Mongolia**
HABITAT: **desert, steppe**
SIZE: **body: about 3 m (9¾ ft)**
    tail: about 53 cm (20¾ in) Ⓥ

The bactrian, or two-humped, camel has been domesticated but has not spread outside its native range to the extent that the dromedary has. Only a small number of bactrian camels lives wild in the Gobi Desert, and even these may be part domestic stock. It is thought, however, that Mongolian stocks may be slowly increasing.

Apart from its two humps, the main characteristic of the bactrian camel is its long, shaggy hair, which keeps it warm in winter but is shed in summer, leaving the body almost naked. Docile, slow-moving animals, these camels move with a rolling gait, which is the result of their ability to raise both legs on one side at the same time. They feed on virtually any vegetation, such as grass, the foliage of trees and bushes, and small plants.

After a gestation of 370 to 440 days, the female gives birth to 1 young, which is active within only 24 hours. It is suckled for about a year and fully grown when about 5 years old.

NAME: **Dromedary,** *Camelus dromedarius*
RANGE: **N. Africa, Middle East; introduced in Australia**
HABITAT: **semi-arid and arid grassland, desert, plains**
SIZE: **body: 2.2–3.4 m (7¼–11 ft)**
    tail: 50 cm (19¾ in)

The dromedary, or one-humped camel, now exists only as a domesticated animal, which it has been, so it is thought, since 4000 B.C. Before then it probably lived in North Africa and Arabia. Today there are two main types: a heavily built, slow-moving animal, used as a beast of burden, and a light, graceful, fast-running racer, used for riding. Both have short, coarse hair, longest on the crown, neck, throat and hump. Dromedaries feed on grass and any other available plants and can survive in areas of sparse, tough vegetation.

Certain adaptations fit the dromedary for life in hot, dry climates; the most significant of these is its ability to go for long periods without drinking, linked with its ability to conserve water in the body. Its hump is its most important specialization, for it gives protection from the sun by absorbing heat and carries fat stores, which are metabolized to provide energy and water. The camel does not store water in the hump but can do so in the stomach lining. The kidneys are able to concentrate urine, thus avoiding water loss, and moisture can be absorbed from faecal material. Moreover, the body temperature of the camel drops at night and rises so slowly during the day that the animal does not need to sweat for a long time to cool itself. During an extended period without water, the camel is able to lose up to 27 per cent of its body weight without detrimental effect. This loss can be recovered in 10 minutes by drinking. In one experiment, a thirsty camel drank 104 litres (23 UK gal/27 US gal) in a few minutes.

Females breed only every other year. After a gestation of 365 to 440 days, the female moves away from the herd to give birth to her single young. When it is able to walk, after a day or so, the mother rejoins the herd with her young. Although the calf is suckled for almost a year, it starts to nibble plants as soon as it is born, and by 2 months old is regularly feeding on vegetation. Mother and young keep in touch by calling, the calf giving distress calls if it becomes separated from her.

DROMEDARY

GUANACO

VICUNA

BACTRIAN CAMEL

Chevrotains, Musk Deer, Deer I

TRAGULIDAE: Chevrotain Family

There are 4 species of chevrotain, or mouse deer, found in tropical forest and mangrove swamps in Africa and Asia. They are tiny, delicate creatures, which look like minute deer with mouselike heads but are probably related to both camels and pigs. They stand only 20 to 35 cm (7¾ to 13¾ in) high at the shoulder and weigh only 2.3 to 4.6 kg (5 to 10¼ lb). Active at night, they feed largely on plants and fruit.

NAME: **Water Chevrotain**, *Hyemoschus aquaticus*
RANGE: **Africa: Guinea to Cameroon, Zaire, Gabon, Central African Republic**
HABITAT: **forest, near water**
SIZE: **body: 75–85 cm (29½–33½ in)**
    tail: 10–15 cm (4–6 in)

About the size of a hare, with a hunched back, small head and short, slender legs, the water chevrotain has a variable pattern of white spots on its back and up to three white stripes along its flanks. It rests during the day in thick undergrowth or in a hole in a river bank and emerges at night to forage for grass, leaves and fruit, as well as some insects, crabs, fish, worms and small mammals. Water chevrotains are solitary except in the breeding season, each individual occupying its own territory. Chevrotains always live near water and are good swimmers; if danger threatens, they often escape by diving deeply.

At breeding time, the male simply finds the female by scent, and they mate without aggression. The female gives birth to 1 young after a gestation of about 4 months. The young is suckled for 8 months but takes some solid food at 2 weeks old.

NAME: **Lesser Malay Chevrotain,** *Tragulus javanicus*
RANGE: **S.E. Asia, Indonesia**
HABITAT: **lowland forest, usually near water**
SIZE: **body: 40–47 cm (15¾–18½ in)**
    tail: 5–8 cm (2–3¼ in)

The tiny, deerlike Malay chevrotain has a robust body on extremely slender legs. It has no horns, but in males, the canine teeth in the upper jaw are enlarged into tusks. A nocturnal creature, it inhabits the dense undergrowth, making little tunnel-like trails through which it moves; it feeds on grass, leaves, fallen fruit and berries. It lives alone except when breeding.

The female gives birth to 1 young after a gestation of about 5 months.

MOSCHIDAE: Musk Deer Family

The 3 species of musk deer, all in the genus *Moschus*, occur in central and northeastern Asia. Sometimes classified with the chevrotains or with the true deer, musk deer are in several respects intermediate between these two groups. They stand about 50 to 60 cm (19¾ to 23½ in) high at the shoulder and have no horns, but they do possess large tusks, formed from the upper canine teeth. The name musk deer comes from the waxy secretions produced by a gland on the abdomen of the male.

NAME: **Forest Musk Deer,** *Moschus chrysogaster*
RANGE: **Himalayas to central China**
HABITAT: **forest, brushland at 2,600–3,600 m (8,500–11,800 ft)**
SIZE: **body: about 1 m (3¼ ft)**
    tail: 4–5 cm (1½–2 in)

Long, thick, bristly hairs cover the body of the forest musk deer and help protect the animal from the often harsh weather conditions of its habitat. Male and female look more or less alike, but the male has larger tusks, developed from the upper canine teeth, and a gland on the abdomen, which secretes musk in the breeding season. Only mature males have these glands. Usually solitary, musk deer may occasionally gather in groups of up to 3. They are active in the morning and evening, feeding on grass, moss and shoots in summer and lichens, twigs and buds in winter.

At the onset of the breeding season, males fight to establish dominance and thus access to the largest number of females. They wrestle with their necks, trying to push one another to the ground, and may inflict deep wounds with their tusks. The female gives birth to a single young after a gestation of about 160 days.

CERVIDAE: Deer Family

There are about 34 species of true deer, distributed over North and South America, Europe, northwest Africa and Asia. Found in habitats ranging from the Arctic to the tropics, deer are slim, long-legged, elegant herbivores. Their most obvious characteristic is the pair of antlers, possessed by males of all species except the Chinese water deer. Most deer shed and regrow their antlers in an annual cycle, shedding them in late winter or early spring and growing them in summer, before the autumn rutting contests for dominance. The cycle is primarily under hormonal control and is influenced by changes in daylength.

NAME: **Chinese Water Deer,** *Hydropotes inermis*
RANGE: **China, Korea; introduced in England**
HABITAT: **river banks with reedbeds and rushes, grassland, fields**
SIZE: **body: 77.5 cm–1 m (30½ in–3¼ ft)**
    tail: 6–7.5 cm (2¼–3 in)

The only true deer to lack antlers in the male, the Chinese water deer has tusks, formed from enlarged upper canine teeth; these are larger in males than in females. Both male and female have small scent glands on each side of the groin and are the only deer to possess such glands. A nocturnal animal, this deer usually lives alone or in pairs and rarely gathers in herds. It feeds on reeds, coarse grass and other vegetation.

Males contest in fierce fights for dominance, in the rutting season before breeding. After a gestation of about 6 months, the female gives birth to 4 young — this is the largest litter produced by any deer.

NAME: **Chinese Muntjac,** *Muntiacus reevesi*
RANGE: **S.E. China, Taiwan; introduced in England and France**
HABITAT: **dense vegetation, hillsides; parkland in introduced range**
SIZE: **body: 80 cm–1 m (31½ in–3¼ ft)**
    tail: 11–18 cm (4¼–7 in)

The antlers of the male Chinese muntjac are small, rarely exceeding 15 cm (6 in) in length, but this deer also has tusks, formed from the upper canine teeth; females have smaller tusks than males. The Chinese muntjac lives in a territory, which it rarely leaves, and prefers to stay in cover of vegetation. It lives alone or in pairs and seldom forms herds. Primarily nocturnal, it may be active in the morning in quiet, undisturbed areas. It feeds on grass, low-growing leaves and shoots.

In dominance contests, males fight with their tusks, rather than their antlers, and make doglike barking noises. The female usually gives birth to 1 young after a gestation of about 6 months.

NAME: **Tufted Deer,** *Elaphodus cephalophus*
RANGE: **S. China, N. Burma**
HABITAT: **dense undergrowth, near water**
SIZE: **body: about 1.6 m (5¼ ft)**
    tail: 7–12 cm (2¾–4¾ in)

The male tufted deer is characterized by the tuft of hair on the forehead at the base of the antlers; the antlers themselves are short and often almost hidden by the tuft. A nocturnal, normally solitary deer, this species feeds on grass and other plant material.

The female gives birth to 1 young after a gestation of about 6 months.

FOREST MUSK DEER
(male)

WATER CHEVROTAIN

CHINESE WATER DEER
(male)

LESSER MALAY CHEVROTAIN

TUFTED DEER
(male)

CHINESE MUNTJAC

Deer 2

NAME: **Père David's Deer,** *Elaphurus davidianus*
RANGE: **originally China; now in wildlife parks and reintroduced in China**
HABITAT: **wildlife parks**
SIZE: **body: 1.5 m (5 ft)**
 tail: 50 cm (19¾ in)

This interesting deer became extinct in the wild in about 1920. However, at the beginning of this century, some of the few remaining specimens left in China were brought to England to live in the grounds of Woburn Abbey, in Bedfordshire, where they have thrived. Populations of Père David's deer can now be found in zoos and parks around the world, and they have been reintroduced in China.

Père David's deer has a mane of thick hair around its neck and throat and a rather longer tail than most deer. One tine of each antler usually points backward, while the other points upward and forks. Although they feed mainly on grass, the deer supplement their diet with water plants. For most of the year they live in herds led by a dominant male, but the male lives alone for 2 months before and 2 months after the rutting season.

During the rutting season, the male fights rivals to gain or retain dominance over a harem. Females give birth to 1 or 2 young after a gestation of about 250 days.

NAME: **White-tailed Deer,** *Odocoileus virginianus*
RANGE: **S. Canada, USA, Central and South America to Peru and Brazil**
HABITAT: **forest, swamps, open brushland**
SIZE: **body: 1.5–2 m (5–6½ ft)**
 tail: up to 28 cm (11 in)

One of the most adaptable animals in the world, the white-tailed deer is found from near-Arctic regions to the tropics. Its adaptability is reflected in its feeding habits: it browses and grazes on many kinds of grasses, weeds, shrubs, twigs, fungi, nuts and lichens. A slender, sprightly creature, the white-tailed deer has a long tail, white on its underside, a white band across its nose and a white patch on the throat. White-tailed deer are shy, elusive animals and do not usually congregate in large herds; in severe winter weather, however, they may congregate in a group in a sheltered spot, out of the wind.

It is not certain whether or not males are polygamous, but in the breeding season, they engage in savage battles over mates. The gestation period is 6½ to 7 months; young females usually produce only a single offspring, but older females may have litters of 2 or even 3. Young are able to walk right away and are suckled for about 4 months.

NAME: **Elk/Moose,** *Alces alces*
RANGE: **N. Europe and Asia: Scandinavia to Siberia; Alaska, Canada, N. USA; introduced in New Zealand**
HABITAT: **coniferous forest, often near lakes and rivers**
SIZE: **body: 2.5–3 m (8¼–9¾ ft)**
 tail: 5–7.5 cm (2–3 in)

The largest of the deer, the elk is identified by its size, its broad, overhanging muzzle and the flap of skin, known as the bell, hanging from its throat. The massive antlers of the male are flattened and palmate, with numerous small branches. The elk is less gregarious than other deer and is usually alone outside the breeding season. In winter, it feeds on woody plants, but in summer water plants provide the bulk of its food. It wades into water to feed and swims well.

Bellowing males display to attract females, and they engage in fierce contests with rivals. Following an 8-month gestation, the female gives birth to a single calf, very rarely to twins. The calf is suckled for about 6 months but stays with its mother for a year.

NAME: **Reindeer/Caribou,** *Rangifer tarandus*
RANGE: **N. Europe and Asia: Scandinavia to Siberia; Alaska, Canada, Greenland**
HABITAT: **tundra**
SIZE: **body: 1.2–2.2 m (4–7¼ ft)**
 tail: 10–21 cm (4–8¼ in)

Once divided into several species, all caribou and reindeer, including the domesticated reindeer, are now considered races of a single species. The races vary in coloration from almost black to brown, grey and almost white. The reindeer is the only deer in which both sexes have antlers, although those of the female are smaller. The antlers are unique in that the lowest, forward-pointing tine is itself branched. Females are gregarious and gather in herds with their young, but adult males are often solitary. Some populations migrate hundreds of miles between their breeding grounds on the tundra and winter feeding grounds farther south. Grass and other tundra plants are their main food in summer, but in winter reindeer feed mainly on lichens, scraping away the snow with their hoofs to expose the plants.

In autumn, males fight to gather harems of 5 to 40 or so females. The female produces 1, occasionally 2, young after a gestation of about 240 days. Young reindeer are able to run with the herd within a few hours of birth.

NAME: **Red Deer/Wapiti,** *Cervus elaphus* **(conspecific with** *C. canadensis***)**
RANGE: **W. Europe, N.W. Africa, Asia to W. China, N.W. America; introduced in New Zealand**
HABITAT: **open deciduous woodland, mountains, plains, moorland**
SIZE: **body: 1.6–2.5 m (5¼–8¼ ft)**
 tail: 12–15 cm (4¾–6 in)

Known as the wapiti in North America and the red deer in Britain, this deer is reddish-brown in summer but greyish-brown in winter. Most older males have antlers with two forward-pointing tines near the base, while young males usually have one tine. In autumn and winter, the male has a mane of longer hair on the neck. A gregarious species, red deer live in herds and are active in the morning and late afternoon or evening, feeding on grass, heather, leaves and buds.

In the autumn, males take part in fierce, antler-clashing fights in order to obtain territories and harems. They defend their females throughout the breeding season and then return to all-male herds in the winter. Females give birth to 1 calf, rarely 2, after a gestation of about 8 months. The young deer is able to walk a few minutes after birth.

NAME: **Roe Deer,** *Capreolus capreolus*
RANGE: **Europe and Asia: Britain to S.E. Siberia, S. China**
HABITAT: **woodland**
SIZE: **body: 95 cm–1.3 m (37½ in–4¼ ft)**
 tail: 2–4 cm (¾–1½ in)

The smallest of the native European deer, the roe deer is unique in having almost no tail. It has a pale rump, and the rest of its coat is reddish-brown in summer and greyish-brown in winter; fawns are spotted. The antlers of the male never have more than three points apiece. These shy, graceful deer are generally solitary, except in the breeding season, but may gather in small groups in the winter. Active at night, they browse on shrubs and broad-leafed trees.

In the breeding season, the male takes a territory and marks its boundaries by rubbing the trunks of trees with his antlers until the bark is frayed and the wood exposed. He has only 1 mate and defends her and his territory against rivals. The period until birth is 9 or 10 months, which is much longer than that of most deer and includes a period of delayed implantation — once the egg is fertilized, it lies dormant in the uterus for about 4½ months before true gestation begins. This ensures that both mating and birth take place at the optimum time. Before giving birth to her 1 or 2 young, the female roe deer chases away her offspring of the previous year.

ELK (male)

RED DEER (male)

WHITE-TAILED DEER
(male)

REINDEER (male)

ROE DEER (male)

PERE DAVID'S DEER (male)

Deer 3
Giraffes, Pronghorn

NAME: **Pampas Deer,** *Ozotoceros*
   bezoarticus
RANGE: **South America: Brazil,**
   Paraguay, Uruguay, N. Argentina
HABITAT: **grassland, open plains**
SIZE: **body: 1.1–1.3 m (3½–4¼ ft)**
   tail: 10–15 cm (4–6 in)   Ⓔ

The slender, long-legged pampas deer once lived only among tall pampas grass, but now that much of this land has been turned over to grain, the deer may even frequent wooded country. As with most deer, only the male has antlers; the front prong of these is not divided, while the hind portion branches once or twice. Males also have glands in the feet which give off a strong, garlicky smell, noticeable more than 1 km (0.6 ml) away. In winter, pampas deer live alone or in pairs, but in spring they may form larger groups. They rest in cover during the day and emerge in the evening to feed on grass. Some races of this deer are now extremely rare, due to uncontrolled hunting and the loss of their habitat to agriculture.

Unlike most deer, the male pampas deer stays with the female after her single offspring is born and helps her to guard it from predators.

NAME: **Northern Pudu,** *Pudu*
   mephistophiles
RANGE: **South America: Colombia to**
   N. Peru
HABITAT: **forest, swampy savanna at**
   2,000–4,000 m (6,600–13,000 ft)
SIZE: **body: 65 cm (25½ in)**
   tail: 2.5–3.5 cm (1–1¼ in)   Ⓘ

The smallest of the native New World deer, the northern pudu has short, delicate legs, a rounded back and small, simple antlers. Its dark-brown hair is thick and dense. A shy, inconspicuous creature, little is known of its habits, and its high, remote habitat has made it hard to observe. It is thought to live in small groups or alone and to feed on leaves, shoots and fruit.

Females produce a single young, sometimes twins, normally between November and January.

GIRAFFIDAE: Giraffe Family
The giraffe family is an interesting and specialized offshoot of the deer family. It appears to have originated in the Old World and is now reduced to only 2 species: the giraffe itself and the okapi, both found in Africa.

Both animals have unique, skin-covered, blunt horns, which are never shed. The giraffe has an extraordinarily elongate neck and legs, which make it the tallest terrestrial animal.

NAME: **Giraffe,** *Giraffa camelopardalis*
RANGE: **Africa, south of the Sahara**
HABITAT: **savanna**
SIZE: **body: 3–4 m (9¾–13 ft)**
   tail: 90 cm–1.1 m (35½ in–3½ ft)

The giraffe, with its long legs and its amazingly long neck, when erect stands up to 3.3 m (11 ft) at the shoulder and nearly 6 m (19½ ft) at the crown. Its characteristic coloration of a light body and irregular dark spots is very variable, both geographically and between individuals; some animals may be almost white or black, or even unspotted. Both male and female have skin-covered horns, one pair on the forehead and sometimes a smaller pair farther back, on the crown. Some animals have yet another small horn, or bump, between these pairs. The tail ends in a tuft of long hairs.

Gregarious animals, giraffes usually live in troops of up to 6, sometimes 12, and may occasionally gather in larger herds. A troop consists of females and their offspring, led by a male. Males fight for possession of females, wrestling with their heads and necks. The troop ambles around its territory, feeding mostly in the early morning and afternoon on the foliage, buds and fruits on the top of acacia and thorn trees. They may also eat grass, plants and grain crops. At midday, giraffes rest in shade and at night lie down for a couple of hours or rest standing.

Females give birth to a single offspring, rarely twins, after a gestation of over a year — usually 400 to 468 days. Births invariably occur at first light. The young is suckled for 6 to 12 months and continues to grow for 10 years.

NAME: **Okapi,** *Okapia johnstoni*
RANGE: **Zaire**
HABITAT: **rain forest**
SIZE: **body: 1.2–2 m (4–6½ ft)**
   tail: 30–42 cm (11¾–16½ in)

An inhabitant of dense forest, the okapi, although long hunted by the local pygmy tribes, was only made known to the outside world in 1901, when it was discovered by the then Governor of Uganda. He thought it was related to the zebra because of its stripes, but, in fact, it bears a remarkable resemblance to primitive ancestors of the giraffe, known only from fossils. The okapi has a compact body, which slopes down toward the hindquarters, and distinctive stripes on its legs. Only males possess short, skin-covered horns, similar to those of the giraffe. The tongue is so long that the okapi can use it to clean its own eyes and eyelids.

Okapis live alone, each in its own home range, and meet only in the breeding season. They feed on leaves, buds and shoots of trees, which they can reach with their long tongues, and on grass, ferns, fruit, fungi and manioc.

Pairing usually takes place between May and June or November and December but may occur at any time. The female gives birth to 1 young after a gestation of 421 to 457 days. The young okapi suckles for up to 10 months and is not fully developed until 4 or 5 years of age.

ANTILOCAPRIDAE: Pronghorn Family
The North American pronghorn, found in Canada, USA and northern Mexico, is the sole living representative of a New World group of antelopelike ruminants. There is also, however, a body of opinion that suggests this animal should be included in the cattle family, Bovidae, but it is kept apart on account of its curious horn structure.

NAME: **Pronghorn,** *Antilocapra*
   americana
RANGE: **central Canada, W. USA, Mexico**
HABITAT: **open prairie, desert**
SIZE: **body: 1–1.5 m (3¼–5 ft)**
   tail: 7.5–10 cm (3–4 in)   Ⓔ

Both male and female pronghorns have true, bony horns, although those of females are small and inconspicuous. The horns are covered with sheaths of specialized, fused hairs, and pronghorns are unique in that these sheaths are shed annually. The small, forward-pointing branch on each horn, the prong, is in fact part of this sheath.

One of the fastest running mammals in North America, the pronghorn can achieve speeds of up to 65 km/h (40 mph). It is also a good swimmer. In summer, it moves in small, scattered groups but congregates in larger herds of up to 100 animals in winter. Pronghorns are active during the day, but feed mostly in the morning and evening, taking grasses, weeds and shrubs such as sagebrush. White hairs on the pronghorn's rump become erect if the animal is alarmed and act as a warning signal to other pronghorns.

Some males collect harems, fighting rival males for the privilege. The female gives birth to her young after a gestation of 230 to 240 days; there is usually only 1 in a female's first litter, but in subsequent years, she produces 2, or even 3, young. Only 4 days after birth, pronghorns can outrun man.

Pronghorns are now rare, due to overhunting, competition for food from domestic livestock and the destruction of their natural habitat.

NORTHERN PUDU (male)

OKAPI (male)

PRONGHORN (male)

GIRAFFE

PAMPAS DEER (male)

Bovids 1

NAME: **Greater Kudu,** *Tragelaphus*
    strepsiceros
RANGE: **Africa: Lake Chad to Eritrea,**
    Tanzania; Zambia to Angola and
    South Africa; introduced in N. Mexico
HABITAT: **thick acacia bush; rocky, hilly**
    country; dry river beds, near water.
SIZE: **body: 1.8–2.45 m (6–8 ft)**
    tail: 35–55 cm (13¾–21½ in)

The most elegant of antelope, the large,
slender male kudu has long horns,
which spread widely in two to three
open spirals; the female occasionally has
small horns. When running, the bull
lays his horns flat along his back. Over
the kudu's wide range, there are variations in coloration and the number of
stripes on the sides. Kudu are browsers,
feeding early and late on leaves, shoots
and seeds and, in dry areas, wild
melons. They also make night raids on
cultivated fields and can jump over a
2 m (6½ ft) fence. Their senses of hearing
and smell are good, although their sight
seems poor.

Kudu live mostly in small herds of 6
to 12 females with young, sometimes
with 1 or 2 older bulls. Otherwise males
are solitary or form bachelor herds.
After a gestation of about 7 months, the
female produces 1 calf, which suckles
for about 6 months.

NAME: **Eland,** *Tragelaphus oryx*
RANGE: **Africa: Ethiopia, E. Africa to**
    Angola and South Africa; mostly in
    game parks in Namibia, N. Cape
    Province, Natal, Mozambique
HABITAT: **open plains, savanna, mopane**
    bush, montane forest to 4,500 m
    (14,750 ft), semi-desert
SIZE: **body: 2.1–3.5 m (6¾–11½ ft)**
    tail: 50–90 cm (19¾–35½ in)

The size of an ox, the eland is the largest
of the antelopes, and a fully grown bull
may weigh as much as 900 kg (2,000 lb).
The cow is smaller and more slightly
built, with lighter horns and no mat of
hair on the forehead. Eland live in
troops of 6 to 24 animals and are always
on the move, depending on the
availability of food and water. In times
of drought, they wander widely and
form large herds. Old solitary bulls are
common; young bulls form single-sex
troops. Eland are browsers, feeding in
the morning and at dusk, and even on
moonlit nights, on leaves, shoots,
melons, tubers, bulbs, onions and
thick-leaved plants. They have a good
sense of smell and eyesight and will
move upwind of danger.

There is usually 1 calf, born after a
gestation of 8½ to 9 months; it lies hidden for a week, then follows the female,
who suckles it for about 6 months.

BOVIDAE: Bovid Family

This biologically and economically important family of herbivorous ungulates
contains about 123 species, of which
domesticated cattle, sheep and goats
must be the best-known members. The
family probably originated in Eurasia
and moved only recently to North
America; it is absent from South
America and most diverse in Africa.

Over their wide range, bovids utilize
almost all types of habitat, from grassland, desert and tundra to dense forest.
Coupled with this diversity of habitat is
great diversity of body form and size,
and bovids range between buffaloes and
tiny antelope. There are, however,
common features within the group.
Fore and hind toes are reduced to split,
or artiodactyl (even-toed), hoofs, based
on digits 3 and 4. There is a complex
four-chambered stomach in which
vegetable food is degraded by micro-
organism symbiosis. Linked with this
digestive system, bovids chew the cud,
bringing up food from the first stomach
and rechewing it. Normally both male
and female have defensive hollow
horns, which are larger in the male.

NAME: **Bongo,** *Tragelaphus euryceros*
RANGE: **Africa: Sierra Leone to Sudan**
    (not Nigeria), Kenya, Tanzania
HABITAT: **forest, bush, bamboo jungle**
SIZE: **body: 1.7–2.5 m (5½–8¼ ft)**
    tail: 45–65 cm (17¾–25½ in)

The adult male bongo is the largest of
the forest-dwelling antelopes and may
weigh up to 227 kg (500 lb); the chestnut-
coloured coat darkens with age in the
male. Both sexes have narrow, lyre-
shaped horns, which they lay along their
slightly humped backs when running, to
prevent them catching up in branches.
Shy animals, bongos rest up in dense
cover during the day, browsing at dawn
and dusk on leaves, shoots, bark, rotten
wood and fruit; they also dig for roots
with their horns. At night they will venture into clearings and plantations to
feed on grass.

They live in pairs or small groups of
females and young with a single male;
old males are solitary. One young is
born after a gestation of 9½ months.

NAME: **Nyala,** *Tragelaphus angasi*
RANGE: **Africa: Malawi to South Africa:**
    Natal
HABITAT: **dense lowland forest, thickets**
    in savanna, near water
SIZE: **body: 1.35–2 m (4½–6½ ft)**
    tail: 40–55 cm (15¾–21½ in)

These antelope live in the densest cover,
emerging only at dusk and dawn. The
males are large and slenderly built, with
big ears and shaggy coats. Females and
juveniles are reddish-brown and lack
the long fringe of hair underneath the
body, the horns and the white facial
chevron that distinguish the male;
females are much smaller than males.
Nyala live in parties of 8 to 16 cows and
young, alone or with one or more bulls.
Solitary bulls and parties of bulls are
also found, and toward the end of the
dry season, herds of up to 50 animals
may form. They browse on leaves,
shoots, bark and fruit of trees, standing
on their hind legs to reach high leaves.
They also eat new, tender grass.

The single young is born after a gestation of 8½ months; females mate again
a week after the birth.

NAME: **Nilgai,** *Boselaphus tragocamelus*
RANGE: **peninsular India (not Sri Lanka)**
HABITAT: **forest, low jungle**
SIZE: **body: 2–2.1 m (6½–6¾ ft)**
    tail: 46–54 cm (18–21¼ in)

The nilgai is the only member of its
genus and is the largest antelope native
to India. It has slightly longer front legs
than hind ones and a long, pointed head.
The male has short horns and a tuft of
hair on the throat; both sexes have short
wiry coats, reddish-brown in the male
and lighter in the female. Females and
calves live in herds; males are usually
solitary or form small parties. Nilgai are
browsers but also like fruit and sugar-
cane and can do considerable damage to
the crop.

Females commonly produce 2 calves
after a gestation of about 9 months.
Bulls fight each other on their knees for
available females, which mate again
immediately after calving.

NAME: **Four-horned Antelope,** *Tetracerus*
    quadricornis
RANGE: **peninsular India (not Sri Lanka)**
HABITAT: **open forest**
SIZE: **body: 1 m (3¼ ft)**
    tail: 12.5 cm (5 in)

This little antelope is the only one in its
genus. The male is unique among
Bovidae in having two pairs of short,
unringed, conical horns: the back pair 8
to 10 cm (3¼ to 4 in) long, the front pair
2.5 to 4 cm (1 to 1½ in) long; these may
be merely black, hairless skin. These are
not gregarious antelope — normally
only two are found together, or a female
with her young. They graze on grasses
and plants and drink often, running for
cover at the least hint of danger with a
peculiar, jerky motion.

Four-horned antelope mate during
the rainy season and usually produce 1
to 3 young after a gestation of about 6
months.

NYALA (male)

BONGO

NILGAI (male)

FOUR-HORNED ANTELOPE (male)

GREATER KUDU (male)

ELAND (male)

Bovids 2

NAME: **Gaur, *Bos gaurus***
RANGE: **India, S.E. Asia**
HABITAT: **hill forest**
SIZE: **body: 2–2.5 m (6½–8¼ ft)**
    tail: 60–80 cm (23½–31½ in) Ⓥ

Once common in hilly, forested areas throughout their range, gaur now only occur in scattered herds in remote areas and in parks and reserves. Gaur are legally protected, but this is hard to enforce except in reserves, and the population is still threatened.

The gaur is a strong, heavily built animal, with a massive head, thick horns and a prominent muscular ridge on its shoulders that slopes down to the middle of its back. Females are smaller than males and have shorter, lighter horns. Gaur range in colour from reddish to dark brown or almost black, with white hair on the lower half of the legs. In small herds of up to 12 animals, they take shelter in the shade and seclusion of forest in the heat of the day and at night, but they venture into the open to feed in the early morning and late afternoon, when they graze and also sometimes browse on the leaves and bark of trees.

During the breeding season, the timing of which varies from area to area, bulls roam through the forest searching for females on heat. When a male finds a mate, he defends her from other males. The female moves slightly away from the herd to give birth to her offspring in a safe, secluded spot; they rejoin the herd a few days later.

NAME: **Banteng, *Bos javanicus***
RANGE: **Bali, Burma to Java, Borneo**
HABITAT: **forested, hilly country to 2,000 m (6,600 ft)**
SIZE: **body: 2 m (6½ ft)**
    tail: 85 cm (33½ in) Ⓥ

The banteng is blue-black, with white stockings and rump, and is quite cow-like in its appearance; females and young are a bright reddish-brown. Bulls may reach 1.5 m (5 ft) at the shoulder, and they have a hairless shield on the crown between the horns. Wary and shy, bantengs are found in thickly forested areas, where there are glades and clearings in which they can graze during the night. In the monsoon season, they move up the mountains and browse on bamboo shoots.

Gregarious animals, bantengs live in herds of 10 to 30 animals, although occasionally large bulls may become solitary. They mate during the dry season, and females produce 1 or 2 calves after a gestation of 9½ to 10 months. Small populations of two subspecies are found: *B. j. biarmicus* in Burma, Thailand and parts of Indo-China; *B. j. lowi* in Borneo.

NAME: **Water Buffalo, *Bubalus arnee***
RANGE: **India, S.E. Asia; introduced in Europe, Africa, Philippines, Japan, Hawaii, Central and South America, Australia**
HABITAT: **dense growth, reed grass in wet areas**
SIZE: **body: 2.5–3 m (8¼–9¾ ft)**
    tail: 60 cm–1 m (23½ in–3¼ ft)

A large, thickset, clumsy creature, with huge splayed hoofs, the water buffalo stands 1.5 to 1.8 m (5 to 6 ft) at the shoulder. It has a long, narrow face, and the span of its flattened, crescent-shaped horns is the largest of all bovids — they can measure as much as 1.2 m (4 ft) along the outer edge. Its bulky body is sparsely covered in quite long, coarse, blackish hair, and there is a tuft of coarse hair in the middle of the forehead. Water buffaloes feed early and late in the day and at night on the lush grass and vegetation that grows near and in lakes and rivers. When not feeding, they spend much of their time submerged, with only their muzzles showing above water, or wallowing in mud, which, when dried and caked, gives them some protection from the insects that plague them.

Water buffaloes are gregarious and live in herds of various sizes. In the breeding season, males detach a few cows from the main herd and form their own harems. Each cow produces 1 or 2 calves after a gestation of 10 months, which it suckles for almost a year. Water buffaloes live for about 18 years.

Tame and docile, these animals have been domesticated and used as beasts of burden in India and Southeast Asia since about 3000 B.C. They also yield milk of good quality and their hides make excellent leather. It is estimated that the domestic population in India and Southeast Asia alone is now at least 75 million, and water buffaloes have been widely introduced in countries where conditions are suitable. Some of these introduced populations have become feral, as in Australia. Truly wild stocks number no more than 2,000.

NAME: **Anoa, *Bubalus depressicornis***
RANGE: **Sulawesi**
HABITAT: **lowland forest**
SIZE: **body: 1.6–1.7 m (5¼–5½ ft)**
    tail: 18–31 cm (7–12¼ in) Ⓔ

The anoa is the smallest of the buffaloes, an adult male standing only 69 to 106 cm (27 to 42 in) at the shoulder. However, it is stockily built, with a thick neck and short, heavy horns, which are at most 38 cm (15 in) long. Although wary, the anoa is aggressive when cornered. Juveniles have thick, woolly, yellow-brown hair, which becomes dark brown or blackish, blotched with white, in adults; old animals may have almost bare skins. Perhaps because of this, anoas appear to enjoy bathing and wallowing in mud. They feed alone during the morning, mainly on water plants and young cane shoots, then spend the rest of the day lying in the shade, generally in pairs. They only form herds just before the females are due to calve. Usually 1 young is born after a gestation of 9½ to 10 months.

When unmolested, anoas have a life span of 20 to 25 years, but destruction of their normal habitat has driven them into inaccessible, swampy forest, and their survival is further threatened by unrelenting hunting, for their horns, meat and thick hides.

NAME: **Yak, *Bos mutus***
RANGE: **W. China, Tibetan plateau, N. India, Kashmir**
HABITAT: **desolate mountain country to 6,100 m (20,000 ft)**
SIZE: **body: up to 3.25 m (10½ ft) (male)**
    tail: 50–80 cm (19¾–31½ in) Ⓔ

Originally these massive animals were found throughout their range, but centuries of hunting and persecution have forced them to retreat into areas of mountain tundra and ice desert, and now they cannot live in warm, lowland areas. Sturdy and sure-footed and covered in long, blackish-brown hairs, which form a fringe reaching almost to the ground, they are, however, well equipped to cope with the rigours of terrain and climate in their habitat. Males may stand up to 2 m (6½ ft) at the shoulder, females are smaller and weigh only about one-third as much as males. All have heavy, forward-curving horns, which they use for defence; when threatened, they form a phalanx, facing outward with horns lowered, with the calves encircled for protection.

Yaks feed morning and evening on whatever vegetation they can find, spending the rest of the time relaxing and chewing the cud. They are usually found in large groups consisting of females and young with a single bull; bachelor bulls roam in groups of 2 or 3. The female produces 1 calf in the autumn, after a gestation of 9½ to 10 months.

Although wild yaks are an endangered species, they have been domesticated for centuries in Tibet, where, as well as being used as pack animals and to pull carts, they provide milk, meat, hair and wool for cloth, and hides. Domestic yaks are usually about half the size of wild ones and are often without horns. Their coats are redder, mottled with brown, black and sometimes white.

BANTENG (male)

ANOA (male)

GAUR (male)

WATER BUFFALO
(male)

YAK (male)

Bovids 3

NAME: **American Bison,** *Bison bison*
RANGE: **N. America**
HABITAT: **prairie, open woodland**
SIZE: **body: 2.1–3.5 m (6¾–11½ ft)**
    tail: 50–60 cm (19¾–23½ in)

Although there were once millions of bison roaming the North American grasslands, wholesale slaughter by the early European settlers brought them almost to extinction by the beginning of the twentieth century. Since then, due largely to the efforts of the American Bison Society, herds have steadily been built up in reserves, where they live in a semi-wild state, and it is estimated that there are now some 20,000 animals. The male may be as much as 2.9 m (9½ ft) at the shoulders, which are humped and covered in the shaggy, brownish-black fur that also grows thickly on the head, neck and forelegs. The female looks similar to the male but is smaller; young are more reddish-brown. Both sexes have short, sharp horns.

Primarily grazers, bison live in herds that vary from a family group to several thousand; huge numbers formerly made seasonal migrations in search of better pasture. They feed morning and evening, and during the day rest up, chewing the cud or wallowing in mud or dust baths to rid themselves of parasites.

During the mating season, bulls fight for cows, which give birth to a single calf, away from the herd, after a gestation of 9 months. Within an hour or two, mother and calf rejoin the herd. The calf is suckled for about a year and remains with its mother until it reaches sexual maturity at about 3 years old.

NAME: **European Bison,** *Bison bonasus*
RANGE: **Poland, USSR**
HABITAT: **open woodland, forest**
SIZE: **body: 2.1–3.5 m (6¾–11½ ft)**
    tail: 50–60 cm (19¾–23½ in)

Like its American counterpart, the European bison, which was formerly found throughout Europe, has been reduced to semi-wild herds in reserves: three in Poland and eleven in USSR, with the largest in the Bialowieza Forest on the border between them. The drop in numbers has been caused by the almost total eradication of forests, for these bison are browsers, living mainly on leaves, ferns, twigs, bark and, in autumn, almost exclusively on acorns.

The European bison closely resembles the American, but is less heavily built, with longer hind legs. It has more scanty, shorter hair on the front of the body and head, and the horns, too, are lighter and much longer, reaching as much as 51 cm (20 in) in the male. The female produces a single calf after a gestation of 9 months, and it remains with her until it is 2 to 3 years old.

NAME: **African Buffalo,** *Synceros caffer*
RANGE: **Africa, south of the Sahara**
HABITAT: **varied, always near water**
SIZE: **body: 2.1–3 m (6¾–9¾ ft)**
    tail: 75 cm–1.1 m (29½ in–3½ ft)

The powerfully built African buffalo is the only member of its genus, although 2 types exist, the smaller, reddish, forest-dwelling buffalo, *S. c. nanus*, and *S. c. caffer*, described here, which lives in savanna and open country. It has a huge head with a broad, moist muzzle, large drooping ears and heavy horns, the bases of which may meet across the forehead. An aggressive animal and a formidable fighter, it is extremely dangerous to hunt, since it may charge without provocation or, if wounded, wait in thick bush and attack a pursuing hunter. Apart from man, its enemies are the lion and occasionally the crocodile, both of which usually succeed in killing only young or sick animals.

African buffaloes have adapted to live in a variety of conditions, from forest to semi-desert, wherever there is adequate grazing and plenty of water, for they drink morning and evening and enjoy lying in water and wallowing in mud. They feed mainly at night, on grass, bushes and leaves, resting up in dense cover during the day. Although their eyesight and hearing are poor, they have a strongly developed sense of smell. Buffaloes are gregarious, living in herds which range from a dozen or so to several hundred animals, often led by an old female but dominated by a mature bull. Old bulls are ousted from the herd and live alone in groups of 2 to 5.

Although normally silent, buffaloes bellow and grunt during the mating season that varies throughout the range and appears to be related to climate. A single calf is born after a gestation of 11 months; it is covered in long, blackish-brown hair, most of which is lost as it matures. African buffaloes live for about 16 years.

NAME: **Bay Duiker,** *Cephalophus dorsalis*
RANGE: **Africa: Sierra Leone to E. Zaire**
    and N. Angola
HABITAT: **thick forest and jungle**
SIZE: **body: 70 cm–1 m (27½ in–3¼ ft)**
    tail: 8–15 cm (3–6 in)

The subfamily Cephalophinae contains two groups: the forest duikers, of which the bay duiker is one, and the bush duikers. The bay duiker is typical of its group, with rather slender legs, a slightly hunched back and a smooth glossy coat. Both male and female have small, backward-pointing horns, which are sometimes obscured by the crest of hairs on the forehead. Duikers are timid and when disturbed dash for thick cover — the name duiker means "diving"

buck. They are mainly active at night, when they feed on grass, leaves and fruit, even scrambling up into bushes or on logs to reach them.

Bay duikers live singly or in pairs and produce 1 young after a gestation of 7 to 8 months. The young is independent at about 3 months old.

NAME: **Yellow Duiker,** *Cephalophus*
    sylvicultor
RANGE: **Africa: Senegal to Kenya,**
    Zambia, N. Angola
HABITAT: **moist highland forest**
SIZE: **body: 1.15–1.45 m (3¾–4¾ ft)**
    tail: 11–20.5 cm (4¼–8 in)

Another forest duiker and the largest of its subfamily, the yellow duiker is remarkable for the well-developed crest of hairs on its forehead and for the yellowish-orange patch of coarse, erectile hairs that grow in a wedge shape on its back. Both sexes have long, thin, sharp-pointed horns. The young loses its dark coloration at about 8 months.

Yellow duikers live in pairs or alone, keeping to thick cover. They are active at night and have a varied diet that includes leaves, grass, herbs, berries, termites, snakes, eggs and carrion. They are hunted by man for their meat and have many other enemies, ranging from leopards and jackal to pythons and large birds of prey.

NAME: **Common/Grey Duiker,**
    Sylvicapra grimmia
RANGE: **Africa, south of the Sahara**
HABITAT: **all types except desert and rain**
    forest, up to 4,600 m (15,000 ft)
SIZE: **body: 80 cm–1.15 m (31½ in–3¾ ft)**
    tail: 10–22 cm (4–8½ in)

The common bush duiker is slightly different from the forest duikers, with its straighter back and thicker, grizzled coat. The crest is quite well developed, and the male has sharp horns, which the female does not always have. Common duikers are adaptable and can survive in almost any habitat from scrub country to open grassland, even invading cultivated lands. The male establishes a fiercely defended territory, in which the animals have regular runs, defecating and resting places. They browse at night on leaves and twigs, which they will stand on their hind legs to reach. They also eat fruit, berries, termites, snakes, eggs and, especially, guineafowl chicks.

Usually found alone or in pairs, they may form small groups in the breeding season, which varies throughout the range and appears to be linked to the rains. The female produces 1 young after a gestation of 4 to 4½ months; normally 2 young are born each year.

YELLOW DUIKER

AMERICAN BISON (male)

COMMON DUIKER (male)

EUROPEAN BISON (female)

AFRICAN BUFFALO (male)

BAY DUIKER

Bovids 4

NAME: **Lechwe, *Kobus leche***
RANGE: **Africa: Zaire, Zambia, Angola, Botswana, South Africa**
HABITAT: **flood plains, swamps, lakes**
SIZE: **body: 1.3–1.7 m ($4\frac{1}{4}$–$5\frac{1}{2}$ ft)**
tail: 30–45 cm ($11\frac{3}{4}$–$17\frac{3}{4}$ in) Ⓥ

There are 3 races of lechwe, whose coloration varies from bright chestnut to greyish-brown. In all races, the male has thin, lyre-shaped horns, up to 91 cm (3 ft) long, which form a double curve; they are particularly fine in *K. l. kafuensis.* With their long, pointed, wide-spreading hoofs, lechwe are perfectly adapted to an aquatic way of life and cannot move quickly on dry ground. They come out of the water only to rest and calve, spending most of their time wading in water up to about 50 cm (20 in) deep, where they feed on grasses and water plants. They swim well and will even submerge, with only the nostrils showing, if threatened. Apart from man, they are preyed on mainly by lion, cheetah, hyena and hunting dogs.

Lechwe are sociable and form herds of several hundred during the breeding season, when males fight fiercely, even though they are not territorial. At other times, young males form large, single-sex herds. After a gestation of 7 to 8 months, the female produces a single calf, which suckles for 3 to 4 months.

NAME: **Uganda Kob, *Kobus kob thomasi***
RANGE: **Africa: Uganda to south of Lake Victoria**
HABITAT: **open grassy plains, lightly wooded savanna, near permanent water**
SIZE: **body: 1.2–1.8 m (4–6 ft)**
tail: 18–40 cm (7–$15\frac{3}{4}$ in)

A graceful, sturdily built, medium-sized antelope, the Uganda kob is a subspecies of the nominate race, which it closely resembles except that the white on the face completely encircles the eyes. Male and female look alike, but only the male has horns, which are lyre shaped, with an S-curve when seen from the side. Kob usually live in single sex herds of 20 to 40, sometimes up to 100. They are grazers, feeding usually in the morning and at dusk, although during the day they will go into the water and eat water plants. They are preyed on mainly by lion, leopard, spotted hyena and hunting dogs. Solitary animals often lie flat and hide when threatened.

In the breeding season, each rutting male has an area 9 to 15 m (30 to 50 ft) in diameter, which he defends against other males. Females move freely through these rutting areas, mating with several males. One young is born after a gestation of $8\frac{1}{2}$ to 9 months, and, since the female mates again almost at once, two births are possible in a year.

NAME: **Common/Defassa Waterbuck, *Kobus ellipsiprymnus***
RANGE: **Africa, south of the Sahara to the Zambezi, east to Ethiopia**
HABITAT: **savanna, woodland, stony hills, near water**
SIZE: **body: 1.8–2.2 m (6–$7\frac{1}{4}$ ft)**
tail: 22–45 cm ($8\frac{1}{2}$–$17\frac{3}{4}$ in)

The many races of this waterbuck vary in coloration from yellowish- or reddish-brown to grey and greyish-black; some have a white ring or patch on the rump. It has large, hairy ears, which are white inside and tipped with black, and the male has heavy, much-ringed horns, which sweep back in a crescent shape. A large animal, standing 1.2 to 1.4 m (4 to $4\frac{1}{2}$ ft) at the shoulder, it weighs 159 to 227 kg (350 to 500 lb), but glands in the skin exude a musky-smelling oily secretion, and the meat is easily tainted when the animal is skinned, so it is not much hunted. The chief predators are lion, leopard and hunting dogs. True to their name, waterbuck spend much time near water and drink often; they take refuge in reedbeds when threatened. They are grazers, feeding on tender young grass shoots.

Common waterbuck move in small herds of up to 25, usually females and young with a master bull; young bulls form bachelor herds. The female produces 1 young after a gestation of about 9 months.

NAME: **Reedbuck, *Redunca arundinum***
RANGE: **Africa: Zaire, Tanzania, south to South Africa**
HABITAT: **open plains, hilly country with light cover, near water**
SIZE: **body: 1.2–1.4 m (4–$4\frac{1}{2}$ ft)**
tail: 18–30 cm (7–$11\frac{3}{4}$ in)

A medium-sized antelope — about 91 cm (3 ft) at the shoulder — the common reedbuck is a graceful animal, with distinctive movements. It runs with a rocking motion, flicking its thick, hairy tail, and the male marks and defends his territory by displaying his white throat patch and making bouncing leaps with his head raised. Reedbuck also make a characteristic clicking sound when running and whistle through their noses when alarmed or on the defence. The female resembles the male but is smaller and lacks his ridged, curved horns; juveniles are a greyish-brown. As their name suggests, reedbuck are always found near water, although not in it, and they spend much time lying up in reedbeds or tall grass. They graze on grass and shoots and will raid crops.

Reedbuck are usually found alone or in pairs or small family groups. A single young is born after a gestation of $7\frac{3}{4}$ months and reaches maturity, acquiring adult coloration, at about a year old.

NAME: **Roan Antelope, *Hippotragus equinus***
RANGE: **Africa, south of the Sahara**
HABITAT: **open woodland, dry bush, savanna, near water**
SIZE: **body: 2.4–2.6 m (8–$8\frac{1}{2}$ ft)**
tail: 60–70 cm ($23\frac{1}{2}$–$27\frac{1}{2}$ in)

There are approximately 6 races of roan antelope, which vary in coloration from grey to reddish-brown. The roan is a large antelope, the largest in Africa after the eland and kudu, and as the name suggests, it superficially resembles a horse, with its long face and stiff, well-developed mane. The male's backward-curving horns are short but strong; the female's are lighter. Roan antelope usually live in herds of up to 20 females and young, led by a master bull, often alongside oryx, impala, wildebeest, buffalo, zebra and ostriches. Young males form bachelor herds. They are preyed on mainly by lion, leopard, hunting dogs and hyena. At least 90 per cent of the roan's food is grass, and they rarely eat leaves or fruit, so need to drink often.

Roan antelope are aggressive, and males will fight on their knees with vicious, backward sweeps of their horns. In the breeding season, the bull detaches a cow from the herd and they live alone for a while. The female produces 1 calf after a gestation of $8\frac{1}{2}$ to 9 months; it attains sexual maturity at $2\frac{1}{2}$ to 3 years old.

NAME: **Rhebok, *Pelea capreolus***
RANGE: **South Africa**
HABITAT: **grassy hills and plateaux with low bush and scattered trees**
SIZE: **body: 1–1.2 m ($3\frac{1}{4}$–4 ft)**
tail: 10–20 cm (4–$7\frac{3}{4}$ in)

A small, graceful antelope, weighing 22.5 kg (50 lb) at most, the rhebok is covered in soft, woolly hair. The male has upright, almost straight horns, 15 to 27 cm (6 to $10\frac{1}{2}$ in) long. Rhebok feed on grass and the leaves of shrubs and are very wary, bouncing off the moment they are disturbed, with a run that jerks up their hindquarters. They are found in family parties, consisting of a master ram with a dozen or more females and young; immature males are normally solitary. The male is highly territorial and marks out his fairly extensive range by tongue clicking, display and urination. Despite his small size, the ram is extremely pugnacious and is known to attack and even kill sheep, goats and mountain reedbuck; he will also attack smaller predators.

In the breeding season, males stage fierce mock battles without actually doing any harm, and they will also chase each other. One, sometimes 2, young are born after a $9\frac{1}{2}$-month gestation.

LECHWE (male)

RHEBOK (male)

UGANDA KOB (male)

REEDBUCK (male)

ROAN ANTELOPE (male)

COMMON WATERBUCK (male)

Bovids 5

NAME: **Arabian Oryx,** *Oryx leucoryx*
RANGE: **S.E. Saudi Arabia: Rub' al Khali area**
HABITAT: **desert**
SIZE: **body: 1.6 m (5¼ ft)**
    tail: 45 cm (17¾ in)   Ⓔ

This is the smallest and rarest of the oryx and the only one found outside Africa. It lives in extreme desert conditions, feeding on grass and shrubs and travelling widely to find food. It is well adapted to its arid habitat, for it can live without drinking, obtaining the moisture it needs from its food; it also uses its hoofs and horns to scrape out a hollow under a bush or alongside a dune in which to shelter from the sun.

Although generally sociable, male oryx fight among themselves in the breeding season, and if cornered, will attack. One young is born after an 8-month gestation.

The present endangered status of the Arabian oryx is the result of overhunting, for its slender horns, its hide and meat are all prized. Although protected by law, it may already be extinct in the wild, since none has been seen since 1972. It is hoped, however, to breed from captive animals and so reintroduce this oryx to the wild.

NAME: **Blue Wildebeest,** *Connochaetes taurinus*
RANGE: **Africa: S. Kenya to N. South Africa**
HABITAT: **open grassland, bush savanna**
SIZE: **1.7–2.4 m (5½–8 ft)**
    tail: 60 cm–1 m (23½ in–3¼ ft)

The clumsy appearance of the blue wildebeest, the lugubrious expression given by its black face and tufty beard, its rocking-horse gait and its constant snorts and grunts have earned it a reputation as a "clown". Nevertheless, it is a most successful species.

Wildebeest are extremely gregarious, and herds numbering tens of thousands may be seen in East Africa during the dry season, when they make migrations of as much as 1,600 km (1,000 mls) in search of water and grazing. Breeding herds usually consist of up to 150 females and young, with 1 to 3 males. The bulls patrol the outside of their herd, keeping it closely grouped and defending a zone around it, even when migrating. Wildebeest feed almost exclusively on grass and need to drink often. They are frequently seen in association with zebra and ostrich; perhaps the wariness of the former offers them some protection against their common predators: lion, cheetah, hunting dog and hyena.

The female looks like the male but is smaller. After a gestation of 8½ months, she produces 1 calf, which can stand within 3 to 5 minutes of birth.

NAME: **Addax,** *Addax nasomaculatus*
RANGE: **Africa: E. Mauritania, W. Mali; patchy distribution in Algeria, Chad, Niger and Sudan**
HABITAT: **sandy and stony desert**
SIZE: **body: 1.3 m (4¼ ft)**
    tail: 25–35 cm (9¾–13¾ in)   Ⓥ

With its heavy head and shoulders and slender hindquarters, the addax is a clumsy-looking animal. Coloration varies widely between individuals, but there is always a mat of dark-brown hair on the forehead, and both sexes have thin, spiral horns. Addax are typical desert-dwellers, with their large, wide-spreading hoofs, adapted to walking on soft sand, and they never drink, obtaining all the moisture they need from their food, which includes succulents. Their nomadic habits are closely linked to the sporadic rains, for addax appear to have a special ability to find the patches of desert vegetation that suddenly sprout after a downpour. They are normally found in herds of 20 to 200.

The female produces 1 young after a gestation of 8½ months.

NAME: **Haartebeest,** *Alcelaphus buselaphus*
RANGE: **Africa, south of the Sahara**
HABITAT: **grassy plains**
SIZE: **body: 1.7–2.4 m (5½–8 ft)**
    tail: 45–70 cm (17¾–27½ in)

The nominate race, the bubal haartebeest, is extinct, but there are 12 subspecies, a further 2 of which (*A. b. swayne* and *A. b. tora*) are endangered, due to disease, hunting and destruction of their habitat.

Haartebeest are strange-looking animals, with backs that slope down slightly from high shoulders and long heads, with a pedicle on top from which spring the horns. Both male and female have horns, which show great intra-specific variation in both size and shape. Coloration also varies from deep chocolate to sandy fawn; females are paler than males.

These sociable antelope are found in herds of from 4 to 30, consisting of females and young with a master bull. He watches over his herd, often from a vantage point on top of a termite mound. Although haartebeest can go for long periods without water, they drink when they can and enjoy wallowing; they also use salt licks with avidity. They are partial to the young grass that grows after burning and often graze with zebra, wildebeest and gazelle and, like these, are preyed on largely by lion.

A single calf is born after a gestation of 8 months. It remains with its mother for about 3 years, at which time young males form a troop of their own.

NAME: **Bontebok,** *Damaliscus dorcas*
RANGE: **South Africa: W. Cape Province**
HABITAT: **open grassland**
SIZE: **body: 1.4–1.6 m (4½–5¼ ft)**
    tail: 30–45 cm (11¾–17¾ in)   Ⓞ

The strikingly marked bontebok was at one time almost extinct but is now fully protected and out of danger, and the population in game reserves numbers several thousand. The very similar blesbok, *D. d. phillipsi* (= *albifrons*), was also endangered, but it, too, is now flourishing, since it has been established in many reserves.

The sexes look alike, but females and juveniles are paler. Bontebok are grazers, active morning and evening. When disturbed, they move off swiftly upwind in single file. They are remarkably agile and can scale fences or wriggle under or through them.

The female produces a single calf after a 7½-month gestation, and young remain with their mothers until they are about 2 years old, when young males form bachelor herds. Outside the breeding season, bontebok live in mixed herds of from 20 to 500 animals.

NAME: **Sassaby/Tsessebi,** *Damaliscus lunatus*
RANGE: **Africa, south of the Sahara, east to Ethiopia, Somalia**
HABITAT: **open plains, flood plains, grassland with scattered bush**
SIZE: **body: 1.5–2 m (5–6½ ft)**
    tail: 40–60 cm (15¾–23½ in)

The nominate race of sassaby is found from Zambia to northern South Africa; it is probably conspecific with *D. korrigum*, known as the tiang or topi, which is found elsewhere in its range. Together, sassabys and topis are the most numerous of all antelope in Africa. In shape they are similar to the true haartebeests, but neither the slope of the back nor the length of the head are so exaggerated. Coloration and horns vary from race to race and between sexes; females are usually paler than males. Sassabys are active early and late in the day, when they feed on grass and herbage and also drink. They can, however, go without water for as long as 30 days.

Sassabys are not as gregarious as the haartebeests and generally move in parties of 8 to 10, which may join to form herds of up to 200 in the dry season. The mature male is highly territorial and marks his central stamping ground with dung and with scent, by rubbing his face and neck on bushes, grass stems and the ground. He watches over his territory and his harem and defends them from rivals and predators.

The female produces a single calf after a gestation of 7½ to 8 months.

BONTEBOK

BLUE WILDEBEEST (male)

SASSABY

HAARTEBEEST

ADDAX

ARABIAN ORYX (male)

Bovids 6

NAME: **Klipspringer,** *Oreotragus*
oreotragus
RANGE: **Africa: N. Nigeria, east to**
Somalia, south to South Africa
HABITAT: **rocky outcrops, hillsides,**
mountains to 4,000 m (13,000 ft)
SIZE: **body: 75 cm-1.15 m (29½ in-3¾ ft)**
tail: 7–23 cm (2¾–9 in)

The klipspringer always occurs where there are rocky outcrops interspersed with grassy patches and clumps of bush. It is a fairly small antelope, with strong legs and blunt-tipped hoofs the consistency of hard rubber; it fills a niche similar to that of the chamois. As it leaps about among the rocks, it is cushioned from bumps by its long, thick, bristly coat. The female is slightly heavier than the male and except in 1 race, *O. o. schillingsi*, does not have horns.

Klipspringers are sometimes found in small parties, more often in pairs, in a territory marked out by glandular secretions and defended against interlopers. They feed morning and evening and on moonlit nights, and will stand on their hind legs to reach the leaves, flowers and fruit that form the bulk of their diet. They also eat succulents, moss and some grass and drink when water is available.

Klipspringers probably mate for life. The female produces 1 young after a gestation of about 7 months, and there may be 2 young born in a year.

NAME: **Beira Antelope,** *Dorcatragus*
megalotis
RANGE: **Africa: Somalia**
HABITAT: **dry, bush-clad mountains,**
stony hills
SIZE: **body: 80–90 cm (31½–35½ in)**
tail: 6–7.5 cm (2¼–3 in) Ⓥ

A rare antelope, the beira is often mistaken for the klipspringer, although it has a slightly longer head, much bigger ears and longer, slimmer legs. The hind legs, especially, are long, with the result that the rump is higher than the shoulders. There is no crest, and only the male has horns; the female is larger than the male.

Beiras live in pairs or small family parties on extremely stony hillsides close to a grassy plain. Their highly specialized hoofs have elastic pads underneath which give a good grip on the stones. Beiras feed in the early morning and late afternoon on leaves of bushes, particularly mimosa, grass and herbage and do not need to drink.

Little is known of their habits or biology, for not only are they rare but their coloration blends so well with the background that they are impossible to spot unless they move. The female gives birth to a single young.

NAME: **Oribi,** *Ourebia ourebi*
RANGE: **Africa: Sierra Leone to Ethiopia,**
Tanzania, Zambia, South Africa
HABITAT: **wide, grassy plains with low**
bush, near water
SIZE: **body: 92 cm–1.1 m (3–3½ ft)**
tail: 6–10.5 cm (2¼–4¼ in)

The oribi is small and graceful, with a long neck and slender legs, longer behind than in front. The silky coat has a sleek, rippled look, and the black-tipped tail is conspicuous when the animal runs. Below each large, oval ear there is a patch of bare skin that appears as a black spot. The female has no horns and is larger than the male.

Pairs or small parties of up to 5 animals live in a territory, which the male marks out by rubbing glandular secretions on twigs and grass stems. Here the oribis have regular runs, resting and defecating places. They are active early and later in the day and on moonlit nights, when they feed on grass, plants and leaves. During the day and when danger threatens, they lie quietly in long grass or by a bush or rock.

The female gives birth to 1 young after a gestation of 6½ to 7 months.

NAME: **Royal Antelope,** *Neotragus*
pygmaeus
RANGE: **Africa: Sierra Leone, Liberia,**
Ivory Coast, Ghana
HABITAT: **forest, forest clearings**
SIZE: **body: 35–41 cm (13¾–16 in)**
tail: 5–6 cm (2–2¼ in)

This dainty, compact little animal, the smallest African antelope, weighs 3 to 4.5 kg (7 to 10 lb) — not much more than a rabbit. Indeed, it is called "king of the hares" by local tribespeople and so "royal" antelope by Europeans. It has a rounded back and a short tail, which it holds tightly against its rump. The male has tiny, sharp horns, which the female lacks; young are darker in colour than adults. Royal antelopes live in pairs or alone in a small territory, which they usually mark out with dung heaps. They are timid and secretive and are mainly active at night, when quite large numbers may feed together on leaves, buds, shoots, fungi, fallen fruit, grass and weeds. They sometimes venture into vegetable plots and cocoa and peanut plantations. Although they are preyed on by a wide range of mammals, birds and even large snakes, their small size often enables them to slip away unseen from danger, with their bellies almost on the ground. Their vulnerability is also compensated for by their astounding abillity to leap, like springboks, as much as 3 m (10 ft) up into the air.

Royal antelopes probably pair for life; the female produces 1 young.

NAME: **Kirk's Dik-dik,** *Madoqua kirki*
RANGE: **Africa: Somalia to Tanzania;**
S.W. Angola, Namibia
HABITAT: **bush country with thick**
undergrowth and scattered trees
SIZE: **body: 55–57 cm (21½–22½ in)**
tail: 4–6 cm (1½–2¼ in)

There are 7 races of this small, dainty antelope, which occur in two widely separated regions. Coloration of the soft coat varies from pale grey-brown in dry areas to a much darker shade in wet areas. The nose is slightly elongated and the legs long and thin, with the hind legs always bent, so the hindquarters slope downward. Males have tiny horns, with a crest of long hair between them; the slightly larger female lacks horns.

Dik-diks are found alone or in pairs, often with their 2 most recent young. Males scent-mark and fiercely defend the boundaries of a clearly defined territory, within which there are regularly used paths and places for resting and defecating. Shy, secretive animals, dik-diks browse at sunset and at night on leaves, shoots, buds and flowers, especially those of the *Acacia*. They also eat fallen fruit, dig up roots and tubers with their horns and hoofs and frequent salt licks; they do not need to drink. Their many enemies include leopard, caracal, serval, wild cats, eagles and man.

Dik-diks pair for life and produce 1 young after a gestation of 6 months; there are two litters a year.

NAME: **Grysbok,** *Raphiceros melanotis*
RANGE: **Africa: Tanzania to South Africa**
HABITAT: **grassy plains, bush savanna at**
the foot of hills
SIZE: **body: 60–75 cm (23½–29½ in)**
tail: 5–8 cm (2–3 in)

This rough-coated, stocky little antelope has relatively short legs, slightly longer behind, which gives it a sloping back. Males have short, sharply pointed horns and generally darker coloration than females. Solitary outside the breeding season, the grysbok establishes a fairly small territory, which is marked out by means of scent and dropping sites. It feeds in the morning and late in the afternoon on the foliage of trees and bushes and is particularly fond of grapevine leaves. During the day, the grysbok rests in the shade of a bush or rock or in long grass. Its main predators are leopard, caracal and crowned eagle, and when threatened, the antelope lies flat, darting away with a zigzag gallop when the enemy is near, only to dive suddenly for cover and disappear again.

The breeding biology of the grysbok is not well known, but it is thought to be similar to that of the steenbok, which produces 1 young after a gestation of 5½ months.

ROYAL ANTELOPE (male)

GRYSBOK (male)

ORIBI
(male)

KIRK'S DIK-DIK (male)

KLIPSPRINGER (male)

BEIRA ANTELOPE
(male)

Bovids 7

NAME: **Impala**, *Aepyceros melampus*
RANGE: **Africa: Kenya, Uganda, south to N. South Africa**
HABITAT: **light mopane woodland, acacia savanna**
SIZE: **body: 1.2–1.6 m (4–5¼ ft)**
tail: 30–45 cm (11¾–17¾ in)

A graceful, medium-sized antelope, with a glossy coat, the impala is identified by the unique bushy tuft of dark hairs above the hind "heels", the vertical dark stripes on white on the back of the thighs and tail and, in the male, the long, elegant, lyre-shaped horns. The impala is remarkable also for its fleetness and for the amazing leaps it makes — as far as 10 m (33 ft) and as high as 3 m (10 ft) — seemingly for enjoyment, as well as to escape from predators.

Impalas are extremely gregarious, and in the dry season troops may join to form herds of 200 or so. They are active day and night and eat quantities of grass and leaves, flowers and fruit. In the breeding season, the ram establishes a territory and a harem of 15 to 20 females, which he defends fiercely; immature males form separate troops. After a gestation of 6½ to 7 months, the female produces 1 young, which is born at midday, when many predators are somnolent, and remains hidden until it is strong enough to join the herd.

NAME: **Blackbuck**, *Antilope cervicapra*
RANGE: **India and Pakistan**
HABITAT: **open grassy plains**
SIZE: **body: 1.2 m (4 ft)**
tail: 18 cm (7 in)

The only species in its genus, the blackbuck is one of the few antelope in which the coloration of male and female is dissimilar. The dominant male in the herd is dark, almost black on back and sides and has long, spirally twisted horns; the female is yellowish-fawn and lacks horns. Subordinate males have smaller horns and retain female coloration. They only darken and develop large horns if they assume the dominant position in a herd, following the death of the leading male.

Blackbucks feed largely on grass and are active morning and evening, resting in the heat of the day. The female is unusually alert, and it is she who first gives warning of danger. When alarmed, blackbucks flee with leaps and bounds that soon settle into a swift gallop.

Blackbucks are normally found in herds of 15 to 50 — smaller groups consisting of a dominant male with females and young; as males mature, they are driven out and form their own small parties. The breeding male sets up a territory and defends it and his harem against rivals. One young, sometimes 2, is born after a gestation of 6 months.

NAME: **Springbok**, *Antidorcas marsupialis*
RANGE: **Africa: Angola, South Africa, Botswana (Kalahari Desert)**
HABITAT: **treeless grassland (veld)**
SIZE: **body: 1.2–1.4 m (4–4½ ft)**
tail: 19–27 cm (7½–10½ in)

The brightly coloured, strikingly marked springbok has a most unusual glandular pouch in its skin that stretches from the middle of the back to the base of the tail. When the animal is excited or alarmed, the pouch opens and reveals a crest of long, stiff, white hairs. Male and female look alike, both with ridged, strong horns. The springbok's name is derived from its ability to bound as high as 3.5 m (11½ ft) into the air, half a dozen times in succession, either in alarm or play, with back curved, legs stiff and crest displayed.

Springboks eat the leaves of shrubs and bushes and grass and are independent of water. They were once exceedingly numerous, and in times of drought, herds of up to a million would make long migrations in search of fresh grazing, devastating the pasture and farmland that lay in their path. As a result, thousands were slaughtered, but they have now been reintroduced throughout their range and are again thriving. In the breeding season, the male may establish a territory and a harem of 10 to 30, but large mixed herds are the norm. The female produces 1 young after a gestation of about 6 months.

NAME: **Dibatag**, *Ammodorcas clarkei*
RANGE: **Africa: Somalia, E. Ethiopia**
HABITAT: **sandy or grassy plains with scattered bushes**
SIZE: **body: 1.5–1.6 m (5–5¼ ft)**
tail: 30–36 cm (11¾–14¼ in) Ⓥ

Although superficially it resembles the gerenuk, the dibatag is much greyer, and the male has shorter and quite different horns. The long, thin, black-tufted tail is generally held upright when the animal is running and gives it its name, which derives from the Somali words *dabu* (tail) and *tag* (erect). These animals live in pairs or family parties, consisting of an adult male and 3 to 5 females with young, in a seasonal territory, moving with the rains wherever food supplies are plentiful. They are active morning and evening, browsing on leaves and young shoots of bushes, which, like the gerenuk, they stand on their hind legs to reach. They also eat flowers, berries and new grass; they do not need to drink.

As a rule, 1 calf is born in the rainy season after a gestation of 6 to 7 months, but it is possible for a female to produce 2 young in a year.

NAME: **Gerenuk**, *Litocranius walleri*
RANGE: **Africa: Somalia, Ethiopia to Kenya, Tanzania**
HABITAT: **dry thorn-bush country, desert**
SIZE: **1.4–1.6 m (4½–5¼ ft)**
tail: 23–35 cm (9–13¾ in)

This large, graceful-looking gazelle is remarkable chiefly for its long neck (gerenuk means giraffe-necked in Somali) and for its long legs. It has a small, narrow head, large eyes and mobile lips; there are tufts of hair on the knees, and the short, almost naked tail is held close against the body except in flight, when it is curled up over the animal's back. The male has horns and is larger than the female.

Gerenuks are usually found in pairs or family parties consisting of a male and 2 to 5 females with young. They are browsers, living almost entirely on leaves and young shoots of thorny bushes and trees, which they reach by standing against the trunk on their hind legs, using a foreleg to pull down the branches, and stretching their necks. They feed morning and evening, standing still in the shade at midday. They are quite independent of water. Their main predators are cheetah, leopard, lion, hyena and hunting dog.

A single young is born, usually in the rainy season, after a gestation of about 6½ months.

NAME: **Thomson's Gazelle**, *Gazella thomsoni*
RANGE: **Africa: Sudan, Kenya, N. Tanzania**
HABITAT: **open plains with short grass**
SIZE: **body: 80 cm–1.1 m (31½ in–3½ ft)**
tail: 19–27 cm (7½–10½ in)

This graceful, small gazelle has a distinctive broad, dark stripe along its sides, in marked contrast to the white underparts. The male is larger than the female and has much stronger horns. There are about 15 races of Thompson's gazelle which show only minor variations of colouring or horn size. They feed morning and evening, mainly on short grass and a small amount of foliage; they need to drink only when the grazing is dry. Their chief predators are cheetah, lion, leopard, hyena and hunting dog.

These gazelles live in loosely structured groups, which may vary between 1 old ram with 5 to 65 females and young, herds of young males with 5 to 500 members and groups of pregnant and recently calved females. When grazing is good, a mature male may establish a territory, which he marks by urination and droppings and by scraping and smearing of ground and bushes with horns and glandular secretions. Females calve at any time of year after a 6-month gestation and may produce 2 calves a year.

THOMSON'S GAZELLE
(male)

BLACKBUCK (male)

GERENUK
(male)

IMPALA
(male)

SPRINGBOK

DIBATAG
(male)

Bovids 8

NAME: **Saiga,** *Saiga tatarica*
RANGE: **River Volga to central Asia**
HABITAT: **treeless plains**
SIZE: **body: 1.2–1.7 m (4–5½ ft)**
  tail: 7.5–10 cm (3–4 in)

The saiga is a migratory species, well adapted to its cold, windswept habitat. It has a heavy fawnish-cinnamon coat, with a fringe of long hairs from chin to chest, which in the winter changes to a uniformly creamy-white and becomes exceedingly thick and woolly. It is thought, too, that the saiga's enlarged, proboscislike nose, with downward-pointing nostrils, may be an adaptation for warming and moisturizing inhaled air. The nasal passages are lined with hairs, glands and mucous tracts, and in each nostril there is a sac, lined with mucous membranes that appear in no other mammal but the whale. The male has horns, regarded as being of medicinal value by the Chinese; this led to overhunting, but saigas have been protected since 1920, and there are now over a million of them.

Saigas feed on low-growing shrubs and grass, and in autumn large herds gather and move off southward to warmer, lusher pastures. When spring comes, groups of 2 to 6 males begin to return northward, followed by the females. In May, after a gestation of about 5 months, the female gives birth to 1 to 3 young, which are suckled until the autumn.

NAME: **Serow,** *Capricornis sumatraensis*
RANGE: **N. India to central and S. China; S.E. Asia to Sumatra**
HABITAT: **bush and forest at 600–2,700 m (2,000–9,000 ft)**
SIZE: **body: 1.4–1.5 m (4½–5 ft)**
  tail: 8–21 cm (3–8¼ in)   Ⓔ

The slow, but extremely sure-footed, serow, with its short, solid hoofs, is found on rocky slopes and ridges of thickly vegetated mountains. It is active early and late in the day, feeding on grass and leaves; it lies up in the shelter of an overhanging rock for the rest of the time. The hairs on its back and sides are light coloured at the base and black at the tip, giving the coat an overall dark appearance; there is a completely black stripe along the centre of the back, and the mane varies from white to black on different individuals. Both male and female have horns, which they use to defend themselves, particularly against the dogs with which they are hunted by the Chinese, who believe that different parts of the serow have great healing properties.

Little is known of their breeding habits except that 1, more often 2, young are born after a gestation of about 8 months.

NAME: **Common Goral,** *Nemorhaedus goral*
RANGE: **S.E. Siberia to Manchuria, Korea**
HABITAT: **mountains at 1,000–2,000 m (3,300–6,600 ft)**
SIZE: **body: 90 cm–1.3 m (35½ in–4¼ ft)**
  tail: 7.5–20 cm (3–7¾ in)

These animals are mountain-dwellers, found generally where there are grassy hills and rocky outcrops near forests. They have long, sturdy legs and their long coats, of guard-hairs overlaying a short, woolly undercoat, give them a shaggy appearance. Male and female look alike, both with horns, but the male has a short, semi-erect mane.

Apart from old bucks, which live alone for most of the year, gorals normally live in family groups of 4 to 8. They feed on grass in the early morning and late afternoon, resting on a rocky ledge in the middle of the day. The female gives birth to 1, rarely 2, young after a gestation of about 6 months.

NAME: **Mountain Goat,** *Oreamnos americanus*
RANGE: **N. America: Rocky Mountains from Alaska to Montana, Idaho, Oregon; introduced in South Dakota**
HABITAT: **rocky mountains above the tree-line**
SIZE: **body: 1.3–1.6 m (4¼–5¼ ft)**
  tail: 15–20 cm (6–7¾ in)

This splendid-looking animal is not a true goat, but a goat-antelope, and is the only one in its genus. It is found among boulders and rocky screes above the tree-line and is well adapted to its cold, harsh habitat. It has thick, woolly underfur and a long, hairy, white coat, which is particularly thick and stiff on the neck and shoulders, forming a ridge, or hump. The hoofs have a hard, sharp rim, enclosing a soft, spongy inner pad, which gives the mountain goat a good grip on rocks and ice. Both sexes have beards and black, conical horns.

Mountain goats are slow-moving, but amazingly sure-footed, climbing to dizzying heights and seemingly inaccessible ledges in their search for grass, sedges and lichens to eat. They also browse on the leaves and shoots of trees and will travel considerable distances to search out salt licks. In winter, they come down to areas where the snow is not too deep, even to the coast in places. In really severe weather, they may take refuge under overhanging rocks or in caves.

These goats are probably monogamous, and the female produces 1 or 2 kids in the spring after a gestation of about 7 months. The young kids are remarkably active and within half an hour of birth are able to jump about among the rocks.

NAME: **Chamois,** *Rupicapra rupicapra*
RANGE: **Europe to Middle East**
HABITAT: **mountains**
SIZE: **body: 90 cm–1.3 m (35½ in–4¼ ft)**
  tail: 3–4 cm (1¼–1½ in)

For nimbleness, audacity and endurance, the chamois is unparalleled among mountain-dwellers. It thrives in wild and inhospitable surroundings, where weather conditions may be savage, and has been known to survive as long as 2 weeks without food. It is the only species in its genus. It is slimly built, with distinctive horns that rise almost vertically, then sweep sharply backward to form a hook. The legs are sturdy, and the hoofs have a resilient, spongy pad underneath, which gives the chamois a good grip on slippery or uneven surfaces. The coat is stiff and coarse, with a thick, woolly underfleece.

Chamois graze on the tops of mountains in summer, on herbs and flowers; in winter, they come farther down the slopes and browse on young pine shoots, lichens and mosses. They are wary animals and a sentinel is always posted to warn of danger. Females and young live together in herds of 15 to 30; old males are solitary except in the rutting season in the autumn. Fighting is common among males, with older rams locking horns with young animals in a struggle to assert their supremacy. The female usually has 1 kid, but 2 or 3 are fairly common.

NAME: **Ibex,** *Capra ibex*
RANGE: **Switzerland, Italy**
HABITAT: **Alps to 3,000 m (10,000 ft)**
SIZE: **body: 1.5 m (5 ft)**
  tail: 12–15 cm (4¾–6 in)

From Roman times, different parts of these animals have been regarded as possessing miraculous healing powers, and ibex were hunted to the point of extinction. Today, however, a few small, protected herds survive in reserves. The most remarkable feature of the male ibex is the long, backward-sweeping horns; the female has shorter horns. The coat is a brownish-grey, with longer hair on the back of the neck forming a mane in old males. The male has a small beard on the chin.

The ibex live above the tree-line, only descending to the upper limits of forest in the harshest winter conditions. In summer, they climb up into alpine meadows, where they graze on grass and flowers. At this time, females are found with young and subadults, and males form their own groups, within which token fights often take place to establish an order of rank. Only in the winter rutting season do males rejoin female herds. The female gives birth to 1 young after a gestation of 5 to 6 months.

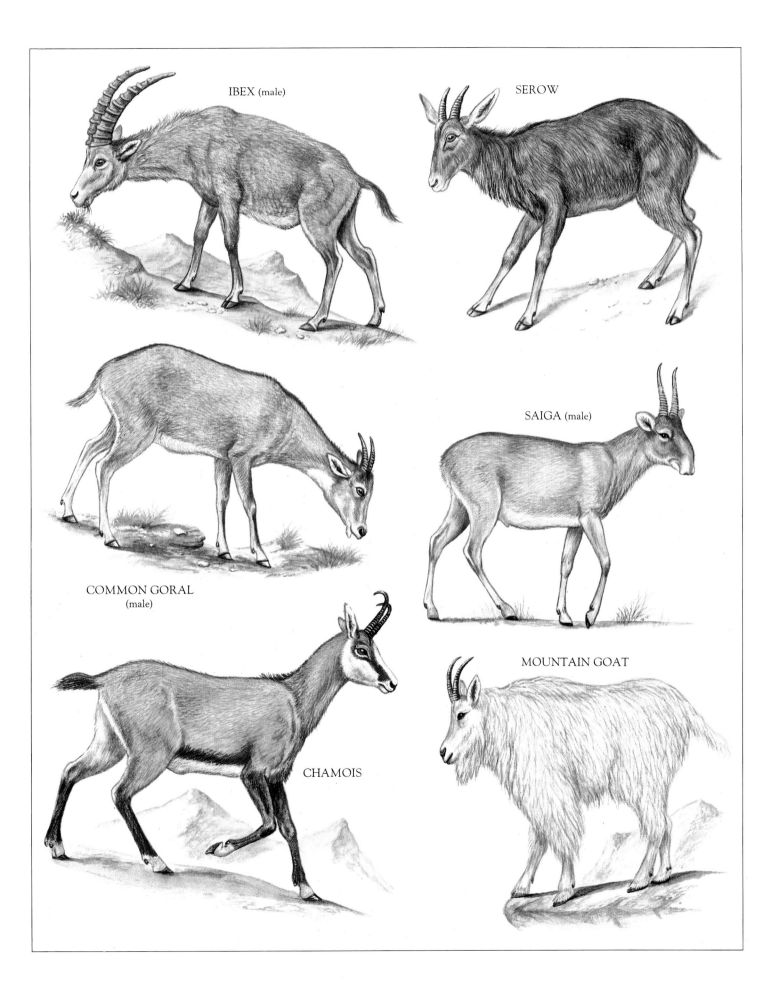

IBEX (male)

SEROW

COMMON GORAL
(male)

SAIGA (male)

CHAMOIS

MOUNTAIN GOAT

Bovids 9

NAME: **Musk Ox,** *Ovibos moschatus*
RANGE: **N. Canada, Greenland**
HABITAT: **tundra**
SIZE: **body: 1.9–2.3 m (6¼–7½ ft)**
 tail: 9–10 cm (3½–4 in)

In prehistoric times, the musk ox occurred throughout northern Europe, Siberia and North America. It was largely exterminated, surviving only in northern Canada and Greenland; however, it has now been successfully reintroduced in Norway and Alaska. It is the only species in its genus. The musk ox is superbly equipped for life in harsh, arctic conditions, for it has a dense undercoat, which neither cold nor water can penetrate, and an outer coat of long, coarse hair that reaches almost to the ground and protects it from snow and rain. The broad hoofs prevent it from sinking in soft snow. Both sexes have heavy horns that almost meet at the base, forming a broad frontal plate. Facial glands in the bull emit a strong, musky odour in the rutting season, hence the animal's name.

Musk oxen are gregarious, living in herds of as many as 100. In the mating season, young bulls are driven out by old, master bulls and form small bachelor groups or remain solitary. The female produces 1 young after a gestation of 8 months. When threatened, musk oxen form a circle, facing outward with horns lowered, with the young in the middle; this is an effective defence against wolves, their natural enemies, but not against men with guns. Musk oxen feed mainly on grass, but they also eat mosses, lichens and leaves and will dig through the snow to find food.

NAME: **Himalayan Tahr,** *Hemitragus jemlahicus*
RANGE: **India: Kashmir (Pir Pamjal Mountains), Punjab; Nepal, Sikkim**
HABITAT: **tree-covered mountain slopes**
SIZE: **body: 1.1 m (3½ ft)**
 tail: 9 cm (3½ in)

Although the tahr appears much like a goat, with its heavy, shaggy coat that forms a mane around the shoulders, it differs from true goats in having a naked muzzle and no beard; the horns, too, are long and not twisted, and there are glands on the feet.

Tahrs are goatlike in their habits as well, for they live on the most precipitous mountainsides, where they climb and leap about with supreme ease. They are gregarious, living in herds of 30 to 40 on almost any vegetation they can reach. Wary animals, they always post a sentinel to watch for danger approaching from below.

The breeding season peaks in the winter, when the female gives birth to 1 or 2 young after a 6- to 8-month gestation.

NAME: **Takin,** *Budorcas taxicolor*
RANGE: **Asia: Burma; China: Szechuan and Shensi Provinces**
HABITAT: **dense thickets in mountain forest, 2,400–4,250 m (7,900–14,000 ft)**
SIZE: **body: 1.2 m (4 ft)**
 tail: 10 cm (4 in) ®

The takin lives in dense bamboo and rhododendron thickets near the upper limits of the tree-line in some of the most rugged country in the world. It is a clumsy-looking, solidly built animal, with thick legs and large hoofs with dew claws. The coat ranges from yellowish-white to blackish-brown, always with a dark stripe along the back. Both males and females have horns.

Although old bulls are generally solitary, in summer they join large herds, which graze in the evening on grass and herbage near the tops of mountains; in winter takins move down to the valleys, where they live in smaller groups, eating grass, bamboo and willow shoots. They are shy animals, spending most of their time under cover in dense thickets, in which there are regularly used paths leading to their grazing grounds and salt licks.

The female produces 1 young after a gestation of about 8 months; it is strong and active and able to follow its mother after about 3 days.

NAME: **Barbary Sheep,** *Ammotragus lervia*
RANGE: **N. Africa: Atlantic coast to Red Sea, south to N. Mali and Sudan; introduced in S.W. USA**
HABITAT: **dry, rocky, barren regions**
SIZE: **body: 1.3–1.9 m (4¼–6¼ ft)**
 tail: 25 cm (9¾ in)

The barbary sheep is the only sheep indigenous to Africa, and the only species in the genus. The mane of long, soft, thick hairs on its throat, chest and upper forelegs differentiates it from other wild sheep, but, like them, both sexes have horns, those of the female being almost as heavy as the male's.

Small family parties, consisting of a breeding pair and their offspring of various litters, wander about in search of food: grass, herbaceous plants, and leaves and twigs of low-growing bushes. They obtain all the water they need from this diet and by licking up dew. Barbary sheep, having no cover in which to hide when danger threatens, rely on the camouflage effect of their sandy-coloured coats and remain perfectly still. They are killed by man for their flesh, hides, hair and sinews.

In captivity, barbary sheep produce one litter a year of 1 or 2 young. They have been successfully crossed with domestic goats, and the offspring with chamois.

NAME: **American Bighorn,** *Ovis canadensis*
RANGE: **N.W. America**
HABITAT: **upland and mountainous areas**
SIZE: **body: 1.2–1.8 m (4–6 ft)**
 tail: 15 cm (6 in) Ⓥ

The American bighorn is found on high mountain pastures in summer, when groups of males or females with young graze independently on grass and herbage. In winter, they form mixed herds and move to lower pastures. The high-ranking male bighorn sheep is a most impressive animal, with massive spiral horns up to 1.15 m (3¾ ft) long. Horn size is of great significance in establishing rank order among males; smaller-horned and, therefore, lower-ranking males are treated as females by dominant males, which perhaps prevents them being driven out of the herd. In the rutting season, high-ranking males of comparable horn size have fierce battles, rushing at each other and crashing their horns together; the fighting may go on for hours, and occasionally an animal is killed. Females have very short horns.

The ewe produces 1 or 2 lambs, born after a gestation of about 6 months, and is assiduous in her care of the young.

NAME: **Mouflon,** *Ovis orientalis*
RANGE: **Sardinia, Corsica; introduced in Germany, Hungary, Austria and Czechoslovakia**
HABITAT: **rugged mountains**
SIZE: **body: 1.2 m (4 ft)**
 tail: 7 cm (2¾ in)

The mouflon, the wild sheep of Europe, is now found only in reserves in Sardinia and Corsica, but even there is inadequately protected. The male has long, spiral horns, often with the tips curving inward; those of the female are short. It has an extremely woolly underfleece, covered in winter by a coarse, blackish-brown top coat, with a distinctive white saddle patch in the male; in the summer this patch disappears. The female and young are grey or darker brown, with no patch.

Mouflon are active early and late in the day and do not wander far, even when food is scarce. They appear to be able to eat every type of vegetation: grass, flowers, buds and shoots of bushes and trees, even poisonous plants such as deadly nightshade, and so manage to survive. They live in separate groups composed of females with young or males on their own in the summer. In the rutting season, a mature ram will detach a female from the herd and mate with her. Fierce fighting may take place if an old ram is challenged, but there are seldom casualties. The ewe produces 1 lamb after a gestation of 5 months.

MUSK OX

HIMALAYAN TAHR

MOUFLON
(male)

BARBARY SHEEP

TAKIN

AMERICAN BIGHORN
(male)

Squirrels I

NAME: European Red Squirrel, *Sciurus vulgaris*
RANGE: Europe, east to China, Korea and Japan: Hokkaido.
HABITAT: evergreen forest
SIZE: body: 20–24 cm ($7\frac{3}{4}$–$9\frac{1}{2}$ in)
tail: 15–20 cm (6–$7\frac{3}{4}$ in)

Until the arrival of the North American grey squirrel in Britain at the beginning of this century, the only European species was the red squirrel. Populations are now declining in Britain, but red squirrels are still abundant in Europe and Asia. Conifer cones are their main food, although in summer they also eat fungi and fruit.

The length of the breeding season is dictated by local climate: in a good year a female may produce two litters of about 3 young each. The young are born in a tree nest, called a drey, which also doubles as winter quarters.

NAME: Grey Squirrel, *Sciurus carolinensis*
RANGE: S.E. Canada, E. USA; introduced in Britain and South Africa
HABITAT: hardwood forest
SIZE: 23–30 cm (9–$11\frac{3}{4}$ in)
tail: 21–23 cm ($8\frac{1}{4}$–9 in)

The grey squirrel's natural home is the oak, hickory and walnut forests of eastern North America, where its numbers are controlled by owls, foxes and bobcats. It feeds on seeds and nuts — an adult squirrel takes about 80 g ($2\frac{3}{4}$ oz) shelled nuts each day — and on eggs, young birds and insects. Occasionally grey squirrels strip the bark from young trees to gain access to the nutritious sap beneath.

Two litters are produced each year, in early spring and summer. There are up to 7 young in a litter, but usually only 3 or 4 survive. Males are excluded from the nest and take no part in rearing the young. In the south of England, the introduced grey squirrel is ousting the native red squirrel.

NAME: African Giant Squirrel, *Protoxerus stangeri*
RANGE: W. Africa, east to Kenya; Angola
HABITAT: palm forest
SIZE: body: 22–33 cm ($8\frac{1}{2}$–13 in)
tail: 25–38 cm ($9\frac{3}{4}$–15 in)

Sometimes called the oil-palm squirrel, this species feeds primarily on nuts from the oil palm. In regions where calcium is scarce, it has been observed to gnaw bones and ivory. The African giant squirrel is a secretive creature, and its presence is usually only detected by a booming call, which it utters when disturbed. Little is known of its breeding habits, but it probably breeds throughout the year.

ORDER RODENTIA

The largest of the mammalian orders, Rodentia contains at least 1,591 species in 28 families.

SCIURIDAE: Squirrel Family

The differences between squirrels and other families of rodents are not immediately noticeable; most of them are technical and of interest only to zoologists. There are about 246 species of squirrel and generally they are alert, short-faced animals. Some have taken to burrowing and live in vast subterranean townships (prairie dogs); others run and hop about over logs and stones (chipmunks); and many have taken to life in the trees (tree and flying squirrels). Most forms are active by day and are among the most brightly coloured of all mammals. Their eyes are large and vision, including colour vision, good. The few nocturnal species, such as the flying squirrels, are more drab in appearance. Males and females generally look alike.

Squirrels have a wide distribution, occurring in all parts of the world except for southern South America, Australia, New Zealand, Madagascar and the deserts of the Middle East. In temperate climates they undergo periods of dormancy in cold weather. Dormancy differs from true hibernation in that the creature wakes every few days for food. True hibernation occurs in a limited number of squirrel species. Squirrels are social animals and have evolved a complex system of signalling with their bushy tails. They are also quite vocal, and most can make a variety of sounds.

NAME: Indian Striped Squirrel, *Funambulus palmarum*
RANGE: India, Sri Lanka
HABITAT: palm forest
SIZE: body: 11.5–18 cm ($4\frac{1}{2}$–7 in)
tail: 11.5–18 cm ($4\frac{1}{2}$–7 in)

With their distinctive stripes, these little squirrels superficially resemble chipmunks. They are highly active animals, foraging by day for palm nuts, flowers and buds. They may damage cotton trees by eating the buds, but when they feed on the nectar of the silky oak flowers, they do good by pollinating the flowers they investigate.

Males are aggressive and fight for females, but once mating has taken place, they show no further interest in females or young. About three litters, each containing about 3 young, are born during the year. The gestation period is 40 to 45 days. Young females are sexually mature at 6 to 8 months old.

NAME: Black Giant Squirrel, *Ratufa bicolor*
RANGE: Burma to Indonesia
HABITAT: dense forest
SIZE: body: 30–45 cm ($11\frac{3}{4}$–$17\frac{3}{4}$ in)
tail: 30–50 cm ($11\frac{3}{4}$–$19\frac{3}{4}$ in)

The 4 species of giant squirrel are, as their name implies, very large, and they can weigh up to 3 kg ($6\frac{1}{2}$ lb). Black giant squirrels are extremely agile, despite their size, and can leap 6 m (20 ft) or more through the trees; as they do so, their tails trail down like rudders. They feed on fruit, nuts, bark and a variety of small invertebrate animals. Singly or in pairs, they shelter in nests in tree holes.

In the breeding season a huge nest is made in which the female produces 1, sometimes 2, young after a gestation of about 4 weeks.

NAME: African Palm Squirrel, *Epixerus ebii*
RANGE: Ghana, Sierra Leone
HABITAT: dense forest; near swamps
SIZE: body: 25–30 cm ($9\frac{3}{4}$–$11\frac{3}{4}$ in)
tail: 28–30 cm (11–$11\frac{3}{4}$ in) (I)

The African palm squirrel is one of the rarest rodents on record. Forest clearance and swamp-draining activities present an intolerable disturbance to this species, from which it may not recover. However, the inaccessibility of its habitats affords it a measure of protection in some parts of its range, and this may allow time for a thorough survey to be made of its status, which at present is not really known. It is believed to feed largely on the nuts of the *Raphia* swamp palm, but it probably eats other foods as well. Nothing is known of its breeding habits nor the reasons for its rarity.

NAME: Prevost's Squirrel, *Callosciurus prevosti*
RANGE: S.E. Asia
HABITAT: forest
SIZE: body: 20–28 cm ($7\frac{3}{4}$–11 in)
tail: 15–25 cm (6–$9\frac{3}{4}$ in)

The sharp contrast of colours in the coat of Prevost's squirrel makes it one of the most distinctive members of the family (the generic name means "beautiful squirrel"). These squirrels forage by day for seeds, nuts, buds, shoots and, occasionally, birds' eggs and insects. They live singly or in pairs. Shortly before giving birth, the female leaves her normal nest in a hollow tree and builds a nest of sticks, twigs and leaves high up in the branches. Here, safe from the attentions of ground-living predators, she gives birth to a litter of 3 or 4 young. It is not known exactly how many litters each female produces in a year, but in some parts of the range there may be as many as four.

INDIAN STRIPED SQUIRREL

EUROPEAN RED SQUIRREL

BLACK GIANT SQUIRREL

AFRICAN GIANT SQUIRREL

PREVOST'S SQUIRREL

AFRICAN PALM SQUIRREL

GREY SQUIRREL

Squirrels 2

NAME: African Ground Squirrel, *Xerus erythropus*
RANGE: Africa: Morocco to Kenya
HABITAT: forest, scrub, savanna
SIZE: body: 22–30 cm (8½–11¾ in)
tail: 18–27 cm (7–10½ in)

Rather like that of its North American counterpart, the fur of the African ground squirrel is harsh and smooth, with practically no underfur. It lives in extensive underground burrow systems, which it digs with its strong forepaws. Very tolerant of humans, it carries out its routine of searching for seeds, berries and green shoots by day. These squirrels are a social species and greet one another with a brief "kiss" and a flamboyant flick of the tail.

Mating occurs in March or April and litters of 3 or 4 young are born after a gestation of about 4 weeks. In some areas, African ground squirrels are thought to inflict a poisonous bite; the basis for this mistaken belief is that their salivary glands contain streptobacilli, which cause septicaemia.

NAME: Thirteen-lined Ground Squirrel, *Spermophilus tridecemlineatus*
RANGE: S. central Canada, central USA
HABITAT: short, arid grassland
SIZE: body: 17–29 cm (6¾–11½ in)
tail: 6–14 cm (2¼–5½ in)

These strikingly marked little rodents are active during the day and are often seen in considerable numbers, although their social groups are far looser than those of prairie dogs.

Like other ground squirrels, they have keen eyesight and frequently rear up on their haunches to survey the scene, searching for predators such as hawks, bobcats and foxes. If danger threatens, the squirrels disappear into their burrows. Although some squirrels live among piles of boulders, most dig burrows which vary in size and complexity. Large squirrels may dig tunnels which are 60 m (200 ft) or more in length, with side chambers, but younger individuals make smaller, shallower burrows. Seeds, nuts, fruit, roots and bulbs form the main bulk of the diet of these squirrels, but they sometimes eat insects, birds' eggs and even mice. In winter, they hibernate, having gained layers of body fat to sustain them through the winter. Their body temperature falls to about 2°C (35.6°F) and their hearts beat only about 5 times a minute, compared to 200 to 500 times in an active animal.

After hibernation the squirrels mate; the gestation period is about 4 weeks, and a litter of up to 13 blind, helpless young is born in early summer. The eyes of the young open at about 4 weeks and they are independent of the mother at about 6 weeks.

NAME: Black-tailed Prairie Dog, *Cynomys ludovicianus*
RANGE: central USA
HABITAT: grassland (prairie)
SIZE: body: 28–32 cm (11–12½ in)
tail: 8.5–9.5 cm (3¼–3¾ in)

The prairie dog derives its common name from its stocky, terrierlike appearance and from its sharp, doglike bark, which it utters to herald danger. One of the most social rodent species, prairie dogs live in underground burrows, called townships, containing several thousands of individuals. They emerge by day to graze on grass and other vegetation and can often cause serious damage to cattle ranges. Feeding is regularly interrupted for bouts of socializing, accompanied by much chattering.

Females give birth to litters of up to 10 young during March, April or May, after a 4-week gestation. After being weaned at 7 weeks, the young disperse to the edge of the township. Prairie dogs are commonly preyed on by eagles, foxes and coyotes.

NAME: Eastern Chipmunk, *Tamias striatus*
RANGE: S.E. Canada, E. USA
HABITAT: forest
SIZE: body: 13.5–19 cm (5¼–7½ in)
tail: 7.5–11.5 cm (3–4½ in)

The chipmunk is one of the best-known small mammals in North America, for its lack of fear of man and its natural curiosity make it a frequent sight at camping and picnic sites. Chipmunks dig burrows under logs and boulders, emerging in the early morning to forage for acorns, cherry stones, nuts, berries and all manner of seeds. Occasionally they are sufficiently numerous to cause damage to crops. During the autumn they store food supplies; they do not truly hibernate, but just become somewhat lethargic during winter.

A single litter of up to 8 young is born each spring. Although weaned at 5 weeks, the young stay with their mother for some months. They have a lifespan of about 5 years.

NAME: Woodchuck, *Marmota monax*
RANGE: Alaska, Canada, south to E. USA
HABITAT: forest
SIZE: body: 45–61 cm (17¾–24 in)
tail: 18–25 cm (7–9¾ in)

The woodchuck, or ground hog as it is called in some regions, is a heavily built, rather belligerent rodent. Woodchucks feed in groups and, ever fearful of the stealthy approach of a mountain lion or coyote, one member of the group keeps watch while the others search for edible roots, bulbs, tubers and seeds.

The young — 4 or 5 in a litter — are born in late spring and grow very fast. By autumn they have achieved adult size and are forced away from the parental nest by the aggression of the male. Woodchucks may live for as long as 15 years.

NAME: Red Giant Flying Squirrel, *Petaurista petaurista*
RANGE: Asia: Kashmir to S. China; Sri Lanka, Java, Borneo
HABITAT: dense forest
SIZE: body: 40–58 cm (15¾–22¾ in)
tail: 43–63 cm (17–25 in)

The broad membrane that joins the ankles to the wrists of this handsome creature does not allow true flight, but the squirrel can glide up to 450 m (1,500 ft). Gliding enables the squirrel to move from one tall tree to another without having to descend to the ground each time. By day, these squirrels rest in hollow trees, coming out at dusk to search for nuts, fruit, tender twigs, young leaves and flower buds to eat. They live singly, in pairs or in family groups.

Little is known of the reproductive habits of these squirrels, but they appear to have just 1 or 2 young in each of two or three litters a year. Because the young are not seen to ride on the mother's back, it is assumed that they are deposited in a safe refuge while the mother feeds. These substantial rodents are hunted by local tribespeople for their flesh.

NAME: Northern Flying Squirrel, *Glaucomys sabrinus*
RANGE: S. Canada, W. USA
HABITAT: forest
SIZE: body: 23.5–27 cm (9¼–10½in)
tail: 11–18 cm (4¼–7 in)

By stretching out all four limbs when it jumps, the flying squirrel opens its flight membrane, which extends from wrists to ankles, and is able to glide from one tree to another. Speeds of as much as 110 m/min (360 ft/min) may be achieved. Normally the squirrels forage about in the tree-tops for nuts, living bark, lichens, fungi, fruit and berries, only taking to the air should an owl or other predator appear. In autumn, stocks of nuts and dried berries are laid up in hollow trees, for the flying squirrels do not hibernate in winter.

At any time from April onward, young are born in a softly lined nest in a hollow tree. There are normally between 2 and 6 in a litter, and the young suckle for about 10 weeks, unusually long for small rodents. It is thought that this is because an advanced level of development is necessary before gliding can be attempted.

AFRICAN GROUND SQUIRREL

BLACK-TAILED PRAIRIE DOG

EASTERN CHIPMUNK

THIRTEEN-LINED GROUND SQUIRREL

RED GIANT FLYING SQUIRREL

WOODCHUCK

NORTHERN FLYING SQUIRREL

Pocket Gophers, Pocket Mice

GEOMYIDAE:
Pocket Gopher Family

There are some 30 species of pocket gopher distributed throughout North America, from 54° North to Panama and from coast to coast. They spend most of their lives underground, in complex and wide-ranging burrow systems which they dig with their chisel-like incisor teeth and strong, broad paws. Gophers occur wherever the soil is soft and supports rich vegetation — roots and tubers form the main diet.

The burrowing activities of these animals tend to be detrimental to grassland and great efforts are made to exterminate pocket gophers, which, however, reproduce at a prodigious rate.

NAME: **Plains Pocket Gopher,** *Geomys bursarius*
RANGE: **central USA: Canadian border to Mexico**
HABITAT: **sandy soil in sparsely wooded areas**
SIZE: **body: 18–24 cm (7–9½ in)**
tail: 10–12.5 cm (4–5 in)

Pocket gophers get their common name from the two deep, fur-lined cheek pouches, which can be crammed full of tubers or shoots for transport back to the nest. They lead solitary lives, the male leaving its burrow only to seek out a female during the breeding season; after mating, he returns to his burrow. A litter of 2 or 3 young is born after a gestation of 18 or 19 days. The young are weaned at 10 days but remain in their mother's burrow until they are about 2 months old. They are sexually mature at 3 months.

Although ranchers consider pocket gophers pests, their burrowing does aerate the soil and thus, in the long term, improves the productivity of the pastureland.

NAME: **Northern Pocket Gopher,** *Thomomys talpoides*
RANGE: **S.W. Canada to Colorado, USA**
HABITAT: **grassland and open forest to altitudes of 4,000 m (13,000 ft)**
SIZE: **body: 25–30 cm (9¾–11¾ in)**
tail: 6–9.5 cm (2¼–3¾ in)

The northern pocket gopher often lives in areas which experience intense winter weather, but it does not hibernate. It builds up huge piles of roots and bulbs in underground larders and survives during the winter on these food stores. Pairs mate in early spring and litters of up to 10 young are born after a gestation of 18 days. Females may mate again almost immediately, giving birth a few days after the first litter is weaned.

HETEROMYIDAE:
Pocket Mouse Family

There are about 70 species of pocket mouse and kangaroo rat, with considerable variations in external appearance. Some are mouselike and live in dense forest, others have long hind legs, bound along like kangaroos and live in deserts and arid plains. They eat seeds, as well as insects and other invertebrates, and have deep, fur-lined cheek pouches in which food can be transported. They are fertile animals, some species producing three or four litters a year. The family occurs in North, Central and South America.

NAME: **Silky Pocket Mouse,** *Perognathus flavus*
RANGE: **USA: Wyoming, south to Texas; Mexico**
HABITAT: **low arid plains**
SIZE: **body: 10–12 cm (4–4¾ in)**
tail: 4.5–6 cm (1¾–2¼ in)

The silky pocket mouse has dense soft fur which it keeps in immaculate condition. Although nocturnal, when it emerges from its burrows the sand is still very hot, and to prevent the soles of its feet from burning, they are covered by thick pads of soft fur which also act like snowshoes, spreading the body weight more evenly. It moves on all-fours or on its hind legs only.

Breeding seasons are from April to June and August to September, and there are normally 4 young in a litter. They are born deep in the burrow system and first emerge to forage for seeds at about 3 weeks old.

NAME: **Californian Pocket Mouse,** *Perognathus californicus*
RANGE: **USA: California; south to Mexico: Baja California**
HABITAT: **arid sandy plains**
SIZE: **body: 19–23 cm (7½–9¼ in)**
tail: 10–14.5 cm (4–5¾ in)

Californian pocket mice dig extensive burrow systems in the sandy soil of their habitat. The entrance is usually sited underneath a small shrub or bush to provide some shade and protection from predators. The mice feed largely on seeds, transporting them back to the burrow in their cheek pouches. They also eat green plants on occasion but seldom, if ever, drink; their bodies are adapted to survive without drinking. The breeding season lasts from April to September, but there is a marked decline in activity during the hottest part of the summer. Up to 7 young are born in each litter, after a gestation of about 25 days.

NAME: **Pale Kangaroo-mouse,** *Microdipodops pallidus*
RANGE: **USA: W. central Nevada**
HABITAT: **wind-swept sand-dunes**
SIZE: **body: 6.5–8 cm (2½–3¼ in)**
tail: 6.5–10 cm (2½–4 in)

To help it cover the great distances it must travel to find the sparse food supplies in its barren habitat, this rodent has powerful hind legs for bounding, with broad flat feet, fringed with stiff hairs. As the mouse hops, its long tail streams out, counterbalancing it and giving it the appearance of a small kangaroo.

Kangaroo-mice are long-lived and breed more slowly than other members of their family. In particularly hot, dry summers, they do not breed at all.

NAME: **Desert Kangaroo-rat,** *Dipodomys deserti*
RANGE: **USA: Nevada, south to Mexico**
HABITAT: **arid brush and grassland**
SIZE: **body: 30.5–38 cm (12–15 in)**
tail: 18–21.5 cm (7–8½ in)

Desert kangaroo rats dig their burrows in well-drained, easily dug soils. They are nocturnal and travel great distances in search of food. Since their kidneys are four times more efficient than a human's, they can live their whole lives without ever drinking.

Breeding occurs in any month of the year. Litters of up to 5 young are born after about 30 days' gestation.

NAME: **Mexican Spiny Pocket Mouse,** *Liomys irroratus*
RANGE: **USA: S.W. tip of Texas; Mexico**
HABITAT: **arid woodland**
SIZE: **body: 19–30 cm (7½–11¾ in)**
tail: 9.5–17 cm (3¾–6¾ in)

The coat of this species bears stiff, grooved hairs which form a protective shield around the body, helping to deter some predators. It prefers lush, succulent vegetation to eat, but also forages for seeds and roots which it carries back to its burrow in its cheek pouches. Breeding takes place at any time of year, and litters usually contain about 4 young.

NAME: **Forest Spiny Pocket Mouse,** *Heteromys anomalus*
RANGE: **Colombia, Venezuela; Trinidad**
HABITAT: **tropical rain forest**
SIZE: **body: 12.5–16 cm (5–6¼ in)**
tail: 13–20 cm (5–7¾ in)

This shy, nocturnal rodent lives in burrows on the forest floor. It collects seeds, buds, fruit, leaves and shoots and carries them back to the burrows for eating. Litters of about 4 young are born at any time of the year, but mostly in spring and early summer.

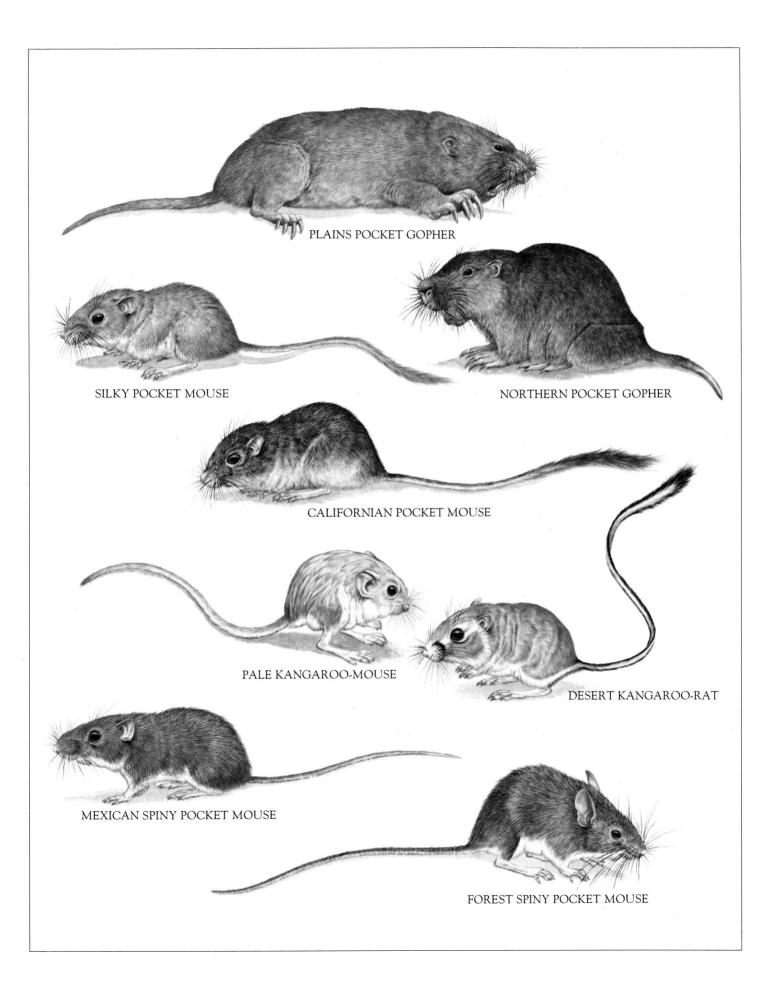

PLAINS POCKET GOPHER

SILKY POCKET MOUSE

NORTHERN POCKET GOPHER

CALIFORNIAN POCKET MOUSE

PALE KANGAROO-MOUSE

DESERT KANGAROO-RAT

MEXICAN SPINY POCKET MOUSE

FOREST SPINY POCKET MOUSE

Mountain Beaver, Beavers, Spring Hare, Scaly-tailed Squirrels

APLODONTIDAE: Mountain Beaver Family

This rodent family contains a single species, the mountain beaver. Its common name is particularly inappropriate because this heavy-bodied, burrowing animal is neither a beaver nor associated with high country. Indeed, its ancestry and evolution are poorly understood, despite the fact that the fossil record suggests that the family is an ancient one.

NAME: **Mountain Beaver/Aplodontia,** *Aplodontia rufa*
RANGE: **N.W. USA**
HABITAT: **moist forest**
SIZE: **body: 30–43 cm (11¾–17 in)
tail: 2.5 cm (1 in)**

The mountain beaver is a solitary creature, and each adult digs its own burrow system of underground nest and tunnels. Although it climbs poorly, rarely going up trees, almost any plant material, including bark and twigs, is eaten by the mountain beaver, and it makes stores of food for the winter months. In spring, beavers produce a litter of 2 or 3 young, which are born in a nest lined with dry vegetation.

CASTORIDAE: Beaver Family

The 2 species of beaver both lead semi-aquatic lives and are excellent swimmers. Their hind feet are webbed and they have broad, flat, hairless tails. Males and females look alike, but males tend to have larger anal scent glands. These glands produce a musky-smelling secretion, probably used for marking territory boundaries.

Beavers are always found near waterways surrounded by dense growths of trees such as willow, poplar, alder and birch. They feed on the bark, twigs, roots and leaves of these trees and use their enormous incisor teeth to fell trees for use in the construction of their complex dams and lodges.

Beavers make dams to create their desired living conditions. A pair starts by damming a stream with branches and mud to create a lake, deep enough not to freeze to the bottom in winter, in which to hoard a winter food supply of branches. A shelter with sleeping quarters is made of branches, by the dam or on an island or bank, or a burrow is dug in the river bank. When beavers have felled all the available trees in their territory, they dig canals into the woods to float back trees from farther afield. Most of the beavers' activity takes place at night.

NAME: **American Beaver,** *Castor canadensis*
RANGE: **N. America: Alaska to Texas**
HABITAT: **rivers, lakes, with wooded banks**
SIZE: **body: 73 cm–1.3 m (28¾ in–4¼ ft)
tail: 21–30 cm (8¼–11¾ in)**

One of the largest rodents, the American beaver weighs up to 27 kg (60 lb) or more. It is well adapted for its aquatic habits: the dense fur provides both waterproofing and insulation and its ears and nostrils can be closed off by special muscles when it is under water, allowing it to stay submerged for up to 15 minutes.

A beaver colony normally consists of an adult pair and their young of the present and previous years. Two-year-old young are driven out to form their own colonies. Autumn is a busy time for the beavers, when they must make repairs to the lodge and dam and stockpile food for the winter. They mate in midwinter and the young, usually 2 to 4, are born in the spring. The young are well developed at birth and are able to swim and feed themselves after about a month.

NAME: **Eurasian Beaver,** *Castor fiber*
RANGE: **now only in parts of Scandinavia, Poland, France, S. Germany, Austria and USSR**
HABITAT: **rivers, lakes, with wooded banks**
SIZE: **body: 73 cm–1.3 m (28¾ in–4¼ ft)
tail: 21–30 cm (8¼–11¾ in)**

The largest European rodent, the beaver has the same habits and much the same appearance as the American beaver, and they are considered by some experts to be only one species. Like its American counterpart, this beaver builds complex dams and lodges but, where conditions are right, may simply dig a burrow in the river bank which it enters under water. It feeds on bark and twigs in the winter and on all kinds of vegetation in summer.

Beavers are monogamous animals and females are believed to mate for life: the male may mate with females other than his partner. Pairs produce litters of up to 8, usually 2 to 4, young in the spring.

PEDETIDAE: Spring Hare Family

This African rodent family contains a single species, the spring hare. Its forelegs are short but the hind legs are relatively long and powerful, and it leaps along in hops of at least 3 m (10 ft). The long bushy tail acts as a counterbalance when the animal is travelling at high speed.

NAME: **Spring Hare,** *Pedetes capensis*
RANGE: **Kenya to South Africa**
HABITAT: **dry open country**
SIZE: **35–43 cm (13¾–17 in)
tail: 37–47 cm (14½–18½ in)**

When alarmed or travelling distances, spring hares bound along like kangaroos, but when feeding, they move on all fours. A nocturnal animal as a rule, the spring hare spends the day in its burrow, emerging at night to feed on bulbs, roots, grain and sometimes a few insects. Several burrows occur in the same area, some occupied by individuals, others by families. There is probably only one litter a year of 1, sometimes 2, young, born in the burrow.

ANOMALURIDAE: Scaly-tailed Squirrel Family

The 7 species of scaly-tailed squirrel are all tree-dwelling rodents, found in the forests of west and central Africa. Apart from a single "non-flying" species, they all have broad membranes at the sides of the body which can be stretched out to form a parachutelike structure, allowing the animal to glide through the air. Scaly-tailed squirrels are not closely related to true squirrels (Sciuridae).

NAME: **Beecroft's Flying Squirrel,** *Anomalurus beecrofti*
RANGE: **W. and central Africa**
HABITAT: **forest**
SIZE: **body: 30–40.5 cm (11¾–16 in)
tail: 23–43 cm (9–17 in)**

Beecroft's flying squirrel travels from tree to tree, rarely descending to the ground. With flight membranes extended, it leaps off one branch and glides up to 90 m (300 ft), to land on another tree. It finds all its food up in the trees and feeds on berries, seeds and fruit, as well as on some green plant material. Most of its activity takes place at night. These rodents generally live singly or in pairs and make dens in tree holes. They produce two litters a year of 2 or 3 young each.

NAME: **Zenker's Flying Squirrel,** *Idiurus zenkeri*
RANGE: **Cameroon, Zaire**
HABITAT: **forest**
SIZE: **body: 6–10 cm (2¼–4 in)
tail: 7.5–13 cm (3–5 in)**

This small flying squirrel has an unusual tail with long hairs projecting from each side, giving it a feathery appearance. Like its relatives, Zenker's squirrel is mainly nocturnal and feeds on berries, seeds and fruit. A gregarious species, it lives in holes in trees in groups of up to a dozen.

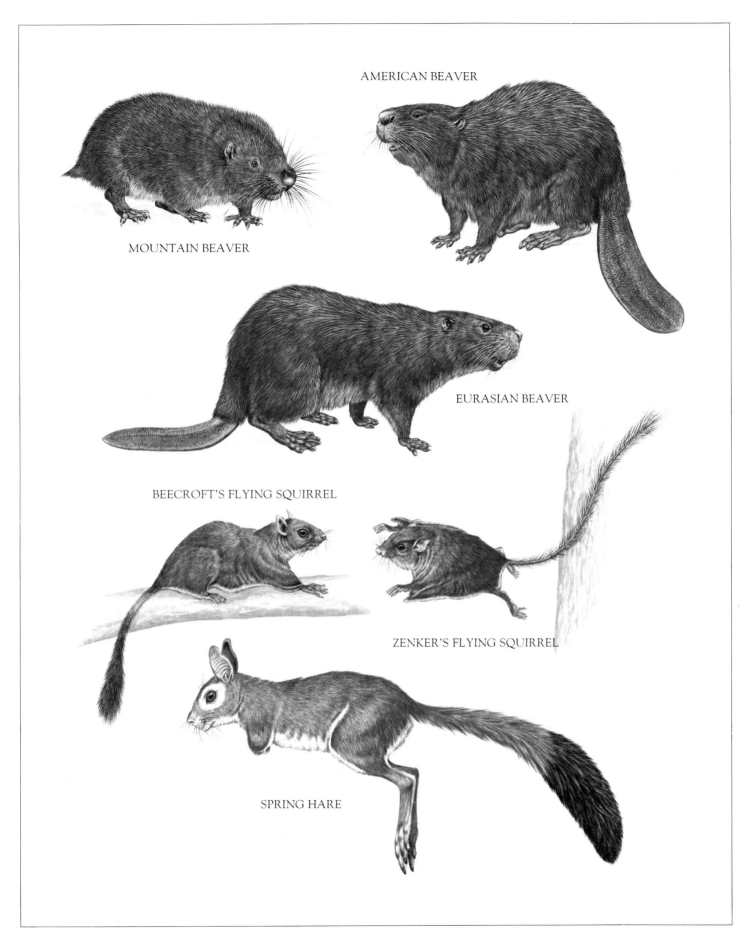

AMERICAN BEAVER

MOUNTAIN BEAVER

EURASIAN BEAVER

BEECROFT'S FLYING SQUIRREL

ZENKER'S FLYING SQUIRREL

SPRING HARE

New World Rats and Mice

NAME: Baja California Rice Rat,
Oryzomys peninsulae
RANGE: Mexico: tip of Baja California
HABITAT: damp land in dense cover
SIZE: body: 22.5–33 cm (9–13 in)
tail: 11–18 cm (4¼–7 in)

There are 100 species of rice rat, all looking much like the Baja California species and leading similar lives. They feed mostly on green vegetation, such as reeds and sedges, but also eat fish and invertebrates. They can become serious pests in rice fields and can cause severe damage to the plants. They weave grassy nests on reed platforms above water level or, in drier habitats, excavate burrows in which they breed throughout the year, producing up to 7 young in each litter.

NAME: Spiny Rice Rat, *Neacomys guianae*
RANGE: Colombia, east to Guyana
HABITAT: dense humid forest
SIZE: body: 7–10 cm (2¾–4 in)
tail: 7–10 cm (2¾–4 in)

This species is distinguished by the spiny coat, which grows thickly on the rat's back but sparsely on its flanks. Little is known of these rodents, which live on the floors of the most impenetrable forests, but it is thought that they breed the year round, producing litters of 2 to 4 young.

NAME: American Climbing Mouse,
Rhipidomys venezuelae
RANGE: N. Brazil, Venezuela, Guyana
HABITAT: dense forest
SIZE: body: 8–15 cm (3¼–6 in)
tail: 18–25 cm (7–9¾ in)

This secretive, nocturnal mouse lives in the deepest forest. Although it makes its nest in a burrow beneath the roots of a tree, it spends much of its life high in the tree-tops, feeding on lichens, small invertebrates and plants such as bromeliads. Equipped with strong, broad feet and long, sharp claws, it is an agile climber, and its long tail acts as a counterbalance when it jumps from one branch to the next. Climbing mice breed throughout the year, and the usual litter is 2 to 5 young.

NAME: Western Harvest Mouse,
Reithrodontomys megalotis
RANGE: USA: Oregon, south to Panama
HABITAT: grassland
SIZE: body: 5–14 cm (2–5½ in)
tail: 6–9 cm (2¼–3½ in)

Harvest mice tend to prefer overgrown pastures to cultivated farmland. In summer, they weave globular nests, up to 17.5 cm (7 in) in diameter, attached to stalks of vegetation. Litters of about 4 young are born in these nests after a gestation period of about 23 days.

HESPEROMYINAE:
New World Rats and Mice Subfamily

This is one of the 14 subfamilies of the huge rodent family Muridae, which contains rats, mice, voles, gerbils, hamsters and others. There are about 350 species in this subfamily and it is the largest mammalian group. Members occur in all habitats, from deserts to humid forests. These undistinguished but abundant little rodents are of immense importance as the primary consumers in their range and occupy a basic position in a number of food chains.

NAME: Deer Mouse, *Peromyscus maniculatus*
RANGE: Canada to Mexico
HABITAT: forest, grassland, scrub
SIZE: body: 12–22 cm (4¾–8½ in)
tail: 8–18 cm (3¼–7 in)

Deer mice are agile animals, running and hopping with ease through what sometimes seems quite impenetrable bush. They construct underground nests of dry vegetation and may move house several times a year. They have a catholic diet, consisting almost equally of plant and animal matter. Young deer mice start to breed at 7 weeks, and litters of up to 9 young are born after a gestation of between 3 and 4 weeks.

NAME: Golden Mouse, *Ochrotomys nuttalli*
RANGE: S.E. USA
HABITAT: brushy and thicketed scrub
SIZE: body: 8–9.5 cm (3¼–3¾ in)
tail: 7–9.5 cm (2¾–3¾ in)

This little mouse spends most of its life among the vines of wild honeysuckle and greenbrier. Here it weaves a solid-looking nest, which may house a whole family or just a single mouse. The mice also build rough feeding platforms in other spots, where they sit to consume seeds and nuts. Golden mice breed from spring to early autumn. The gestation period is about 4 weeks, and there are usually 2 or 3 young in a litter.

NAME: Northern Grasshopper Mouse,
Onychomys leucogaster
RANGE: S. Canada to N. Mexico
HABITAT: semi-arid scrub and desert
SIZE: body: 9–13 cm (3½–5 in)
tail: 3–6 cm (1¼–2¼ in)

The grasshopper mouse is largely carnivorous: grasshoppers and scorpions are its main prey, but it may even overpower and eat one of its own kind. These mice nest in burrows, which they dig themselves or find abandoned, and breed in spring and summer, producing litters of 2 to 6 young after a 33-day gestation.

NAME: South American Field Mouse,
Akodon arviculoides
RANGE: Brazil
HABITAT: woodland, cultivated land
SIZE: body: 11.5–14.5 cm (4½–5¾ in)
tail: 4.5–6.5 cm (1¾–2½ in)

There are more than 60 species of South American field mouse. These mice are active day and night, although most above-ground activity takes place at night; they feed on a wide range of plant matter. There are usually two litters each year, in November and March, with up to 7 young in a litter. Special breeding chambers in the burrows are set aside for pregnant and nursing females.

NAME: Arizona Cotton Rat, *Sigmodon arizonae*
RANGE: USA: Arizona, south to Mexico
HABITAT: dry grassland
SIZE: 12.5–20 cm (5–7¾ in)
tail: 7.5–12.5 cm (3–5 in)

Cotton rats are so abundant that they are sometimes declared a plague. They normally feed on plants and small insects, but when populations are high, they take the eggs and chicks of bobwhite quail as well as crayfish and fiddler crabs. They breed prolifically, producing their first litter of up to 12 young when just 10 weeks old.

NAME: White-throated Woodrat,
Neotoma albigula
RANGE: USA: California to Texas, south to Mexico
HABITAT: scrub, lightly forested land
SIZE: body: 28–40 cm (11–15¾ in)
tail: 7.5–18.5 cm (3–7¼ in)

A group of up to 100 of these rats will build a large nest of whatever material is available, such as twigs and pieces of cactus. The nest may be up to 2 m (6½ ft) in diameter and is usually situated in a pile of rocks or at the base of a tree. The rats dash in and out of the nest, foraging for shoots, fruit and other plant food. Litters contain 1 to 4 young.

NAME: Fish-eating Rat, *Ichthyomys stolzmanni*
RANGE: Colombia, south to Peru
HABITAT: near rivers and lakes
SIZE: body: 14.5–21 cm (5¾–8¼ in)
tail: 14.5–19 cm (5¾–7½ in)

Fish-eating rats are highly specialized rodents, well adapted to feeding on fish and some aquatic invertebrates, for they have partially webbed feet and swim strongly. The upper incisor teeth are simple, spikelike structures, used to spear the fish, which is then dragged ashore for consumption. The rats breed in burrows, which they dig in a river bank, and produce one or two litters of young each year.

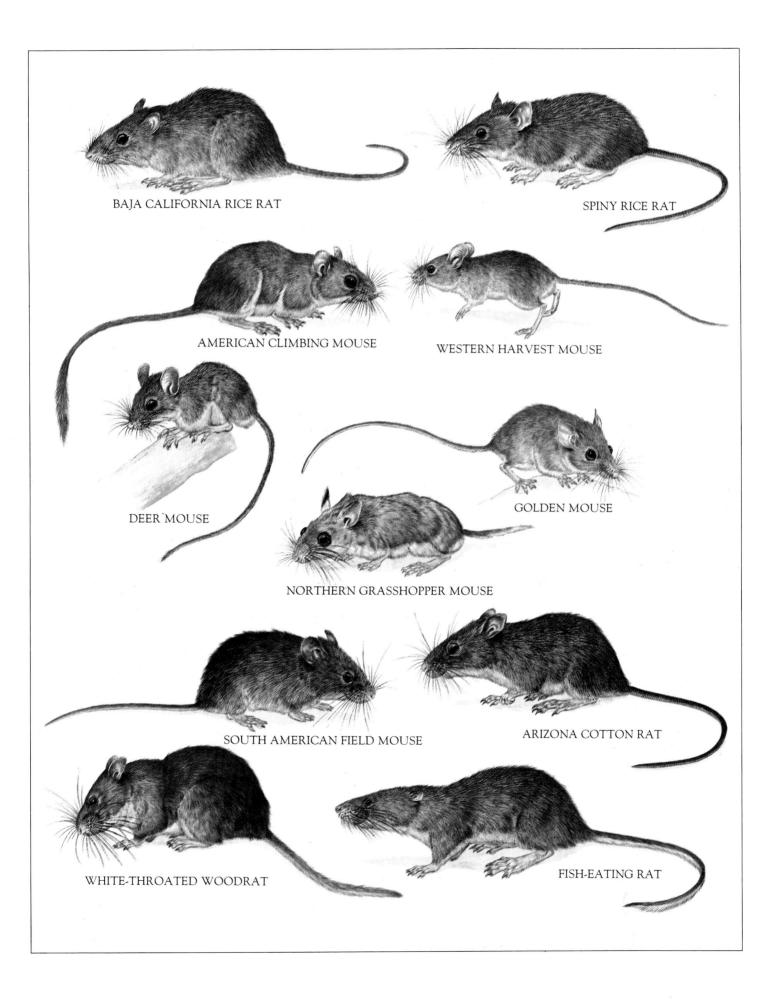

BAJA CALIFORNIA RICE RAT

SPINY RICE RAT

AMERICAN CLIMBING MOUSE

WESTERN HARVEST MOUSE

DEER MOUSE

GOLDEN MOUSE

NORTHERN GRASSHOPPER MOUSE

SOUTH AMERICAN FIELD MOUSE

ARIZONA COTTON RAT

WHITE-THROATED WOODRAT

FISH-EATING RAT

Hamsters, Mole-rats

CRICETINAE: Hamster Subfamily

The true hamsters are small burrowing rodents, found in the Old World from Europe eastward through the Middle East and central Asia. There are 15 species known, all of which are characterized by a body shape like that of a thick-set rat with a short tail. All species have capacious cheek pouches, used for carrying food back to the burrow; when full, the pouches may extend back beyond the level of the shoulder blades.

NAME: **Common Hamster,** *Cricetus cricetus*
RANGE: **W. Europe, USSR**
HABITAT: **grassland, cultivated land**
SIZE: **body: 22–30 cm (8½–11¾ in)**
tail: 3–6 cm (1¼–2¼ in)

The common hamster occupies a burrow system with separate chambers for sleeping and food storage. It feeds on seeds, grain, roots, potatoes, green plants and insect larvae. In late summer, it collects food for its winter stores. Grain is particularly favoured, and up to 10 kg (22 lb) may be hoarded. From October to March or April, the hamster hibernates, waking periodically to feed on its stores.

During the summer, females usually produce two litters, each of 6 to 12 young, which themselves bear when they are just 2 months old.

NAME: **Golden Hamster,** *Mesocricetus auratus*
RANGE: **Middle East**
HABITAT: **steppe**
SIZE: **body: 17–18 cm (6¾–7 in)**
tail: 1.25 cm (½ in)

Golden hamsters are primarily nocturnal creatures, but they may be active at times during the day. Adults live alone in burrow systems which they dig. They are omnivorous, feeding on vegetation, seeds, fruit and even small animals. Their cheek pouches are large and, when filled, double the width of head and shoulders.

Golden hamsters are highly aggressive and solitary creatures, and females must advertise clearly when they are sexually receptive. They do this by applying a specific vaginal secretion to rocks and sticks in their territories, and cease to mark as soon as the receptive phase of the oestrous cycle is over. The usual litter contains 6 or 7 young, and there may be several litters a year.

The domesticated strain of the species makes a popular pet.

NAME: **Striped Hamster,** *Phodopus sungorus*
RANGE: **USSR, N. China**
HABITAT: **arid plains, sand-dunes**
SIZE: **body: 5–10 cm (2–4 in)**
tail: absent

This small, yet robust, hamster is active at night and at dawn and dusk. Little is known about it, but its habits seem to be similar to those of other hamsters. It feeds on seeds and plant material and fills its cheek pouches with food to carry back to its burrow.

Striped hamsters are more sociable than their golden cousins and may occur in quite large colonies. They are so well adapted to their cool habitat that they do not breed well at room temperature. Litters contain 2 to 6 young, which are weaned at 21 days. Females mate again immediately after giving birth, so can produce litters at three-weekly intervals.

SPALACINAE: Blind Mole-rat Subfamily

This small subfamily of highly specialized burrowing rodents probably contains only the 3 species found in the eastern Mediterranean area, the Middle East, southern USSR and northern Africa. They are heavy-bodied rodents, with short legs, small feet and a remarkable absence of external projections: there is no tail, external ears are not apparent, and there are no external openings for the tiny eyes that lie buried under the skin.

NAME: **Lesser Mole-rat,** *Spalax leucodon*
RANGE: **S.E. Europe; Africa: Libya**
HABITAT: **grassland and cultivated land**
SIZE: **body: 15–30.5 cm (6–12 in)**
tail: absent

Mole-rats live in complex burrow systems, with many chambers and connecting tunnels, which they dig with their teeth and heads, rather than with their feet. They feed underground on roots, bulbs and tubers but may occasionally venture above ground at night to feed on grasses, seeds and even insects. One litter of 2 to 4 young is born in early spring, after a gestation of about 4 weeks.

MYOSPALACINAE: Eastern Asiatic Mole-rat Subfamily

There are 4 species in this group, found in USSR and China. These mole-rats are stocky, burrowing rodents, equipped with heavily clawed limbs for digging. They do not have external ears, but their tiny eyes are apparent, and there is a short, tapering tail.

NAME: **Common Chinese Zokor,** *Myospalax fontanieri*
RANGE: **China: Szechuan to Hopei Provinces**
HABITAT: **grassland, steppe**
SIZE: **body: 15–27 cm (6–10½ in)**
tail: 3–7 cm (1¼–2¾ in)

The zokor lives in long burrows, which it digs with amazing speed among the roots of trees and bushes, using its long, sharp claws. As it goes, it leaves behind a trail of "mole hills" on the surface. Grain, roots and other underground parts of plants are its main foods, and the zokor ventures above ground only occasionally at night, when it runs the risk of being caught by an owl.

RHIZOMYINAE: Mole- and Bamboo Rat Subfamily

There are 6 species in this subfamily which fall into two closely related groups: 2 species of East African mole-rats and 4 species of Southeast Asian bamboo rats. All are plant-eating, burrowing rodents.

NAME: **Giant Mole-rat,** *Tachyoryctes macrocephalus*
RANGE: **Africa: Ethiopia**
HABITAT: **montane grassland**
SIZE: **body: 18–25.5 cm (7–10 in)**
tail: 5–8 cm (2–3 in)

The giant mole-rat has a stout, molelike body and small, yet functional, eyes. It is a powerful burrower, equipped with short, strong limbs and claws. When a pile of soil has built up behind it, the animal turns and pushes the soil to the surface with the side of its head and one forefoot. It is active both day and night and feeds on plant material above and below ground.

Surprisingly for such a common species, nothing is known of the breeding habits of this mole-rat.

NAME: **Bamboo Rat,** *Rhizomys sumatrensis*
RANGE: **Indo-China, Malaysia, Sumatra**
HABITAT: **bamboo forest**
SIZE: **body: 35–48 cm (13¾–18¾ in)**
tail: 10–15 cm (4–6 in)

The bamboo rat has a heavy body, short legs and a short, almost hairless tail. Its incisor teeth are large and strong, and it uses these and its claws for digging. It burrows underground near clumps of bamboo, the roots of which are its staple diet. It also comes out of its burrow to feed on bamboo above ground and on other plants, seeds and fruit.

The usual litter is believed to contain 3 to 5 young, and there may be more than one litter a year.

COMMON HAMSTER

GOLDEN HAMSTER

STRIPED HAMSTER

LESSER MOLE-RAT

COMMON CHINESE ZOKOR

GIANT MOLE-RAT

BAMBOO RAT

Crested Rat,
Spiny Dormice and relatives

LOPHIOMYINAE:
Crested Rat Subfamily
A single species is known in this group: the crested or maned rat, found in dense mountain forests. The animal's general appearance is not at all ratlike, for it is about the size of a guinea pig and has long, soft fur and a thick, bushy tail. The female is generally larger than the male.

NAME: **Crested Rat**, *Lophiomys imhausi*
RANGE: E. Africa
HABITAT: forest
SIZE: body: 25.5–36 cm (10–14¼ in)
tail: 14–18 cm (5½–7 in)

Crested rats are skilled climbers, leaving their daytime burrows at night to collect leaves, buds and shoots among the trees. Along the neck, back and part of the tail is a prominent mane of coarse hairs, which can be erected when the animal is excited or alarmed. The raised crest exposes a long scent gland, which produces a stifling odour to deter predators. The hairs lining the gland have a unique, wicklike structure to help broadcast the odour. Another curious feature of these rodents is the reinforced skull, the significance of which is not known.

NESOMYINAE:
Madagascan Rat Subfamily
This subfamily includes 11 species, of which 10 are found only in Madagascar, and 1 occurs in South Africa. It is a varied group, and the species seem to have become adapted to fill different ecological niches, but they do have some common features. It has been suggested, however, that they are not all of common ancestry and that a considerable amount of evolutionary convergence has occurred.

NAME: **Madagascan Rat**, *Nesomys rufus*
RANGE: Madagascar
HABITAT: forest
SIZE: body: 19–23 cm (7½–9 in)
tail: 16–19 cm (6¼–7½ in)

This species of Madagascan rat is mouselike in appearance, with long, soft fur and a light-coloured belly. Its hind feet are long and powerful, and the middle three toes are elongated. It seems likely that the Madagascan rat is an adept climber, using its sharp claws to grip smooth bark.

Although little is known of this creature's habits, it is probable that its diet is made up of small invertebrates, buds, fruit and seeds. Nothing is known of its breeding cycle.

NAME: **White-tailed Rat**, *Mystromys albicaudatus*
RANGE: South Africa
HABITAT: grassland, arid plains
SIZE: body: 14–18 cm (5½–7 in)
tail: 5–8 cm (2–3 in)

The white-tailed rat is the only member of its group to occur outside Madagascar. It is a nocturnal creature and spends the day in an underground hole, emerging at dusk to feed on seeds and other plant material. It is said that the strong smell of these rodents repels mammalian predators, such as suricates and mongooses, but they are caught by barn, eagle and grass owls.

White-tailed rats appear to breed throughout the year, producing litters of 4 or 5 young. A curious feature of the early development of these rats is that the young become firmly attached to the female's nipples and are carried about by her. They only detach themselves when about 3 weeks old.

PLATACANTHOMYINAE:
Spiny Dormouse Subfamily
There are 2 species in this subfamily. Their common name originates from the flat, pointed spines that are intermixed with the fur, particularly on the back. Spiny dormice are well adapted for tree-climbing, for their feet are equipped with sharp claws and padded soles, and the digits spread widely.

NAME: **Malabar Spiny Dormouse**, *Platacanthomys lasiurus*
RANGE: S. India
HABITAT: forest, rocks
SIZE: body: 13–21 cm (5–8¼ in)
tail: 7.5–10 cm (3–4 in)

The Malabar spiny dormouse lives in trees and feeds on seeds, grain and fruit. The long tail, with its dense, bushy tip, is used as a balancing aid when the animal is moving in trees. A nest of leaves and moss is made for shelter in a hole in a tree or among rocks. This rodent sometimes becomes an agricultural pest because of its diet, and it has been known to destroy quantities of crops such as ripe peppers.

Although spiny dormice are not uncommon, nothing is known of the breeding habits of these secretive rodents. The other species of the subfamily, the Chinese pygmy dormouse, *Typhlomys cinereus*, lives in forest in southeast China.

OTOMYINAE:
African Swamp Rat Subfamily
There are about 12 species in this group, all found in Africa, south of the Sahara to Cape Province. They inhabit a range of habitats and climatic zones, including mountains, arid regions and swampy areas, and are all competent swimmers. A characteristic of the group is that the members have only two pairs of mammary glands, situated in the lower abdomen.

NAME: **Swamp Rat**, *Otomys irroratus*
RANGE: Zimbabwe to South Africa
HABITAT: damp grassland, swamps
SIZE: 13–20 cm (5–7¾ in)
tail: 5–17 cm (2–6¾ in)

A plump-bodied rodent, the African swamp rat has a rounded, volelike head and small ears. Characteristic features are the grooves on each side of the incisor teeth. Active day and night, this rat feeds on seeds, berries, shoots and grasses. It will enter water readily and even dive to escape danger. Its nest is usually above ground and is made of plant material, although in some areas swamp rats make use of burrows discarded by other species.

Young females reach sexual maturity at 10 weeks of age, males about 3 weeks later. Although swamp rats seldom damage man's crops, their parasites do transmit tick-bite fever and possibly bubonic plague. Swamp rats are an important food source for many larger predators.

NAME: **Karroo Rat**, *Parotomys brantsi*
RANGE: South Africa: Cape Province
HABITAT: sandy plains
SIZE: body: 13.5–17 cm (5¼–6¾ in)
tail: 7.5–12 cm (3–4¾ in)

Karroo rats are gregarious animals which live in colonies. They dig burrows and sometimes also build nests of sticks and grass above the burrows. They are nervous, wary animals and stay near their shelters most of the time. Leaves of the saltbush tree provide the bulk of their diet, and they generally feed during the day. Every 3 or 4 years, their population increases dramatically, and they feed on agricultural crops, causing considerable damage.

Karroo rats are believed to breed about four times a year, producing litters of 2 to 4 young.

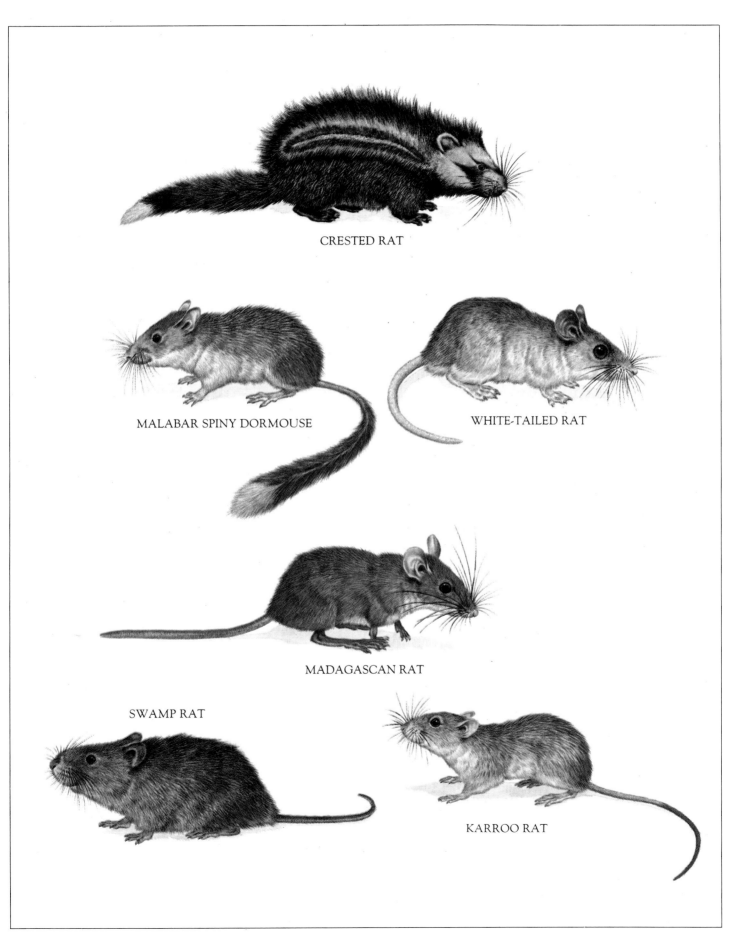

CRESTED RAT

MALABAR SPINY DORMOUSE

WHITE-TAILED RAT

MADAGASCAN RAT

SWAMP RAT

KARROO RAT

Voles and Lemmings

NAME: **Norway Lemming,** *Lemmus lemmus*
RANGE: **Scandinavia**
HABITAT: **tundra, grassland**
SIZE: **body: 13–15 cm (5–6 in)**
 tail: 2 cm ($\frac{3}{4}$ in)

The boldly patterned Norway lemming is active day and night, alternating periods of activity with short spells of rest. Grasses, shrubs and particularly mosses make up its diet; in winter it clears runways under the snow on the ground surface in its search for food. These lemmings start to breed in spring, under the snow, and may produce as many as eight litters of 6 young each throughout the summer.

Lemmings are fabled for their dramatic population explosions, which occur approximately every three or four years. It is still not known what causes these, but a fine, warm spring following two or three years of low population usually triggers an explosion that year or the next. As local populations swell, lemmings are forced into surrounding areas. Gradually more and more are driven out, down the mountains and into the valleys. Many are eaten by predators, and more lose their lives crossing rivers and lakes. Lemmings do not deliberately commit suicide.

NAME: **Southern Bog Lemming,** *Synaptomys cooperi*
RANGE: **N.E. USA; S.E. Canada**
HABITAT: **bogs, meadows**
SIZE: **body: 8.5–11 cm (3$\frac{1}{4}$–4$\frac{1}{4}$ in)**
 tail: 2 cm ($\frac{3}{4}$ in)

Sociable animals, southern bog lemmings live in colonies of up to about 30 or so. They make burrows just under the ground surface and also clear a network of paths, or runways, on the surface, where they keep the grass cut and trimmed; little piles of cuttings punctuate these runways. Bog lemmings are active day and night. They have powerful jaws and teeth and feed largely on plant material. Breeding continues throughout the spring and summer, and females produce two or three litters a year of 1 to 4 young each.

NAME: **Sagebrush Vole,** *Lagurus curtatus*
RANGE: **W. USA**
HABITAT: **arid plains**
SIZE: **body: 9.5–11 cm (3$\frac{3}{4}$–4$\frac{1}{4}$ in)**
 tail: 1.5–3 cm ($\frac{1}{2}$–1$\frac{1}{4}$ in)

As its common name suggests, this pale-coloured vole is particularly common in some areas of arid plains where the sagebrush is abundant. It makes shallow burrows near the ground surface and is active at any time of the day and night. It feeds on the sagebrush and on other green vegetation and produces several litters a year of 4 to 6 young each.

MICROTINAE:
Vole and Lemming Subfamily

There are about 110 species in this group of rodents, found in North America, northern Europe and Asia; lemmings tend to be confined to the more northerly regions. All species are largely herbivorous and they usually live in groups or colonies. Many dig shallow tunnels, close to the ground surface, and clear paths through the grass within their range. Others are partly aquatic and some species climb among low bushes.

Local populations of voles and lemmings are subject to cyclical variation in density, with boom years of greatly increased population occurring every few years.

NAME: **Southern Mole-vole,** *Ellobius fuscocapillus*
RANGE: **central Asia**
HABITAT: **grassy plains**
SIZE: **body: 10–15 cm (4–6 in)**
 tail: 0.5–2 cm ($\frac{1}{4}$–$\frac{3}{4}$ in)

The mole-vole is a more habitual burrower than other voles and lemmings and is more specifically adapted. Its snout is blunt, its eyes and ears small to minimize damage from the soil, and it has short, strong legs. It probably uses its teeth to loosen the soil when burrowing, for while its incisors are stronger than is usual in voles, its claws, although adequate, are not as strong as those of other burrowing rodents.

Mole-voles feed on roots and other underground parts of plants. Like moles, they make shallow tunnels in which to search for food and deeper, more permanent tunnels for nesting. They breed at any time of year, probably according to the availability of food, and produce 3 or 4 young at a time.

NAME: **Bank Vole,** *Clethrionomys glareolus*
RANGE: **Europe (not extreme north or south), east to USSR**
HABITAT: **woodland**
SIZE: **body: 8–11 cm (3$\frac{1}{4}$–4$\frac{1}{4}$ in)**
 tail: 3–6.5 cm (1$\frac{1}{4}$–2$\frac{1}{2}$ in)

The bank vole feeds on softer plant material than most voles. It will climb on bushes to find its food, eating buds, leaves and fruit, as well as some insects. It is active night and day, with several rest periods, and, like other voles, it clears well-defined runways in the grass and makes shallow tunnels. Nests are usually made under logs or among tree-roots, and in summer females produce several litters of 3 to 5 young each.

NAME: **Meadow Vole,** *Microtus pennsylvanicus*
RANGE: **Canada, N. USA**
HABITAT: **grassland, woodland, often near water**
SIZE: **body: 9–12.5 cm (3$\frac{1}{2}$–5 in)**
 tail: 3.5–6.5 cm (1$\frac{1}{4}$–2$\frac{1}{2}$ in)

The meadow vole is a highly adaptable species, found in a wide range of habitats. It is a social animal, but each adult has its own territory. The voles clear runways in the grass, which they keep trimmed, and feed on plant material such as grass, seeds, roots and bark. A nest of grass is made in the ground or in a shallow burrow under the runways.

The female is a prolific breeder, producing at least three and often as many as twelve or thirteen litters a year of up to 10 young each. The gestation period is 3 weeks, and females start to breed at only 3 weeks old.

NAME: **Muskrat,** *Ondatra zibethicus*
RANGE: **Canada, USA; introduced in Europe**
HABITAT: **marshes, freshwater banks**
SIZE: **body: 25–36 cm (9$\frac{3}{4}$–14$\frac{1}{4}$ in)**
 tail: 20–28 cm (7$\frac{3}{4}$–11 in)

An excellent swimmer, this large rodent spends much of its life in water and has webbed hind feet and a long, naked, vertically flattened tail which it uses as a rudder. It feeds on aquatic and land vegetation and occasionally on some mussels, frogs and fish. It usually digs a burrow in the bank of a river, but where conditions permit, it builds a lodge from plant debris in shallow water and inside it constructs a dry sleeping platform above water level. The lodge may shelter as many as 10 animals.

At the onset of the breeding season, the muskrats' groin glands enlarge and produce a musky secretion, believed to attract male and female to one another. They breed from April to August in the north, and throughout the winter in the south of their range. Two or three litters of 3 or 4 young are born after a gestation of 29 or 30 days.

NAME: **European Water Vole,** *Arvicola terrestris*
RANGE: **Europe, east to E. Siberia**
HABITAT: **freshwater banks, grassland**
SIZE: **body: 14–19 cm (5$\frac{1}{2}$–7$\frac{1}{2}$ in)**
 tail: 4–10 cm (1$\frac{1}{2}$–4 in)

Although competent in water, the European water vole is less agile than the more specialized muskrats and beavers. It makes a burrow in the bank of a river or stream or burrows into the ground, if far from water. Grasses and other plant material are its main food. Water voles breed in summer, producing several litters of 4 to 6 young.

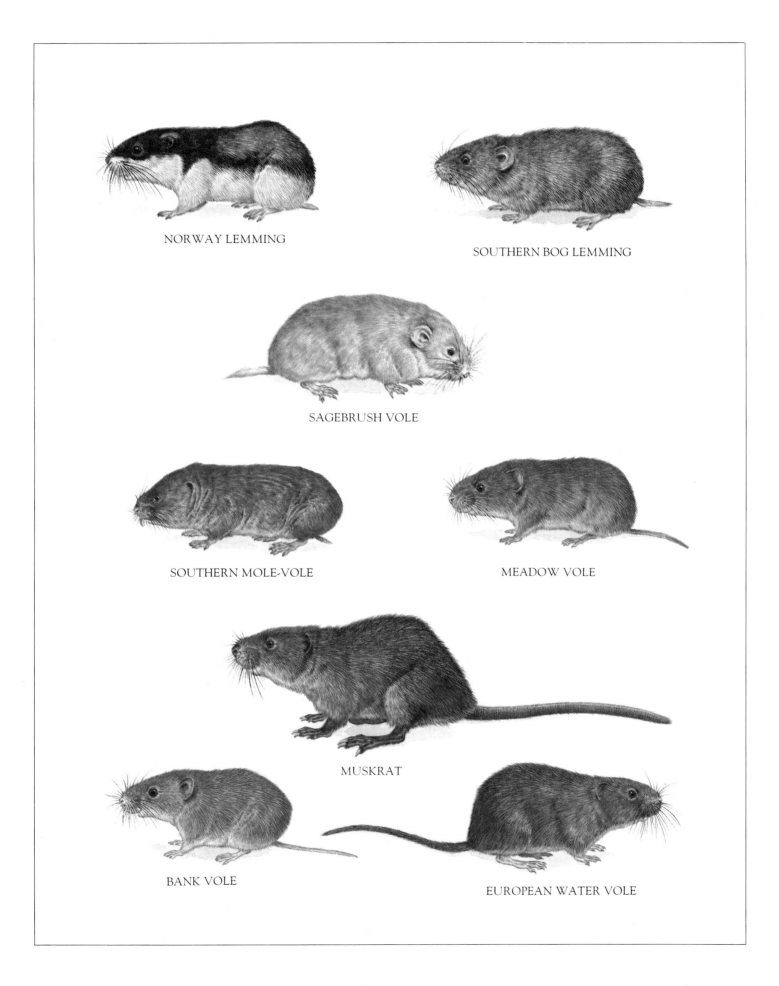

NORWAY LEMMING

SOUTHERN BOG LEMMING

SAGEBRUSH VOLE

SOUTHERN MOLE-VOLE

MEADOW VOLE

MUSKRAT

BANK VOLE

EUROPEAN WATER VOLE

Gerbils

NAME: Large North African Gerbil,
Gerbillus campestris
RANGE: Africa: Morocco to Somalia
HABITAT: sandy desert
SIZE: body: 10–14 cm (4–5½ in)
   tail: 11–12 cm (4¼–4¾ in)

There are about 25 species in the genus *Gerbillus*, all found from Morocco eastward to Pakistan. All occupy the driest deserts and eke out their existence in the most inhospitable environments. The large North African gerbil occurs in groups of 12 or more, living in simple, poorly made burrows, dug in the sand. The gerbils remain hidden by day, emerging at dusk to search for insects, seeds and windblown vegetable matter. They never have to drink but derive all the water they need from the fats contained in seeds. They breed throughout the year, producing litters of up to 7 young after a gestation period of 20 or 21 days.

NAME: South African Pygmy Gerbil,
Gerbillurus paeba
RANGE: S. Africa: S.W. Angola to Cape Province
HABITAT: desert
SIZE: body: 8–9 cm (3¼–3½ in)
   tail: 6.5–7 cm (2½–2¾ in)

Of the 4 species of pygmy gerbil which occur in the southwestern corner of Africa, *G. paeba* is the most widespread. It constructs simple burrows in sandy or gravelly soil, usually with one entrance higher than the other to improve ventilation. Pygmy gerbils feed on whatever plant and animal food is available. When the desert plants bloom, the gerbils lay in stores of seeds and fruit in underground larders. They usually breed about twice a year, but in times of abundant food supplies may breed up to four times in a year. The young are fed by their mother for up to a month.

NAME: Greater Short-tailed Gerbil,
Dipodillus maghrebi
RANGE: Africa: N. Morocco
HABITAT: upland, arid semi-desert
SIZE: body: 9–12.5 cm (3½–5 in)
   tail: 6.5–7 cm (2½–2¾ in)

Among the many gerbil species in North Africa, the greater short-tailed gerbil occupies a specific and specialized ecological niche. It inhabits the foothills and mid-regions of the Atlas mountains and lives among boulder fields and rock scree. Like all gerbils, it is an active animal, emerging at night to forage among the sparse vegetation for seeds, buds and insects. It may have to travel considerable distances each night to find food.

Litters of about 6 young are born in brood chambers underground.

GERBILLINAE: Gerbil Subfamily

There are some 70 or so members of this subfamily of rodents, all of which come from central and western Asia and Africa. They are all well adapted for arid conditions and many occur only in apparently inhospitable deserts. Chief among their adaptations is a wonderfully efficient kidney, which produces urine several times more concentrated than in most rodents, thus conserving moisture. Water loss from the lungs is a major problem for desert-dwelling animals, and all gerbils have specialized nose bones which act to condense water vapour from the air before it is expired. This essential water is then reabsorbed into the system.

To keep their bodies as far as possible from the burning sand, gerbils have long hind legs and feet, the soles of which are insulated with dense pads of fur. Their bellies are pure white to reflect radiated heat. Finally, gerbils adapt to desert life by being strictly nocturnal, never emerging from their burrows until the heat of the day has passed.

Gerbils are all seed-eaters and make large stores of food during the brief periods when the desert blooms. They never occur in great densities, but they are abundant enough to be an important source of food for predators such as fennec foxes and snakes.

NAME: Great Gerbil, *Rhombomys opimus*
RANGE: Iran, east to Mongolia and China
HABITAT: arid scrubland
SIZE: body: 16–20 cm (6¼–7¾ in)
   tail: 13–16 cm (5–6¼ in)

The great gerbil occupies a wide range of habitats, from the cold central Asiatic mountains to the Gobi desert, with its high summer temperatures. It is an adaptable animal, which changes its behaviour to suit its environment. In winter, its activity is in inverse proportion to the depth of snow, and in some colonies the gerbils come to the surface only rarely. Large colonies can do much damage to crops and irrigation channels and great gerbils are considered pests in parts of the USSR.

They are herbivorous animals and build up stores of 60 kg (130 lb) or more of plant material in their burrows on which to live in winter. During the winter, huge numbers of these gerbils are preyed on by owls, stoats, mink and foxes. In spring, however, their rapid breeding soon replenishes the colonies.

NAME: Indian Gerbil, *Tatera indica*
RANGE: W. India, Sri Lanka
HABITAT: plains, savanna, arid woodland
SIZE: body: 15–19 cm (6–7½ in)
   tail: 20–25 cm (7¾–9¾ in)

The Indian gerbil is a sociable animal, living communally in deep burrow systems with many entrances. Often these entrances are loosely blocked with soil to discourage the entry of predatory snakes and mongooses. Sometimes the populations of this species increase to such an extent that the animals leave their normal habitat and invade fields and gardens in search of bulbs, roots, green vegetation, insects and even eggs and young birds. Indian gerbils breed throughout the year, producing litters of up to 8 young. It is thought that these animals are a reservoir of bubonic plague.

NAME: Fat-tailed Gerbil, *Pachyuromys duprasi*
RANGE: Africa: Algerian Sahara to S.W. Egypt
HABITAT: sandy desert
SIZE: body: 10.5–14 cm (4¼–5¼ in)
   tail: 4.5–6 cm (1¾–2¼ in)

This little gerbil derives its name from its habit of storing fat in its stubby tail. During periods when food is abundant, the tail enlarges in size and may even become too fat to be carried; in lean times, the fat is used up and the tail decreases again. These gerbils spend their days in underground burrows, emerging at night to search for whatever seeds and grubs they can find in the scant vegetation. They have enormous ear bones and very acute hearing, which may help them to locate underground insects. Litters of about 6 young are produced throughout the year and the gestation period is 19 to 22 days. The young gerbils are independent at about 5 weeks old.

NAME: Fat Sand Rat, *Psammomys obesus*
RANGE: Algeria, east to Saudi Arabia
HABITAT: sandy desert
SIZE: body: 14–18.5 cm (5½–7¼ in)
   tail: 12–15 cm (4¾–6 in)

The fat sand rat overcomes the problem of the unpredictability of desert food supplies by laying down a thick layer of fat all over its body when food is abundant. It then lives off this fat when food is short. Active day and night, this gerbil darts about collecting seeds and other vegetation which it carries back to its burrow. In early spring, a brood chamber is made and lined with finely shredded vegetation, and the first litter of the year is born in March. There are usually 3 to 5 young in a litter and the breeding season continues until late summer.

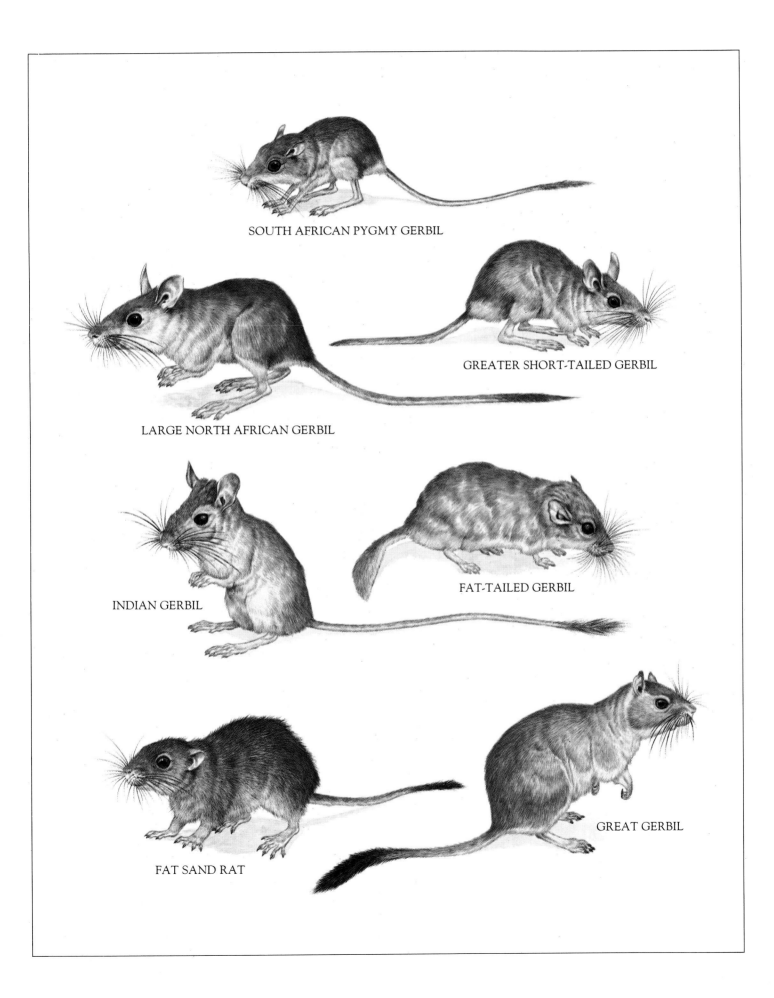

SOUTH AFRICAN PYGMY GERBIL

GREATER SHORT-TAILED GERBIL

LARGE NORTH AFRICAN GERBIL

INDIAN GERBIL

FAT-TAILED GERBIL

FAT SAND RAT

GREAT GERBIL

Climbing Mice, Pouched Rats, Island Water Rats

DENDROMURINAE:
African Climbing Mouse Subfamily

There are about 21 species of climbing mice, all of which occur in Africa, south of the Sahara. They are linked by certain skull and tooth characteristics but otherwise they are quite a varied group. Most are agile climbers and have an affinity for tall vegetation, but some species are ground-living. Many forms have extremely long tails which they wrap around stems and twigs for extra stability. Although abundant, these rodents do not occur in the large groups characteristic of mice and voles. Many species live in extremely dense forest.

One climbing mouse, *Dendroprionomys rousseloti*, is known only from three specimens caught in Zaire, but this may indicate the degree of research effort rather than rarity.

NAME: **African Climbing Mouse, *Dendromus mesomelas***
RANGE: **Cameroon to Ethiopia, south to South Africa**
HABITAT: **swamps**
SIZE: **body: 6–10 cm (2¼–4 in)**
 tail: 7–12 cm (2¾–4¾ in)

The most striking feature of this little rodent is its prehensile tail, which acts as a fifth limb when the mouse climbs the stems of plants. Climbing mice are strictly nocturnal and are always on the alert for owls, mongooses and other predators. They feed on berries, fruits and seeds, and occasionally search for lizards and the eggs and young of small birds. Sometimes they manage to climb into the suspended nests of weaver finches, and there are records of them establishing their own nests in these secure hammocks, although they usually build nests of stripped grass at the base of grass stems.

Climbing mice breed throughout the year, producing litters of 3 to 5 young.

NAME: **Fat Mouse, *Steatomys krebsi***
RANGE: **Angola, Zambia, south to South Africa**
HABITAT: **dry, open, sandy plains**
SIZE: **body: 7–10 cm (2¾–4 in)**
 tail: 4–4.5 cm (1½–1¾ in)

Fat mice are adapted for life in the more seasonal parts of southern Africa. During the rainy season, when seeds, bulbs and insects are abundant, the mice gorge themselves, becoming quite fat. In the dry season, when plant growth stops, they live off their stores of fat. They live singly or in pairs in underground burrows, breeding in the wet season and producing litters of 4 to 6 young.

CRICETOMYINAE:
African Pouched Rat Subfamily

There are about 6 species of African pouched rat, all with a pair of deep cheek pouches for the transportation of food. They occur south of the Sahara and occupy a range of habitats, from sandy plains to the densest forests. One species has taken to living close to human refuse dumps, scavenging for a living.

Normally African pouched rats live alone in underground burrow systems, which they construct or else take over from other species. The burrows include separate chambers for sleeping, excreting and, if the inhabitant is female, breeding. Special larder chambers are used as food stores for the dry season. Despite the complexity of their homes, pouched rats move every few weeks to a new burrow.

NAME: **Giant Pouched Rat, *Cricetomys emini***
RANGE: **Sierra Leone to Malawi**
HABITAT: **dense forest**
SIZE: **body: 25–45 cm (9¾–17¾ in)**
 tail: 36–46 cm (14¼–18 in)

As its name implies, the giant pouched rat is a substantial rodent, weighing up to about 1 kg (2¼ lb). It lives in dark forests or areas of dense scrub, emerging from its burrow at night to forage for roots, tubers, fruit and seeds. Some food is eaten where it is found, but much is taken back to the burrow in the rat's capacious cheek pouches. Giant rats live singly, associating only briefly with others for mating, which occurs at all times of the year. Litters contain 2 or 3 young, born after a gestation of about 6 weeks.

NAME: **Long-tailed Pouched Rat, *Beamys hindei***
RANGE: **S. Kenya**
HABITAT: **forest**
SIZE: **body: 10–11 cm (4–4¼ in)**
 tail: 9.5–10 cm (3¾–4 in)

The long-tailed pouched rat occurs in a restricted part of central Africa, although the related *B. major* extends as far south as northern Zimbabwe. This rat seems to spend much of its life underground in a complex burrow system which it constructs. It feeds mainly on the underground storage organs of plants — tubers and bulbs — although the presence of seeds in its food stores suggests that it spends some time above ground searching for food. Groups of up to two dozen rats live together, breeding at all times of the year. Litters contain 1 to 5 young, which are mature at about 5½ months.

HYDROMYINAE:
Island Water Rat Subfamily

There are about 17 species of these large rats, most of which inhabit rivers and swamps in Australia, New Guinea and its associated islands and the Philippines. All share certain family characteristics of teeth and skull. The origin of the family is uncertain, and it is not known how it spread over such a wide part of the globe. Some species, such as the Australian water rat, *Hydromys chrysogaster*, are common, while others, such as the false water rat of Australia, *Xeromys myoides*, are extremely rare. Little is known about most of these shy, retiring creatures.

NAME: **Australian Water Rat, *Hydromys chrysogaster***
RANGE: **Tasmania, Australia, New Guinea; Aru, Kai and Bruni Islands**
HABITAT: **swamps, streams, marshes**
SIZE: **body: 20–35 cm (7¾–13¾ in)**
 tail: 20–35 cm (7¾–13¾ in)

This imposing rodent has a sleek, streamlined appearance, well suited to its aquatic habits. Its partially webbed hind feet enable it to perform complex manoeuvres in the water. It feeds on small fish, frogs, crustaceans and water birds which it pursues and captures. Catches are often taken to a special food store for later consumption.

In the south of their range, water rats breed in early spring, producing a litter of 4 or 5 young; farther north they may breed all year round.

NAME: **Eastern Shrew Mouse, *Pseudohydromys murinus***
RANGE: **N.E. New Guinea**
HABITAT: **montane forest**
SIZE: **body: 8.5–10 cm (3¼–4 in)**
 tail: 9–9.5 cm (3½–3¾ in)

The Eastern shrew mouse is known from only four specimens taken at between 2,100 and 2,700 m (6,900 and 9,000 ft) in dense forest. However, the species is probably not as rare as this suggests, since there has been little organized survey work in this remote area. Little is known about the natural history of this animal, but since it does not have webbed feet, it is probably less aquatic than its relatives. Its long, strong tail may aid in climbing, and it probably spends much of its time seeking food in the shrub layer of the rain forest.

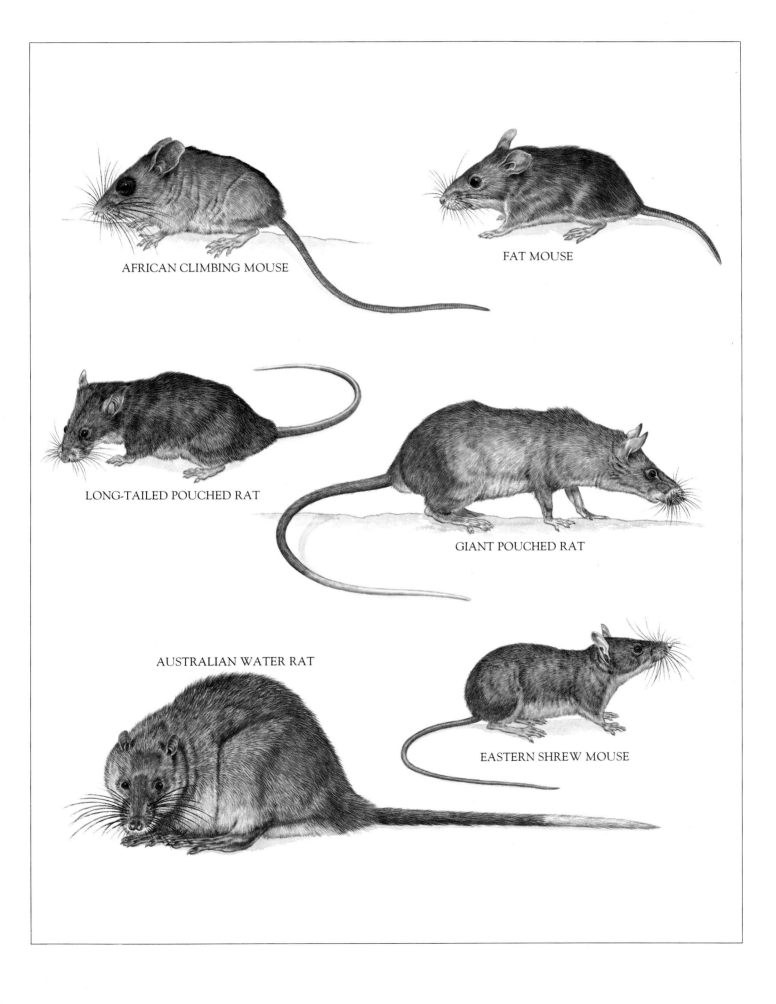

AFRICAN CLIMBING MOUSE

FAT MOUSE

LONG-TAILED POUCHED RAT

GIANT POUCHED RAT

AUSTRALIAN WATER RAT

EASTERN SHREW MOUSE

Old World Rats and Mice

NAME: **Harvest Mouse**, *Micromys minutus*
RANGE: **Europe, east to the Urals**
HABITAT: **hedgerows, reedbeds**
SIZE: **body: 5.5–7.5 cm (2¼–3 in)**
 tail: 5–7.5 cm (2–3 in)

These attractive little mice are among the smallest rodents: a fully grown male weighs about 7 g (¼ oz). Harvest mice build tennis-ball-sized nests of finely stripped grass among reed stems or grass heads. Here, safe from the attentions of owls or weasels, the litter of up to 12 young is born and reared. Adult harvest mice feed on seeds and small insects. They are the only Old World mammals to have truly prehensile tails.

NAME: **Wood Mouse**, *Apodemus sylvaticus*
RANGE: **Ireland, east to central Asia**
HABITAT: **forest edge**
SIZE: **body: 8–13 cm (3¼–5¼ in)**
 tail: 7–9.5 cm (2¾–3¾ in)

The wood mouse is one of the commonest of the European small rodents. Strictly nocturnal, wood mice emerge from their nests under the roots of trees in the evening and often forage in pairs for seeds, insects and seasonal berries. They usually breed between April and November, but may continue throughout the winter, if food is abundant.

NAME: **Rough-tailed Giant Rat**, *Hyomys goliath*
RANGE: **New Guinea**
HABITAT: **forest**
SIZE: **body: 29–39 cm (11½–15¼ in)**
 tail: 25–38 cm (9¾–15 in)

This rat derives its common name from the thick overlapping scales covering the underside of its tail. As these scales are often worn down, it is thought that they act to prevent the tail from slipping when it is used as a brace when the rat is climbing. This shy, secretive species is poorly studied, but it appears to feed on epiphytic plants on the branches of trees and on insects.

NAME: **African Grass Rat**, *Arvicanthis abyssinicus*
RANGE: **W. Africa to Somalia and Zambia**
HABITAT: **savanna, scrub, forest**
SIZE: **body: 12–19 cm (4¾–7½ in)**
 tail: 9–16 cm (3½–6¼ in)

The grass rat is a highly social rodent, living in colonies which sometimes number up to a thousand. The rats dig burrows in the earth or, alternatively, may establish themselves in a pile of rocks. Their staple food is probably grass seeds, but they also eat sweet potatoes and cassava. They breed throughout the year.

MURINAE:
Old World Rats and Mice Subfamily

This subfamily of almost 400 species of rodents contains some of the world's most successful mammals. They are highly adaptable and extremely tolerant of hostile conditions. Many are pests, wreaking havoc on stored grain and root crops, while others act as reservoirs for disease. Rats and mice occur throughout the world and have followed man even to the poles and to the tops of mountains.

NAME: **Four-striped Grass Mouse**, *Rhabdomys pumilio*
RANGE: **central Africa, south to the Cape of Good Hope**
HABITAT: **grass, scrub**
SIZE: **body: 9–13 cm (3½–5¼ in)**
 tail: 8–12.5 cm (3¼–5 in)

This common small rodent lives in a burrow that opens into thick vegetation and feeds on a wide variety of plant and animal food. In central Africa, these mice breed throughout the year, producing up to six litters annually, with 4 to 12 young per litter. In the south, the breeding season is limited to September through to May, and four litters are born.

NAME: **House Rat**, *Rattus rattus*
RANGE: **world-wide (originally native to Asia)**
HABITAT: **associated with man**
SIZE: **body: 20–26 cm (7¾–10¼ in)**
 tail: 20–24 cm (7¾–9½ in)

It has been said that the house, or ship, rat, carrying such diseases as bubonic plague, typhus, rabies and trichinosis, has altered human destiny more than any individual in recorded history. Wherever man has gone, and in all his activities, his unchosen companion has been the house rat. The success of this species is due to its extremely wide-ranging diet and its rapid rate of reproduction. Litters of up to 10 young are born every 6 weeks or so throughout the year.

NAME: **Brown Rat**, *Rattus norvegicus*
RANGE: **world-wide (originally native to E. Asia and Japan)**
HABITAT: **associated with man**
SIZE: **body: 25–30 cm (9¾–11¾ in)**
 tail: 25–32 cm (9¾–12½ in)

The brown rat is a serious pest, living alongside man wherever he lives and feeding on a wide range of food. Brown rats carry *Salmonella* and the bacterial disease tularaemia, but rarely the plague. They breed throughout the year, and the gestation period is 21 days.

NAME: **Stick-nest Rat**, *Leporillus conditor*
RANGE: **S. central Australia; Franklyn Island, South Australia**
HABITAT: **arid grassland**
SIZE: **body: 14–20 cm (5½–7¾ in)**
 tail: 13–18 cm (5¼–7 in)

Stick-nest rats build huge nests of sticks and debris, each inhabited by a pair or by a small colony of rats. The rats are vegetarian but may occasionally eat insects. They breed during the wet season, producing litters of 4 to 6 young.

NAME: **Mosaic-tailed Mouse**, *Melomys cervinipes*
RANGE: **N. E. Australia**
HABITAT: **forest, usually near water**
SIZE: **body: 9–17 cm (3½–6¾ in)**
 tail: 11–17 cm (4¼–6¾ in)

The scales on the tails of most rats and mice are arranged in rings, but in this species they resemble a mosaic. These mice breed during the rainy season — November to April — and produce litters of up to 4 young. They are active climbers and often rest up in pandanus trees in nests made of finely shredded grass and leaves.

NAME: **House Mouse**, *Mus musculus*
RANGE: **world-wide**
HABITAT: **fields; associated with man**
SIZE: **6.5–9.5 cm (2½–3¾ in)**
 tail: 6–10.5 cm (2¼–4¼ in)

Mice eat relatively little, but they spoil vast quantities of stored food such as grain. Wild mice are nocturnal and feed on grass seeds and plant stems and, occasionally, on insects.

NAME: **Hopping Mouse**, *Notomys alexis*
RANGE: **central Australia**
HABITAT: **dry grassland, spinifex scrub**
SIZE: **body: 9–18 cm (3½–7 in)**
 tail: 12–23 cm (4¾–9 in)

The Australian counterpart of the kangaroo rats of North Africa and North America, this rodent emerges from its cool, humid burrows only at night. It feeds on seeds, roots and any green vegetation. Hopping mice breed during the winter months, producing litters of 2 to 5 young.

NAME: **Greater Bandicoot Rat**, *Bandicota indica*
RANGE: **India to S. China; Taiwan, Java**
HABITAT: **forest, scrub; often near man**
SIZE: **body: 16–36 cm (6¼–14¼ in)**
 tail: 16–26 cm (6¼–10¼ in)

These rodents are serious pests in agricultural areas because they not only spoil grain but steal quantities of food for their own underground larders. They breed throughout the year and bear litters of 10 to 12 young.

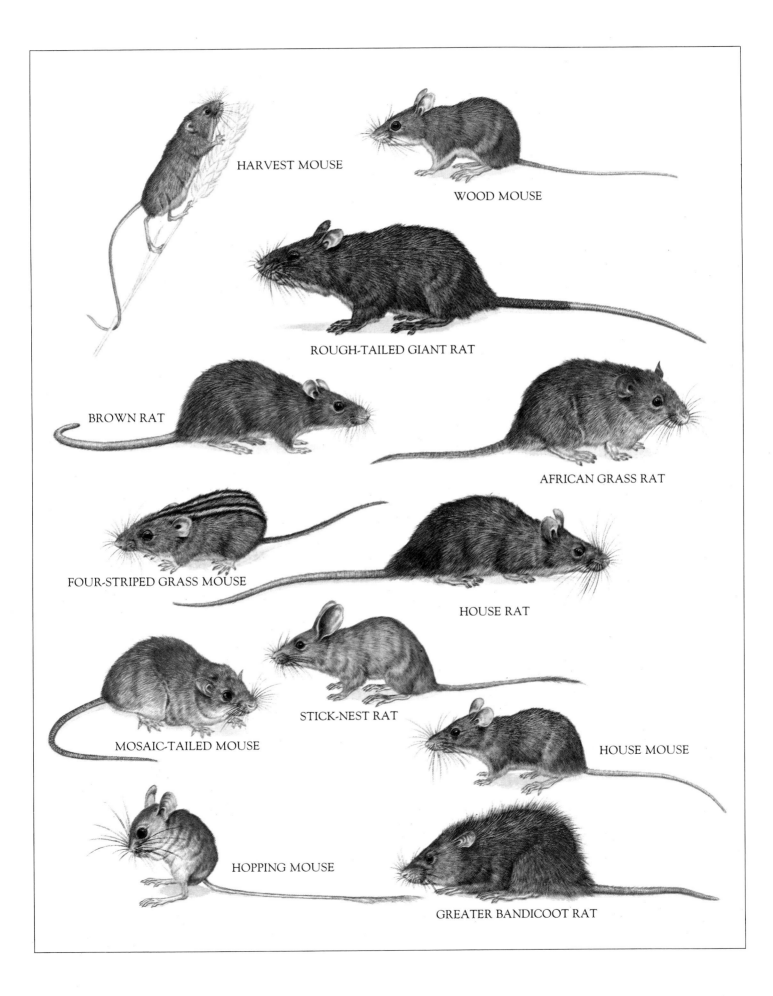

HARVEST MOUSE

WOOD MOUSE

ROUGH-TAILED GIANT RAT

BROWN RAT

AFRICAN GRASS RAT

FOUR-STRIPED GRASS MOUSE

HOUSE RAT

MOSAIC-TAILED MOUSE

STICK-NEST RAT

HOUSE MOUSE

HOPPING MOUSE

GREATER BANDICOOT RAT

Dormice, Jumping Mice, Jerboas

GLIRIDAE: Dormouse Family

There are 14 species of dormouse. Six species, all in the genus *Graphiurus*, occur in Africa, and others are found in Europe, northern Asia and Japan. These nocturnal rodents resemble short, fat squirrels and most have bushy tails. During late summer and autumn, most dormice build up their body fat reserves and then hibernate during the winter. They wake periodically to feed on the fruit and nuts that they store for winter consumption.

NAME: **Fat Dormouse, *Glis glis***
RANGE: **Europe, Asia**
HABITAT: **forest**
SIZE: **body: 15–18 cm (6–7 in)**
  tail: 13–16 cm (5¼–6¼ in)

The largest of its family, the fat dormouse has a long bushy tail and rough pads on its paws to facilitate climbing. It feeds on nuts, seeds, berries and fruit and occasionally catches insects and small birds. In summer, it makes a nest of plant fibre and moss up in a tree, but its winter hibernation nest is usually made nearer the ground in a hollow tree or in an abandoned rabbit burrow. The female produces a litter of 2 to 6 young in early summer.

NAME: **Japanese Dormouse, *Glirurus japonicus***
RANGE: **Japan (except Hokkaido)**
HABITAT: **montane forest**
SIZE: **body: 6.5–8 cm (2½–3¼ in)**
  tail: 4–5 cm (1½–2 in)

A tree-dwelling, nocturnal rodent, the Japanese dormouse spends its days in a tree hollow or in a nest built in the branches. It feeds at night on fruit, seeds, insects and birds' eggs. In winter it hibernates in a hollow tree or even in a man-made shelter such as an attic or nesting box. After hibernation, the dormice mate, and the female gives birth to 3 to 5 young in June or July. Occasionally a second litter is produced in October.

NAME: **African Dormouse, *Graphiurus murinus***
RANGE: **Africa, south of the Sahara**
HABITAT: **varied, forest, woodland**
SIZE: **body: 8–16.5 cm (3¼–6½ in)**
  tail: 8–13 cm (3¼–5¼ in)

African dormice are agile creatures which move swiftly over bushes and vegetation in search of seeds, nuts, fruit and insects. They shelter in trees or rock crevices and, although primarily nocturnal, may be active by day in dense dark forests. Up to three litters of 2 to 5 young are born in summer.

SELEVINIIDAE: Desert Dormouse Family

Only a single species, discovered in 1938, is known from this family.

NAME: **Desert Dormouse, *Selevinia betpakdalensis***
RANGE: **USSR: S.E. Kazakhstan**
HABITAT: **desert**
SIZE: **body: 7–8.5 cm (2¾–3¼ in)**
  tail: 7–9.5 cm (2¾–3¾ in)

The desert dormouse digs a burrow for shelter in which it is believed also to hibernate. It feeds largely on insects but also eats plant food and makes winter food stores of plant material in its burrow. It moves around by making short jumps on its hind legs. In late spring, desert dormice mate and produce a litter of up to 8 young.

ZAPODIDAE: Jumping Mouse Family

The 10 species in this family are all small mouse-shaped rodents, with long hind limbs modified for jumping. Some of these species can leap up to 2 m (6½ ft) when startled. Members of the family are found in open and forested land, as well as in swamps, throughout eastern Europe, Asia and North America.

NAME: **Meadow Jumping Mouse, *Zapus hudsonius***
RANGE: **Canada, N.E. and N. central USA**
HABITAT: **open meadows, woodland**
SIZE: **body: 7–8 cm (2¾–3¼ in)**
  tail: 10–15 cm (4–6 in)

Meadow jumping mice feed on seeds, fruit and insects for which they forage on the ground, bounding along in a series of short jumps. They are primarily nocturnal but in wooded areas may be active day and night under cover of the vegetation. In summer they make nests of grass and leaves on the ground, in grass or under logs; but in winter, they dig burrows or make small nests just above ground, on a bank or mound, in which to hibernate. They do not store food but, prior to hibernation, gain a substantial layer of body fat that sustains them. From October to April, the jumping mouse lies in its quarters in a tight ball, its temperature only just above freezing, and its respiration and heart rates greatly reduced to conserve energy.

Meadow jumping mice produce two or three litters a year. Most mate for the first time shortly after emerging from hibernation and the 4 or 5 young are born after a gestation of about 18 days.

NAME: **Northern Birch Mouse, *Sicista betulina***
RANGE: **N. and central Europe, E. Siberia**
HABITAT: **woodland**
SIZE: **body: 5–7 cm (2–2¾ in)**
  tail: 8–10 cm (3¼–4 in)

This rodent is easily distinguished by the dark stripe down its back and by its tail, which is about one and a half times the length of its body. A nocturnal animal, it spends the day in its burrow, emerging at night to search for insects and small invertebrates to eat. It also feeds on seeds and fruit, particularly during the fattening-up period, prior to hibernating from October to April. Females produce a litter of 3 to 5 young in May or June, after a gestation of 4 or 5 weeks.

DIPODIDAE: Jerboa Family

The 27 species of jerboa form a family of jumping rodents, with hind limbs which are at least four times as long as their manipulative forelimbs. These rodents are found in deserts, semi-arid zones and steppe country in North Africa and Asia, where they construct complex burrow systems.

NAME: **Northern Three-toed Jerboa, *Dipus sagitta***
RANGE: **central Asia: Caucasus to N. China**
HABITAT: **sand-dunes, steppe, pine forest**
SIZE: **body: 10–13 cm (4–5¼ in)**
  tail: 15–19 cm (6–7½ in)

The northern three-toed jerboa feeds on plants, seeds and insects. It needs very little water and is able to survive on the water contained in its food. In summer, it spends its days in a shallow burrow and emerges in the evening to travel to its feeding grounds, leaping along on its powerful hind limbs. In autumn, it digs a deeper burrow and hibernates from November to March. Jerboas mate soon after awakening and may have two litters of 2 to 5 young in a season.

NAME: **Great Jerboa, *Allactaga major***
RANGE: **USSR: Ukraine, east to China**
HABITAT: **steppe, semi-desert**
SIZE: **body: 9–15 cm (3½–6 in)**
  tail: 16–22 cm (6¼–8½ in)

The great jerboa and 8 of the 9 other species in the genus *Allactaga* have five toes on each hind foot. Great jerboas feed on seeds and insects, which they find by combing through the sand with the long slender claws on their front feet. They are nocturnal, spending the day in burrows; they also hibernate in burrows. One or two litters are produced each year.

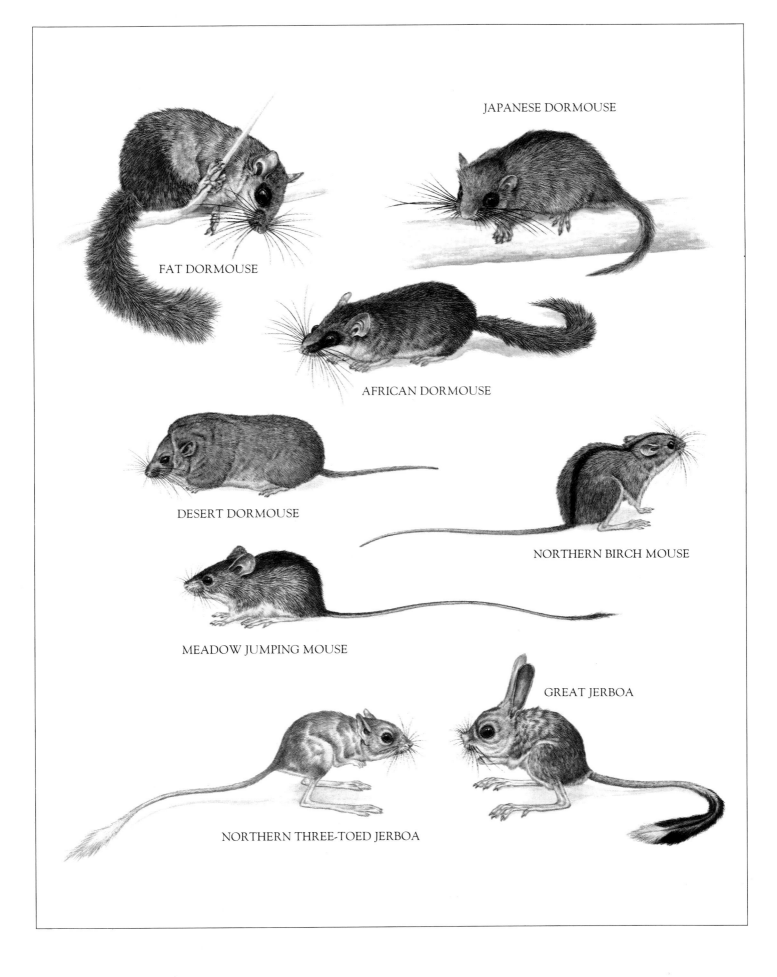

JAPANESE DORMOUSE

FAT DORMOUSE

AFRICAN DORMOUSE

DESERT DORMOUSE

NORTHERN BIRCH MOUSE

MEADOW JUMPING MOUSE

GREAT JERBOA

NORTHERN THREE-TOED JERBOA

Porcupines

HYSTRICIDAE:
Old World Porcupine Family

The 12 species in this family are all large rodents, unmistakable in their appearance, with long spines, derived from hairs, covering back, sides and parts of the tail. Porcupines live in desert, forest and savanna regions of Africa, parts of Asia and Indonesia and the Philippines. Most are primarily ground-living creatures; they move in a clumsy, shuffling manner which rattles their spines. They are generally nocturnal animals and live in burrows, which they dig, or in holes or crevices. They feed on plant material, such as roots, bulbs, tubers, fruit and bark, and on some carrion. Males and females look alike.

NAME: **Indonesian Porcupine, *Thecurus sumatrae***
RANGE: **Sumatra**
HABITAT: **forest**
SIZE: **body: 54 cm (21¼ in)**
tail: 10 cm (4 in)

The Indonesian porcupine's body is covered with flattened spines, interspersed with short hairs. The spines are longest on the back and sides, becoming smaller on the tail; on the underside of the body, the spines are rather more flexible than those elsewhere. The specialized "rattling" quills on the tail expand near the tips; these expanded portions are hollow so that the quills rattle when they are vibrated together as a warning to potential enemies.

NAME: **Crested Porcupine, *Hystrix africaeaustralis***
RANGE: **Africa: Senegal to Cape Province**
HABITAT: **forest, savanna**
SIZE: **body: 71–84 cm (28–33 in)**
tail: up to 2.5 cm (1 in)

The crested porcupine is a stout-bodied rodent, with sharp spines up to 30 cm (11¾ in) long on its back. Specialized hollow quills on the tail can be rattled in warning when the tail is vibrated. If, despite its warnings, a porcupine is still threatened, it will charge backward and drive the sharp, backward-curving spines into its enemy. The spines detach easily from the porcupine but cannot actually be "shot" as was once believed.

Crested porcupines are slow-moving animals that rarely climb trees. They dig burrows in which they spend the day, emerging at night to feed. They are thought to produce two litters a year of 2 to 4 young each. The young are born with soft spines and remain in the nest until their spines harden, when they are about 2 weeks old.

NAME: **Asian Brush-tailed Porcupine, *Atherurus macrourus***
RANGE: **S.E. Asia: Assam to Malaya**
HABITAT: **forest, often near water**
SIZE: **body: 40–55 cm (15¾–21½ in)**
tail: 15–25 cm (6–9¾ in)

A slender, ratlike porcupine, the brush-tailed has a distinctive long tail, tipped with a tuft of bristles. The spines of this species are flattened and grooved and most are short.

During the day, the brush-tailed porcupine shelters in a burrow, among rocks or even in a termite mound and emerges at night to hunt for food, mostly plants, roots, bark and insects. An agile creature, it will climb trees and runs well. It has partially webbed feet and is able to swim. Groups of up to 8 individuals shelter and forage together.

There are 3 other species of brush-tailed porcupine in this genus, all with similar appearance and habits.

NAME: **Long-tailed Porcupine, *Trichys fasciculata***
RANGE: **S.E. Asia: Malaya, Sumatra, Borneo**
HABITAT: **forest**
SIZE: **body: 28–47 cm (11–18½ in)**
tail: 17–23 cm (6¾–9 in)

The spines of the long-tailed porcupine are flattened and flexible, but shorter and less well developed than those of other species. The long tail breaks off easily, and many adults are found in this condition. These porcupines are good climbers and have broad paws, with strong digits and claws for holding on to branches.

ERITHIZONTIDAE:
New World Porcupine Family

The 9 species of New World porcupine are generally similar in appearance to the Old World species and have the same coarse hair and specialized spines. Unlike the Old World porcupines, they are largely tree-living, and the feet are adapted for climbing, with wide soles and strong digits and claws. Six of the species are in the genus *Coendou* and have prehensile tails — a prehensile tail is one equipped with muscles which allow it to be used as a fifth limb, for curling around and grasping branches. Males and females look alike.

New World porcupines are generally, but not exclusively, nocturnal and spend their days in hollows in trees or in crevices in the ground. Both male and female mark their home range with urine. The family occurs in North, Central and South America.

NAME: **North American Porcupine, *Erithizon dorsatum***
RANGE: **Canada; USA: Alaska, W. states, south to New Mexico, some N.E. states.**
HABITAT: **forest**
SIZE: **body: 46–56 cm (18–22 in)**
tail: 18–23 cm (7–9 in)

A thickset animal, this porcupine has spines on neck, back and tail and some longer spines, armed with minute barbs at their tips. It is slow and clumsy, but climbs trees readily to feed on buds, twigs and bark. In summer, it also feeds on roots and stems of flowering plants and even on some crops. It does not hibernate.

The porcupines mate at the beginning of winter, and the courting male often sprays the female with urine before mating, perhaps to prevent other males attempting to court her. After a gestation of 7 months, the young, usually only 1, is born in the late spring. It is well developed at birth, with fur, open eyes and soft quills which harden within an hour. A few hours after birth, the young porcupine can climb trees and feed on solid food.

NAME: **Tree Porcupine, *Coendou prehensilis***
RANGE: **Bolivia, Brazil, Venezuela**
HABITAT: **forest**
SIZE: **body: 30–61 cm (11¾–24 in)**
tail: 33–45 cm (13–17¾ in)

The body of the tree porcupine is covered with short, thick spines. Its major adaptation to arboreal life is its prehensile tail, which it uses to grasp branches when it is feeding. The tail lacks spines and the upper part of its tip is naked, with a calloused pad for extra grip. The hands and feet, too, are highly specialized for climbing, with long curved claws on each digit. Tree porcupines are mainly nocturnal, and slow but sure climbers. They feed on leaves, stems and some fruits. Females seem to produce only 1 young a year.

NAME: **Upper Amazon Porcupine, *Echinoprocta rufescens***
RANGE: **Colombia**
HABITAT: **forest**
SIZE: **body: 46 cm (18 in)**
tail: 10 cm (4 in)

The Upper Amazon porcupine has a short, hairy tail that is not prehensile. Its back and sides are covered with spines, which become thicker and stronger toward the rump. It is an arboreal animal and generally inhabits mountainous areas over 800 m (2,600 ft). Little has been discovered about the breeding habits and biology of this species of porcupine.

CRESTED PORCUPINE

INDONESIAN PORCUPINE

ASIAN BRUSH-TAILED PORCUPINE

LONG-TAILED PORCUPINE

TREE PORCUPINE

UPPER AMAZON PORCUPINE

NORTH AMERICAN PORCUPINE

Guinea Pigs, Capybara, Pacas and Agoutis

CAVIIDAE:
Guinea Pig/Cavy Family

There are about 15 species in this interesting and entirely South American family of ground-living rodents. Within the group are forms known as guinea pigs or cavies, mocos or rock cavies, and the Patagonian "hares", locally called "maras". Cavies and rock cavies have the well-known chunky body shape of domestic guinea pigs, with short ears and limbs, a large head and a tail which is not externally visible. Maras have a more harelike shape, with long legs and upstanding ears.

Cavies feed on plant material, and their cheek teeth continue to grow throughout life to counteract the heavy wear caused by chewing such food. Most live in small social groups of up to 15 or so individuals; sometimes they join in larger groups of as many as 40. They do not hibernate, even in areas which experience low temperatures.

NAME: **Cavy**, *Cavia tschudii*
RANGE: **Peru to N. Argentina**
HABITAT: **grassland, rocky regions**
SIZE: **body: 20–40 cm (7¾–15¾ in)**
  tail: no visible tail

These nocturnal rodents usually live in small family groups of up to 10 individuals, but may form larger colonies in particularly suitable areas. Although their sharp claws are well suited for digging burrows, they often use burrows made by other species or shelter in rock crevices. They feed at dawn and dusk, largely on grass and leaves.

Cavies breed in summer, or throughout the year in mild areas, producing litters of 1 to 4 young after a gestation of 60 to 70 days. The young are well developed at birth and can survive alone at 5 days old. This species is the probable ancestor of the domestic guinea pig, which is kept as a pet and also widely used in scientific research laboratories. Cavies are still kept by upland Indians as a source of fine, delicate meat.

NAME: **Rock Cavy**, *Kerodon rupestris*
RANGE: **N.E. Brazil**
HABITAT: **arid rocky areas**
SIZE: **body: 20–40 cm (7¾–15¾ in)**
  tail: no visible tail

Similar in build to the cavy, the rock cavy has a longer, blunter snout and longer legs. It shelters under rocks or among stones and emerges in the afternoon or evening to search for leaves to eat. It will climb trees to find food.

The female rock cavy is believed to produce two litters a year, each consisting of 1 or 2 young.

NAME: **Mara**, *Dolichotis patagona*
RANGE: **Argentina**
HABITAT: **open arid land**
SIZE: **body: 69–75 cm (27¼–29½ in)**
  tail: 4.5 cm (1¾ in)

The mara has long, slender legs and feet, well adapted for running and bounding along at speeds as great as 30 km/h (18½ mph) in the manner of a hare or jackrabbit. Indeed, the mara fills the niche of the hare in an area where this group is absent.

On each hind foot there are three digits, each with a hooflike claw; each forefoot bears four digits, armed with sharp claws. Maras are active in the daytime and feed on any available plant material. They dig burrows or take them over from other animals, and the litter of 2 to 5 young is born in a nest made in the burrow.

HYDROCHOERIDAE:
Capybara Family

This family contains only 1 species, the capybara, which is the largest living rodent. It resembles a huge guinea pig, with a large head and square muzzle, and lives in dense vegetation near lakes, rivers or marshes.

NAME: **Capybara**, *Hydrochoerus hydrochaeris*
RANGE: **Panama to E. Argentina**
HABITAT: **forest, near water**
SIZE: **body: 1–1.3 m (3¼–4¼ ft)**
  tail: vestigial

The capybara spends much time in water and is an excellent swimmer and diver; it has partial webs between the digits of both its hind and forefeet. When swimming, only its eyes, ears and nostrils show above the water. Capybaras feed on plant material, including aquatic plants, and their cheek teeth grow throughout life to counteract the wear and tear of chewing. They live in family groups and are active at dawn and dusk. In areas where they are frequently disturbed, capybaras may be nocturnal.

Males and females look alike, but there is a scent gland on the nose that is larger in the male. They mate in spring, and a litter of 2 young is born after a gestation of 15 to 18 weeks. The young are well developed at birth.

DINOMYIDAE: Pacarana Family

This South American family contains 1 apparently rare species, the false paca, or pacarana, so called because its striking markings are similar to those of the paca.

NAME: **Pacarana**, *Dinomys branicki*
RANGE: **Colombia to Bolivia**
HABITAT: **forest**
SIZE: **body: 73–79 cm (28¾–31 in)**
  tail: 20 cm (7¾ in)

The pacarana is a slow-moving, docile animal with short, strong limbs and powerful claws. It feeds on plant material, such as leaves, stems and fruit, and sits up on its haunches to examine and eat its food. Its cheek teeth are subject to considerable wear and grow throughout its life. It is most probably nocturnal.

Little is known of its breeding habits, but 2 young seems to be the normal number in a litter.

DASYPROCTIDAE:
Paca and Agouti Family

The 12 species in this family are distributed throughout Central and South America. All are medium to large ground-living rodents, with limbs well adapted for running.

The group splits naturally into three: the nocturnal pacas, the daytime-active agoutis, and the acuchis, about which little is known. All species have been hunted extensively for their flesh; they themselves eat leaves, fruit, roots and stems, all of which may be hoarded in underground stores.

NAME: **Paca**, *Cuniculus paca*
RANGE: **S. Mexico to Surinam, south to Paraguay**
HABITAT: **forest, near water**
SIZE: **body: 60–79 cm (23½–31 in)**
  tail: 2.5 cm (1 in)

The nocturnal paca is usually a solitary animal. It spends its day in a burrow, which it digs in a riverbank, among tree roots or under rocks, emerging after dark to look for food. The paca enters water willingly and will often escape from predators by swimming. It is believed to produce two litters a year of 1, rarely 2, young.

NAME: **Agouti**, *Dasyprocta aguti*
RANGE: **Venezuela, E. Brazil; Lesser Antilles**
HABITAT: **forest, savanna**
SIZE: **body: 41–62 cm (16–24½ in)**
  tail: 1–3 cm (½–1¼ in)

Agoutis are social animals and are active in the daytime. They are good runners and can also jump up to 2 m (6½ ft) vertically, from a standing position. Agoutis dig burrows in a river bank or under a tree or stone and tread well-defined paths from burrows to feeding grounds. They are believed to mate twice a year and bear litters of 2 to 4 young.

CAVY

ROCK CAVY

MARA

CAPYBARA

PACARANA

PACA

AGOUTI

Chinchillas and relatives

CHINCHILLIDAE:
Viscacha and Chinchilla Family

There are about 6 species in this family which is found only in South America. All species have dense, beautiful fur and the importance of the chinchillas, in particular, to the fur trade has led to their becoming relatively endangered as a wild species.

The hind limbs of the Chinchillidae are longer than their forelimbs and they are good at running and leaping; they are also good climbers. They feed on plants, including roots and tubers, and their cheek teeth grow throughout life. Social animals, they live in small family groups which are part of larger colonies. Where the ground is suitable, they dig burrows, but otherwise they shelter under rocks.

NAME: **Plains Viscacha,** *Lagostomus maximus*
RANGE: **Argentina**
HABITAT: **grassland**
SIZE: **body: 47–66 cm (18½–26 in)**
  tail: 15–20 cm (6–7¾ in)

The plains viscacha is a robust rodent with a large head and blunt snout. Males are larger than females. Colonies of plains viscachas live in complex burrows, with networks of tunnels and entrances. They are expert burrowers, digging mainly with their forefeet and pushing the soil with their noses — their nostrils close off to prevent soil from entering them.

Females breed once a year or sometimes twice in mild climates. There are usually 2 young, born after a gestation of 5 months.

NAME: **Chinchilla,** *Chinchilla laniger*
RANGE: **Bolivia, Chile**
HABITAT: **rocky, mountainous areas**
SIZE: **body: 22.5–38 cm (8¾–15 in)**
  tail: 7.5–15 cm (3–6 in) Ⓥ

Chinchillas are attractive animals, with long ears, large eyes and bushy tails. They live in colonies of 100 or more, sheltering in holes and crevices in rocks. They feed on any available vegetation, sitting up to eat and holding their food in their front paws.

Female chinchillas are larger than males and are aggressive toward one another. They breed in winter, usually producing two litters of 1 to 6 young. The gestation period is 111 days, and the young are suckled for 6 to 8 weeks.

The chinchilla's exceedingly soft, dense coat is the cause of its present extreme rarity in the wild, although it is now farmed all over the world for its valuable fur.

CAPROMYIDAE:
Hutia and Coypu Family

There are now about 12 living species in this family; several other species have become extinct relatively recently. The family divides into two groups: the 11 species of hutia, all found on West Indian islands, and the single species of coypu, or nutria, which is semi-aquatic and a native of South America.

NAME: **Hutia,** *Capromys ingrahami*
RANGE: **Bahamas**
HABITAT: **forest**
SIZE: **body 30–50 cm (11¾–19¾ in)**
  tail: 15–30 cm (6–11¾ in) Ⓔ

The hutia feeds mostly on fruit and leaves but occasionally eats small invertebrates and reptiles. It is a good climber and seeks some food in the trees. Active in the daytime, it shelters in a burrow or rock crevice at night. It is believed to breed all year round, provided the temperature stays above 15°C (60°F), and produces litters of 2 to 9 young.

Hutias have been unable to cope with man's introduction of mongooses and dogs to the West Indies, and many species may be heading for extinction.

NAME: **Coypu/Nutria,** *Myocastor coypus*
RANGE: **Bolivia and S. Brazil to Chile and Argentina; introduced in N. America, Europe and Asia.**
HABITAT: **near marshes, lakes, streams**
SIZE: **body: 43–63 cm (17–24¾ in)**
  tail: 25–42 cm (9¾–16½ in)

The semi-aquatic coypu is a skilled swimmer and diver and looks like a beaver with a rat's tail. Its hind feet are webbed and it has dense fur. Coypus feed on aquatic vegetation and possibly on molluscs. They dig burrows in river banks, clear trails in their territory and are extremely destructive to plants and crops. By escaping from farms where they are bred for their fur, these animals have colonized new areas all over the world, and are sometimes considered pests because of the damage they do.

Two to three litters of up to 10 young are produced during the year, and the gestation period is 132 days. Young coypus can swim only a few hours after birth.

OCTODONTIDAE:
Octodont Rodent Family

The 8 species of octodont rodent all occur in South America. Most resemble rats with round noses and long furry tails. They are all good burrowers and feed on plant material.

NAME: **Degu,** *Octodon degus*
RANGE: **W. Peru, Chile**
HABITAT: **mountains, coastal regions**
SIZE: **12.5–19.5 cm (5–7½ in)**
  tail: 10–16 cm (4–6¼ in)

The degu is a stout, short-legged rodent with a large head for its size. Active during the day, it feeds on plants, bulbs and tubers. It is thought to breed all year round and may produce several litters a year of 2 young each.

CTENOMYIDAE:
Tuco-tuco Family

This family of 32 species, all found in South America, is believed by some authorities to be a group of relatives of the octodont rodents which has become highly specialized for a burrowing, underground existence.

NAME: **Tuco-tuco,** *Ctenomys talarum*
RANGE: **E. Argentina**
HABITAT: **grassland**
SIZE: **body: 17–25 cm (6¾–9¾ in)**
  tail: 6–11 cm (2¼–4¼ in)

Tuco-tucos look very much like North American pocket gophers (Geomyidae) and lead similar lives in complex burrow systems. Their front teeth are enormous relative to their body size and are used, when burrowing, to loosen the soil, which is then pushed out of the tunnel with the feet. Tuco-tucos spend nearly all their lives underground and feed on roots, tubers and stems.

In winter and spring, the tuco-tucos mate, and a single litter of 2 to 5 well-developed young is born after a gestation period of about 15 weeks.

ABROCOMIDAE:
Chinchilla-rat Family

There are 2 species only of chinchilla-rat, both native to South America. As their common name suggests, their fur resembles that of the chinchilla, although it is of poorer quality, and they have a ratlike body shape.

NAME: **Chinchilla-rat,** *Abrocoma bennetti*
RANGE: **Chile**
HABITAT: **high coastal plains**
SIZE: **body: 19–25 cm (7½–9¾ in)**
  tail: 13–18 cm (5¼–7 in)

A resident of cold, bleak mountain regions, the chinchilla-rat is a little-known creature. It feeds on plant food and its cheek teeth grow throughout life. Mainly a ground-dweller, it can also climb trees in search of food. Chinchilla-rats live in burrows or rock crevices, apparently in small groups.

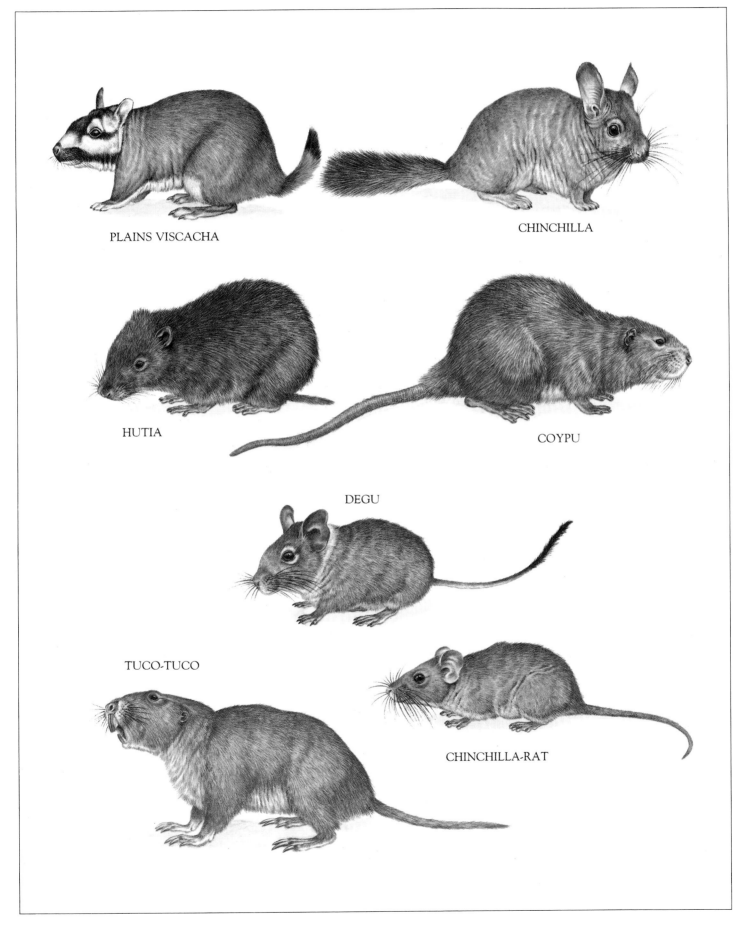

PLAINS VISCACHA

CHINCHILLA

HUTIA

COYPU

DEGU

TUCO-TUCO

CHINCHILLA-RAT

Spiny Rats, Cane Rats and relatives

ECHIMYIDAE:
American Spiny Rat Family

The 45 species of American spiny rat are found from Nicaragua in the north to central Brazil in the south. Most are robust, ratlike creatures with a more or less spiny coat of sharp hairs. They are herbivorous, feeding on a variety of plant material. Most prefer to live close to rivers and streams.

Like many South American rodents, spiny rats have a long gestation and give birth to well-developed young which can run around when only a few hours old.

NAME: **Gliding Spiny Rat,** *Diplomys labilis*
RANGE: **Panama**
HABITAT: **forest**
SIZE: **body: 25–48 cm (9¾–19 in)**
 tail: 20–28 cm (7¾–11 in)

Well adapted for arboreal life, the gliding spiny rat has long, strong toes and sharp, curved claws with which it can grip the smoothest of bark. Its common name derives from its habit of leaping from branch to branch, spreading its limbs to utilize its gliding membrane as it does so.

Gliding spiny rats make their nests in hollows in trees near water. They breed throughout the year, producing litters of 2 young after a gestation of about 60 days. The young are able to clamber around among the branches a few hours after they are born.

NAME: **Armoured Rat,** *Hoplomys gymnurus*
RANGE: **Nicaragua, south to Colombia and Ecuador**
HABITAT: **rain forest, grassy clearings**
SIZE: **body: 22–32 cm (8¾–12½ in)**
 tail: 15–25 cm (6–9¾ in)

The most spiny of all the spiny rats, the armoured rat has a thick coat of needle-sharp hairs along its back and flanks. These rodents live in short, simple burrows in the banks of streams and emerge at night to forage for food. They breed throughout the year, producing litters of 1 to 3 young.

THRYONOMYIDAE:
Cane Rat Family

The 6 species of cane rat are found throughout Africa, south of the Sahara. They are substantial rodents, weighing up to 7 kg (15½ lb), and they are the principal source of animal protein for some tribes. In the cane fields they cause serious damage to the crop, stripping the cane of its outer bark to expose the soft central pith on which they feed.

NAME: **Cane Rat,** *Thryonomys swinderianus*
RANGE: **Africa, south of the Sahara**
HABITAT: **grassy plains, sugar-cane plantations**
SIZE: **body: 35–61 cm (13¾–24 in)**
 tail: 7–25 cm (2¾–9¾ in)

Cane rats do not normally live in burrows, preferring to construct a sleeping platform from chopped-up vegetation when needed. Occasionally, though, they take over disused aardvark or porcupine burrows or seek refuge among a pile of boulders. In southern Africa, cane rats are known to mate from April to June and to give birth to litters of 2 to 4 young after a gestation of 2 months. The young cane rats are born with their eyes open and are able to run around soon after birth.

PETROMYIDAE:
Dassie Rat Family

The single species in this family is an unusual rodent that looks more like a squirrel than a rat. Its common name links it in habit with the dassie, or rock hyrax, for both creatures share a love of the sun and spend much time basking on rocks, moving from place to place to catch the strongest rays. While the others bask, one member of the colony keeps a look-out for predators, such as mongooses, eagles or leopards, uttering a shrill warning call if danger threatens.

NAME: **Dassie Rat,** *Petromus typicus*
RANGE: **Africa: Angola, Namibia**
HABITAT: **rocky, arid hills**
SIZE: **body: 14–20 cm (5½–7¾ in)**
 tail: 13–18 cm (5¼–7 in)

Dassie rats live in colonies and are active in the daytime. They feed on fruit, seeds and berries. They mate in early summer (October) and give birth in late December to a litter of 1 or 2 young.

BATHYERGIDAE:
Mole-rat Family

There are 9 species of mole-rat, all found in Africa, south of the Sahara. They are highly specialized for a subterranean life, for they are virtually blind and have powerful incisor teeth and claws for digging. Their skulls also are heavy and strong, and they use their heads as battering rams. The underground storage organs of plants — tubers and bulbs — are their main food, and worms and insect larvae may be eaten occasionally. Most have thick velvety coats which resemble that of the mole, but the naked mole-rat is quite hairless.

NAME: **Cape Dune Mole-rat,** *Bathyergus suillus*
RANGE: **South Africa to Cape of Good Hope**
HABITAT: **sand-dunes, sandy plains**
SIZE: **body: 17.5–33 cm (6¾–13 in)**
 tail: 4–7 cm (1½–2¾ in)

The largest member of its family, the Cape dune mole-rate may weigh up to 1.5 kg (3 lb). It has relatively huge incisor teeth, each one measuring 2 mm (1/12 in) across, and a fearsome bite. It builds extensive burrow systems, usually close to the surface, which can be a serious menace to root crops.

In November or December females give birth to 3 to 5 well-developed young.

NAME: **Naked Mole-rat,** *Heterocephalus glaber*
RANGE: **Somalia, Ethiopia, N. Kenya**
HABITAT: **arid steppe, light sandy soil**
SIZE: **body: 8–9 cm (3¼–3½ in)**
 tail: 3.5–4 cm (1½ in)

This species, the smallest of its family, is one of the most curious mammals known. Each colony of about 100 rats is ruled by a single queen, who alone breeds. She is tended by a few non-workers of both sexes, which are fatter and more sluggish than the workers. The latter dig the burrows and gather the roots and tubers for the whole colony to eat. The queen appears to be able to inhibit sexual maturation in all other females, but how this is achieved is not known. She breeds throughout the year and may produce up to 20 young in each litter. If she dies or is removed, a nonworker female assumes her role.

CTENODACTYLIDAE:
Gundi Family

The 5 species of gundi are found in Africa. They are extremely agile and can climb almost vertical rock faces. All members of the family are highly vocal and can utter a range of birdlike trills and twitters.

NAME: **Gundi,** *Ctenodactylus gundi*
RANGE: **Africa: Sahara**
HABITAT: **rocky outcrops**
SIZE: **body: 16–20 cm (6¼–7¾ in)**
 tail: 1–2 cm (½–¾ in)

Shy animals, gundis feed only at night on a range of plant material which they usually take back to the safety of a rock crevice to consume. The gestation period of gundis is about 40 days, and the usual litter size is 1 or 2 young, which are able to run about immediately after birth.

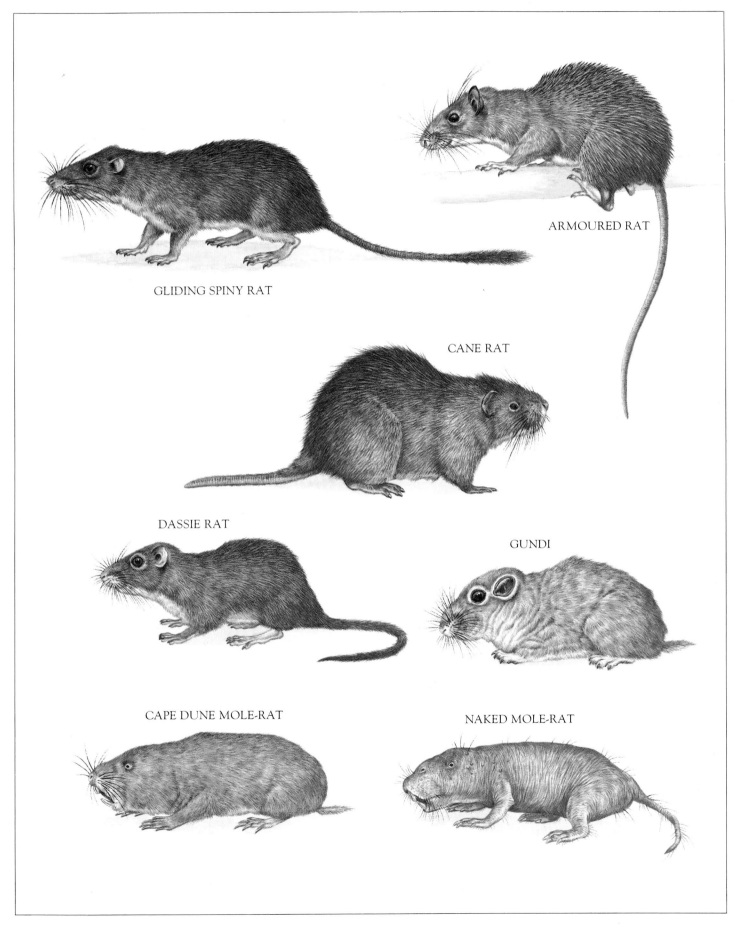

ARMOURED RAT

GLIDING SPINY RAT

CANE RAT

DASSIE RAT

GUNDI

CAPE DUNE MOLE-RAT

NAKED MOLE-RAT

Pikas, Rabbits and Hares I

ORDER LAGOMORPHA

There are 2 families and about 54 species in this order: the pikas, and the rabbits and hares. For many years, these herbivores were regarded as a subgroup of the rodents, but a detailed examination of their structure, dentition and chewing mechanisms suggests that the two groups are distinct but related.

OCHOTONIDAE: Pika Family

There are about 14 species of pika, all in the same genus. They live in north and central Asia and 1 species also occurs in North America. Pikas are smaller than rabbits and have short, rounded ears and no visible tail.

NAME: **Northern Pika, *Ochotona alpina***
RANGE: **Siberia, N.E. China, Japan: Hokkaido; western N. America: Alaska to New Mexico**
HABITAT: **rocky mountain slopes, forest**
SIZE: **body: 20–25 cm ($7\frac{3}{4}$–$9\frac{3}{4}$ in)**

A small, short-legged animal, this pika cannot run fast like a rabbit but moves in small jumps, rarely venturing far. It lives in family groups and takes shelter in a den, made among rocks or tree roots. Grass and slender plant stems are its main food, and, like all pikas, it gathers extra food in late summer and piles it in heaps like little haystacks to use in the winter. If they run short of food in winter, pikas tunnel through the snow to reach their stores.

There may be up to three litters a year, depending on the region; in North America there are two, both in the summer months. Each litter contains 2 to 5 young, born after a gestation of 30 or 31 days.

LEPORIDAE: Rabbit and Hare Family

Found in forest, shrubby vegetation, grassland, tundra and on mountain slopes in the Americas, Europe, Asia and Africa, the rabbits and hares are an extremely successful family of small herbivorous mammals. The common rabbit has been introduced in Australia and New Zealand, where it has proved itself to be remarkably adaptable.

Compared with pikas, the 40 or so species of rabbit and hare have become highly adapted for swift running, with disproportionately well-developed hind limbs. They also have long, narrow ears and small tails. Their teeth are adapted for gnawing vegetation: they have chisel-shaped upper incisors, which grow throughout life, for biting and large cheek teeth for chewing.

NAME: **Brown Hare, *Lepus capensis***
RANGE: **N., E. and S. Africa, Europe (not Iceland or N. Scandinavia), across temperate Asia to China; introduced in N. and South America, Australia and New Zealand**
HABITAT: **open country, farmland, woodland**
SIZE: **body: 44–76 cm ($17\frac{1}{4}$–30 in)**
tail: 7–11 cm ($2\frac{3}{4}$–$4\frac{1}{4}$ in)

A fast-running hare, with long hind limbs, the brown hare is mainly active at dusk and at night. During the day, it remains in a shallow depression in the ground, known as a form, which is concealed among vegetation. It feeds on leaves, buds, roots, berries, fruit, fungi, bark and twigs and is usually solitary.

The female may have several litters a year, each of 1 to 6 young, which are born in the form, fully furred and active, with their eyes open. They are suckled for about 3 weeks and leave their mother about a month after birth.

NAME: **Black-tailed Jack Rabbit, *Lepus californicus***
RANGE: **USA: Oregon, east to South Dakota and Missouri, south to N. Mexico; introduced in some eastern states of USA**
HABITAT: **prairies, cultivated land, arid scrub**
SIZE: **body: 46.5–63 cm ($18\frac{1}{4}$–$24\frac{3}{4}$ in)**
tail: 5–11.5 cm (2–$4\frac{1}{2}$ in)

Identified by its long ears and large, black-striped tail, this jack rabbit has powerful, elongate hind limbs and moves with a fast, bounding gait. For short periods, it may attain speeds of up to 56 km/h (35 mph) and tends to run rather than to take cover if threatened. In summer, it feeds on succulent green plants and grass, turning to more woody vegetation in winter. This jack rabbit, like all lagomorphs, ingests its faecal pellets, and it is thought to obtain additional nutrients when the material passes through the digestive system a second time.

There may be several litters a year, with 1 to 6 young in each; the average gestation period is 43 days. The young are born fully furred, with eyes open, in a shallow depression on the ground.

NAME: **Snowshoe Hare, *Lepus americanus***
RANGE: **Alaska, Canada, N. USA**
HABITAT: **forest, swamps, thickets**
SIZE: **body: 36–52 cm ($14\frac{1}{4}$–$20\frac{1}{2}$ in)**
tail: 2.5–5.5 cm (1–$2\frac{1}{4}$ in)

Also known as the varying hare, this animal has a dark-brown coat in summer that turns white in winter, except for a black edging on the ear tips. This coat is of undoubted camouflage value, although in fact it is a yellowish-buff, with only the visible tips of the hairs pure white. Usually active at night and in the early morning, the snowshoe hare feeds on juicy green plants and grass in summer and twigs, shoots and buds in winter. The population of these hares fluctuates tremendously on a roughly 10-year cycle, due to food availability and predator interactions.

Breeding begins in March, and there may be two or three litters, each of 1 to 7, usually 4, young, which are born well furred, with their eyes open.

NAME: **Natal Red Hare, *Pronolagus crassicaudatus***
RANGE: **Africa: Namibia, South Africa, Botswana, Zimbabwe, Mozambique**
HABITAT: **stony country with scattered vegetation, forest edge**
SIZE: **body: 42–50 cm ($16\frac{1}{2}$–$19\frac{3}{4}$ in)**
tail: 6–14 cm ($2\frac{1}{4}$–$5\frac{1}{2}$ in)

The Natal red hare lives alone in a small territory and is usually active at dusk and at night, feeding on grass and green-leaved plants. During the day, it rests in a shallow depression, or form, near cover of grass or rocks, and if alarmed, it darts into a rock crevice or hole. Like all hares, its hearing is acute and its sight and sense of smell are also good.

The female gives birth to 1 or 2 fully haired young after a gestation of about a month.

NAME: **Volcano Rabbit, *Romerolagus diazi***
RANGE: **Mexico: S.E. of Mexico City**
HABITAT: **slopes of volcanoes**
SIZE: **body: 28.5–31 cm ($11\frac{1}{4}$–$12\frac{1}{4}$ in)**
tail: vestigial Ⓔ

This unusual rabbit, with its extremely restricted distribution, is now rare and strictly protected. It has short, rounded ears and trots, rather than hops, on its short legs as it moves along the runways it makes in the grass. Mainly active at night and at dusk, it feeds on grass and young shoots. Volcano rabbits live in colonies and dig burrows for shelter.

The wild population of this species may be as low as 1,300, but a captive colony exists in the zoo on Jersey in the Channel Islands.

NAME: **Hispid Hare, *Caprolagus hispidus***
RANGE: **Nepal to N.E. India: Assam**
HABITAT: **forest, grassy bamboo thickets**
SIZE: **body: about 47 cm ($18\frac{1}{2}$ in)**
tail: 2.5 cm (1 in) Ⓔ

Also known as the bristly or Assam rabbit, the hispid hare has an unusual coat of coarse, bristly fur; its ears are short and broad and its legs stout. It lives alone or sometimes in pairs and digs a burrow for shelter. Grass shoots, roots and bark are its main foods.

VOLCANO RABBIT

BLACK-TAILED JACK RABBIT

HISPID HARE

NORTHERN PIKA

SNOWSHOE HARE

BROWN HARE

NATAL RED HARE

Rabbits and Hares 2

NAME: **Swamp Rabbit,** *Sylvilagus*
aquaticus
RANGE: **south-central USA: Georgia to**
Texas
HABITAT: **marshland, swamps, wet**
woodland
SIZE: **body: 45–55 cm (17¾–21½ in)**
tail: 6 cm (2¼ in)

The robust, large-headed swamp rabbit takes to water readily and is an expert swimmer and diver. It swims to avoid danger when pursued and also to reach islets or other new feeding areas. With its large, splayed toes, it also moves easily on damp, muddy land. Mainly nocturnal, it emerges from its shelter beneath a log or in a ground hollow at any time after heavy rain and feeds on grass, herbs and aquatic vegetation. It may also forage in grain fields, where these are near swamps. Although usually docile, rival males will attack each other in ferocious, face-to-face fights, sometimes inflicting serious wounds.

A litter of 1 to 6 young, usually 2 or 3, is born, after a gestation of about 40 days, in a shallow depression in the ground lined with grass and fur. The young are born furred, and their eyes open a few days after birth. There are thought to be two litters a year.

NAME: **Brush Rabbit,** *Sylvilagus*
bachmani
RANGE: **W. coast of N. America: British**
Columbia to Baja California
HABITAT: **chaparral, thick brush or scrub**
SIZE: **body: 27–33 cm (10½–13 in)**
tail: 2–4 cm (¾–1½ in)

The small, brown brush rabbit has a tiny tail and rounded ears. The rabbits living in hot inland regions to the south of the range tend to have longer ears than those living on the cooler, humid northern coast. Presumably this is because sound does not travel as well in hot, dry air as in moist, cool air and so they need the larger ears to pick up sounds more effectively. This shy, elusive rabbit stays hidden in cover of dense undergrowth for much of the time, venturing out for only short distances to feed on a wide variety of green plants and taking almost anything within reach. It moves along well-trodden runways through the vegetation but does not make a burrow.

There are three or four litters a year, between January and June. Each litter contains 3 to 5 young, born after a gestation of about 27 days in a shallow hollow in the ground, which is lined with grass and fur. The young are born blind but with a covering of fine fur.

NAME: **Desert Cottontail,** *Sylvilagus*
auduboni
RANGE: **USA: California to Montana,**
south to Arizona and Texas;
N. Mexico
HABITAT: **open plains with scattered**
vegetation, wooded valleys, sagebrush
SIZE: **body: 30–38 cm (11¾–15 in)**
tail: 5–7.5 cm (2–3 in)

The desert cottontail is distinguished from the brush rabbit, with which it overlaps in the south of its range, by its larger size and ears and by its greyish coat. Often abroad at any time of day, it is, however, most active in the late afternoon and at night, when it feeds on grass, leaves of various plants including cultivated plants, and fruit. It can do much damage to gardens and crops. Never far from some form of cover, it darts for safety if alarmed, the white underside of its tail momentarily revealed as it is flicked up. A burrow or a shallow depression in the ground is used for shelter.

Breeding takes place in spring or throughout the year, depending on the area. A litter of 1 to 5 blind, helpless young is born after a gestation of 26 to 30 days.

NAME: **Pygmy Rabbit,** *Sylvilagus*
idahoensis
RANGE: **N.W. USA: Oregon, Idaho,**
Montana, Utah, Nevada, N. California
HABITAT: **arid areas with sagebrush**
SIZE: **body: 23–29 cm (9–11½ in)**
tail: 2–3 cm (¾–1¼ in)

The smallest member of its genus, this species has thick, soft fur and short hind legs. It lives in a burrow that it excavates itself and does not often venture far from its home. During much of the day, it rests up in the burrow, emerging at dusk to feed on sagebrush and any other available plant matter. Its main enemies are coyotes and owls, and if alarmed, the pygmy rabbit takes refuge in its burrow, which usually has 2 or 3 entrances.

Litters of 5 to 8 young, usually 6, are born between May and August.

NAME: **Common Rabbit,** *Oryctolagus*
cuniculus
RANGE: **Europe (except far north and**
east), N.W. Africa; introduced in
many countries including New
Zealand, Australia, Chile
HABITAT: **grassland, cultivated land,**
woodland, grassy coastal cliffs
SIZE: **body: 35–45 cm (13¾–17¾ in)**
tail: 4–7 cm (1½–2¾ in)

The ancestor of the domestic rabbit, this species has been introduced into many areas outside its native range and has been so successful as to become a major pest in some places. Smaller than a hare, with shorter legs and ears, the common rabbit is brownish on the upperparts, with buffy-white underneath. The feet are equipped with large, straight claws.

Gregarious animals, these rabbits live in burrows, which they dig near to one another, and there may be a couple of hundred rabbits in a colony, or warren. They are most active at dusk and during the night but may emerge in the daytime in areas where they are undisturbed. Grass and leafy plants are their main foods, but rabbits also feed on vegetable and grain crops and can cause serious damage to young trees. In winter, they eat bulbs, twigs and bark if more succulent foods are unavailable. As a warning to others of approaching danger, a rabbit may thump the ground with its hind foot.

There may be several litters a year, born in the spring and summer in Europe. There are 3 to 9 young in a litter, and the gestation period is 28 to 33 days. Young are born naked, blind and helpless in a specially constructed burrow, lined with vegetation and fur, which the mother plucks from her belly. They do not emerge from the burrow until about 3 weeks old. The female is on heat again 12 hours after the birth, but not all pregnancies last the term, and many embryos are resorbed. It is estimated that only about 40 per cent of litters conceived are actually born, and the average number of live young produced by a female in a year is 11.

NAME: **Sumatran Rabbit,** *Nesolagus*
netscheri
RANGE: **S.W. Sumatra**
HABITAT: **forested mountain slopes at**
600–1,400 m (2,000–4,600 ft)
SIZE: **body: 36–40 cm (14¼–15¾ in)**
tail: 1.5 cm (½ in) Ⓔ

The only member of its family to have a definitely striped coat, the Sumatran rabbit has buffy-grey upperparts with brown stripes and a line down the middle of its back, from nose to tail. The tiny tail and the rump are reddish, while the limbs are greyish-brown. This unusual rabbit is now extremely rare, even in areas where it was once abundant, because of large-scale clearance of its forest habitat for cultivation.

Primarily nocturnal, the Sumatran rabbit spends the day in a burrow, but it is believed to take over an existing hole, rather than to dig its own. It feeds on leaves and stalks of plants in the forest undergrowth.

COMMON RABBIT

DESERT COTTONTAIL

PYGMY RABBIT

BRUSH RABBIT

SWAMP RABBIT

SUMATRAN RABBIT

Classification

CLASS MAMMALIA: **Mammals**

Subclass Prototheria: Egg-laying Mammals

Order Monotremata: Monotremes
Family Tachyglossidae: Spiny Anteaters
Family Ornithorhynchidae: Platypus

Subclass Theria: Live-bearing Mammals
Infraclass Metatheria: Marsupials

Order Marsupialia: Marsupials
Family Didelphidae: Opossums
Family Microbiotheriidae: Colocolo
Family Caenolestidae: Rat Opossums
Family Dasyuridae: Marsupial Carnivores and
    Insectivores
Family Myrmecobiidae: Numbat
Family Notoryctidae: Marsupial Mole
Family Phascolarctidae: Koala
Family Vombatidae: Wombats
Family Tarsipedidae: Honey Possum
Family Peramelidae: Bandicoots
Family Thylacomyidae: Rabbit-bandicoot
Family Phalangeridae: Phalangers
Family Burramyidae: Pygmy Possums
Family Petauridae: Ringtails
Family Macropodidae: Kangaroos, Wallabies

Infraclass Eutheria: Placental Mammals

Order Edentata: Edentates
Family Myrmecophagidae: American Anteaters
Family Bradypodidae: Sloths
Family Dasypodidae: Armadillos

Order Pholidota
Family Manidae: Pangolins

Order Tubulidentata
Family Orycteropodidae: Aardvark

Order Insectivora: Insectivores
Family Solenodontidae: Solenodons
Family Tenrecidae: Tenrecs, Otter-shrews
Family Chrysochloridae: Golden Moles
Family Erinaceidae: Hedgehogs, Moonrats
Family Soricidae: Shrews
Family Talpidae: Moles, Desmans

Order Macroscelidea
Family Macroscelididae: Elephant Shrews

Order Dermoptera
Family Cynocephalidae: Flying Lemurs

Order Chiroptera: Bats
Family Pteropodidae: Fruit Bats
Family Rhinopomatidae: Mouse-tailed Bats
Family Emballonuridae: Sheath-tailed Bats
Family Craseonycteridae: Hog-nosed Bats
Family Nycteridae: Slit-faced Bats
Family Megadermatidae: False Vampire Bats
Family Rhinolophidae: Horseshoe Bats
Family Hipposideridae: Old World Leaf-nosed Bats
Family Noctilionidae: Fisherman Bats

Family Mormoopidae: Moustached Bats
Family Desmodontidae: Vampire Bats
Family Molossidae: Free-tailed Bats
Family Phyllostomatidae: New World Leaf-nosed
    Bats
Family Vespertilionidae: Evening Bats
Family Natalidae: Funnel-eared Bats
Family Furipteridae: Smoky Bats
Family Thyropteridae: Disc-winged Bats
Family Myzopodidae: Sucker-footed Bat
Family Mystacinidae: New Zealand Short-tailed Bat

Order Scandentia
Family Tupaiidae: Tree Shrews

Order Primates: Primates
Family Cheirogaleidae: Mouse-lemurs
Family Lemuridae: Lemurs
Family Indriidae: Leaping Lemurs
Family Daubentoniidae: Aye-aye
Family Lorisidae: Lorises
Family Tarsiidae: Tarsiers
Family Callitrichidae: Marmosetes and Tamarins
Family Cebidae: New World Monkeys
Family Cercopithecidae: Old World Monkeys
Family Pongidae: Apes
Family Hominidae: Man

Order Carnivora: Carnivores
Family Canidae: Dogs, Foxes
Family Ursidae: Bears
Family Ailuropodidae: Pandas
Family Procyonidae: Raccoons
Family Mustelidae: Mustelids
Family Viverridae: Civets
Family Hyaenidae: Hyenas
Family Felidae: Cats

Order Pinnipedia: Pinnipedes
Family Otariidae: Sea Lions
Family Odobenidae: Walrus
Family Phocidae: Seals

Order Sirenia
Family Dugongidae: Dugong
Family Trichechidae: Manatees

Order Cetacea: Whales
Family Platanistidae: River Dolphins
Family Phocoenidae: Porpoises
Family Delphinidae: Dolphins
Family Monodontidae: White Whales
Family Physeteridae: Sperm Whales.
Family Ziphiidae: Beaked Whales
Family Eschrichtidae: Grey Whale
Family Balaenopteridae: Rorquals
Family Balaenidae: Right Whales

Order Proboscidea
Family Elephantidae: Elephants

Order Hyracoidea
Family Procaviidae: Hyraces

Order Perissodactyla: Odd-toed Ungulates
Family Equidae: Horses

Family Tapiridae: Tapirs
Family Rhinocerotidae: Rhinoceroses

Order Artiodactyla: Even-toed Unguates
Family Suidae: Pigs
Family Tayassuidae: Peccaries
Family Hippopotamidae: Hippopotamuses
Family Camelidae: Camels
Family Tragulidae: Mouse-deer
Family Moschidae: Musk Deer
Family Cervidae: Deer
Family Giraffidae: Giraffes
Family Antilocapridae: Pronghorn
Family Bovidae: Bovids

Order Rodentia: Rodents
Family Sciuridae: Squirrels
Family Geomyidae: Pocket Gophers
Family Heteromyidae: Pocket Mice
Family Aplodontidae: Mountain Beaver
Family Castoridae: Beavers
Family Anomaluridae: Scaly-tailed Squirrels
Family Pedetidae: Spring Hare
Family Muridae:
    Subfamily Hesperomyinae: New World Rats and
        Mice
    Subfamily Cricetinae: Hamsters
    Subfamily Spalacinae: Blind Mole-rats
    Subfamily Myospalacinae: Eastern Asiatic Mole-
        rats
    Subfamily Rhizomyinae: Mole- and Bamboo Rats
    Subfamily Lophiomyinae: Crested Rat
    Subfamily Platacanthomyinae: Spiny Dormice
    Subfamily Nesomyinae: Madagascan Rats
    Subfamily Otomyinae: African Swamp Rats
    Subfamily Microtinae: Voles and Lemmings
    Subfamily Gerbillinae: Gerbils
    Subfamily Dendromurinae: African Climbing
        Mice
    Subfamily Cricetomyinae: African Pouched Rats
    Subfamily Hydromyinae: Island Water Rats
    Subfamily Murinae: Old World Rats and Mice
Family Gliridae: Dormice
Family Seleviniidae: Desert Dormouse
Family Zapodidae: Jumping Mice
Family Dipodidae: Jerboas
Family Hystricidae: Old World Porcupines
Family Erithizontidae: New World Porcupines
Family Caviidae: Guinea Pigs
Family Hydrochoeridae: Capybara
Family Dinomyidae: Pacarana
Family Dasyproctidae: Pacas and Agoutis
Family Chinchillidae: Viscachas and Chinchillas
Family Capromyidae: Hutias and Coypu
Family Octodontidae: Octodont Rodents
Family Ctenomyidae: Tuco-tucos
Family Abrocomidae: Chinchilla-rats
Family Echimyidae: American Spiny Rats
Family Thryonomyidae: Cane Rats
Family Petromyidae: Dassie Rat
Family Bathyergidae: Mole-rats
Family Ctenodactylidae: Gundis

Order Lagomorpha: Lagomorphs
Family Ochotonidae: Pikas
Family Leporidae: Rabbits and Hares

Index

ACKNOWLEDGEMENTS

The Publishers received invaluable help during the preparation of the *Longman World Guide to Mammals* from: Heather Angel, who lent us reference slides; Angus Bellairs, who gave advice; Dr H. G. Cogger, who lent us reference slides; Rosanne Hooper and Zilda Tandy, who assisted with research; Dr Pat Morris of Royal Holloway College, London, and Dr Robert Stebbings of the Institute of Terrestrial Ecology, Huntingdonshire, who both helped with reference; the staff of the Science Reference Library, London; the IUCN Conservation Monitoring Centre, Cambridge, England, for data on threatened species.